Solar Thermal Conversion Technologies for Industrial Process Heating

Solar Thermal Conversion Technologies for Industrial Process Heating presents a comprehensive look at the use of solar thermal energy in industrial applications, such as textiles, chemical processing, and food. The successful projects implemented in a variety of industries are shown in case studies, alongside performance assessment methodologies. The book includes various solar thermal energy conversion technologies and new techniques and applications of solar collectors in industrial sectors.

Features:

- Covers the key designs and novel technologies employed in the processing industries.
- Discusses challenges in the incorporation of the solar thermal system in industrial applications.
- Explores the techno-economic, environmental impact and life cycle analysis with government policies for promoting the system.
- Includes real-world case studies.
- Presents chapters written by global experts in the field.

The book will be useful for researchers, graduate students, and industry professionals with an aim to promote mutual understanding between sectors dealing with solar thermal energy.

Solar Thermal Conversion Technologies for Industrial Process Heating

Edited by
T.V. Arjunan
Vijayan Selvaraj
M.M. Matheswaran

CRC Press
Taylor & Francis Group
Boca Raton New York London

CRC Press is an imprint of the
Taylor & Francis Group, an **informa** business

First edition published 2023
by CRC Press
6000 Broken Sound Parkway NW, Suite 300, Boca Raton, FL 33487–2742

and by CRC Press
4 Park Square, Milton Park, Abingdon, Oxon, OX14 4RN

CRC Press is an imprint of Taylor & Francis Group, LLC

Library of Congress Cataloging-in-Publication Data
Names: Arjunan, T.V., editor. | Selvaraj, Vijayan, editor. | Matheswaran, M.M., editor.
Title: Solar thermal conversion technologies for industrial process heating / edited by T.V. Arjunan, Vijayan Selvaraj, M.M. Matheswaran.
Description: First edition. | Boca Raton: CRC Press, [2023] | Includes bibliographical references and index. |
Identifiers: LCCN 2022046467 (print) | LCCN 2022046468 (ebook) | ISBN 9781032203249 (hbk) | ISBN 9781032203560 (pbk) | ISBN 9781003263326 (ebk)
Subjects: LCSH: Heat engineering. | Manufacturing processes. | Solar heating—Industrial applications.
Classification: LCC TJ255 .S65 2023 (print) | LCC TJ255 (ebook) | DDC 697/.78—dc23/eng/20221202
LC record available at https://lccn.loc.gov/2022046467
LC ebook record available at https://lccn.loc.gov/2022046468

ISBN: 9781032203249 (hbk)
ISBN: 9781032203560 (pbk)
ISBN: 9781003263326 (ebk)

DOI: 10.1201/9781003263326

Typeset in Times
by Apex CoVantage, LLC

Contents

PART I   Solar Thermal Conversion Technologies: Overview

PART II   Solar Energy for Industrial Process Heating

PART III   Sustainability Assessment for Solar Industrial Process Heating

PART IV   Modern Control Systems and Government Polices for Solar Industrial Heat

Editor Biographies

T.V. Arjunan is a professor in the department of Mechanical Engineering, Guru Ghasidas Vishwavidyalaya (Central University), Bilaspur (C.G.). His active research areas include solar thermal applications, heat transfer, desalination, fluid mechanics, energy systems, etc. He has vast experience in teaching and research for approximately 23 years, during which he has developed many innovative solar thermal systems and supervised many research projects of undergraduate, postgraduate, and doctoral theses. He has executed many sponsored projects, workshops/seminars, and research projects more than 3 crore worth from various funding agencies, such as DST, AICTE, CSIR, ISTE, and MHRD. He has carried out many collaborative research projects with researchers working in India and abroad. To his credit, he has 5 patents of which 2 are international applications, authored more than 120 research papers in peer-reviewed international journals, and refereed conferences organized by various professional societies. He has contributed 4 book chapters in the field of solar energy systems published by Springer Nature. In addition, he serves as a reviewer in various highly reputed international journals. He is an active life member in several professional bodies such as ISTE, SAE, Bioenergy Forum, and Solar Energy Society of India. In recognition of his research work, he was awarded the "K F Antia Memorial Award" from the Institution of Engineers (India) and received a certificate of recognition from ZF Windpower Coimbatore Ltd, for contributing in bearing information system proof of concept development. He is an active speaker and has delivered many keynote addresses in international conferences, workshops, and seminars organized by various reputed institutions in India.

Vijayan Selvaraj is working as an associate professor in the department of Mechanical Engineering at Coimbatore Institute of Engineering and Technology, Coimbatore, India. He received his undergraduate, postgraduate, and doctoral degrees from Anna University, Chennai, India in 2006, 2011, and 2018, respectively. He has served as postdoctoral fellow in the department of mechanical engineering at National Institute of Technology, Tiruchirappalli from December 2019 to June 2022. His research interests are development of solar thermal applications such as air heater, dryers, and energy storage system for low and medium temperature applications. He has authored or co-authored more than 30 publications in international refereed journals, conferences, and book chapters. Currently, he is serving as an associate editor in the *Journal of Advanced Mechanical Sciences* (JAMS).

M.M. Matheswaran is a graduate of Mechanical Engineering in 2006 from Anna University, Coimbatore, India, and obtained his Master's degree in Energy Engineering from the same University in 2009 and his Ph.D. degree from Anna University, Chennai, India, in 2020. Presently, he is working as an associate professor in the Mechanical Engineering Department, Jansons Institute of

Technology, Coimbatore, India. His active research areas include solar thermal applications, heat transfer, solar air heaters, fluid mechanics, energy systems, etc. He has published 5 patents and authored 23 research papers in peer-reviewed international journals and refereed conferences organized by various professional societies. Further, he is serving as a reviewer in various highly reputed international journals.

Contributors

Abolfazl Ahmadi
Department of Energy
 Systems Engineering, School of
 New Technologies, Iran University
 of Science & Technology Narmak,
 Tehran, Iran

T.V. Arjunan
Department of Mechanical Engineering
 Guru Ghasidas Vishwavidyalaya
 Bilaspur, Chhattisgarh, India

Thota S.S. Bhaskara Rao
Heat Power Laboratory, Department of
 Mechanical Engineering National
 Institute of Technology Rourkela
 Odisha, India

N.V.V. Krishna Chaitanya
Academy of Scientific and Innovative
 Research, Ghaziabad-201002, India

V.P. Chandramohan
Department of Mechanical Engineering
 National Institute of Technology
 Warangal, Warangal, Telangana, India

Pritam Das
Department of Mechanical Engineering
 National Institute of Technology
 Warangal, Warangal, Telangana,
 India

Jasinta Poonam Ekka
Department of Mechanical Engineering
 Guru Ghasidas Vishwavidyalya
 Bilaspur, Chhattisgarh, India

Mirhamed Hakemzadeh
Solar Energy Research Institute
 Universiti Kebangsaan Malaysia
 Bangi, Selangor, Malaysia

Prashant Kumar Jangde
Department of Mechanical Engineering
 Guru Ghasidas Vishwavidyalaya
 Bilaspur, Chhattisgarh, India

P. Kanagavel
National Institute of Wind Energy
 (NIWE), Ministry of New and
 Renewable Energy, Government of
 India, Chennai, Tamilnadu, India

N. Kannan
Agricultural Engineering, Faculty of
 Agriculture University of Jaffna
 Thirunelvely, Jaffna, Sri Lanka

A. Veera Kumar
Department of Mechanical Engineering
 Coimbatore Institute of Engineering
 and Technology, Coimbatore,
 Tamilnadu, India

K. Ravi Kumar
Department of Energy Science and
 Engineering, Indian Institute of
 Technology Delhi New Delhi, India

M.M. Matheswaran
Department of Mechanical Engineering
 Jansons Institute of Technology
 Coimbatore, Tamilnadu, India

T. Meenakshi
Department of Electronics and
 Communication Engineering
 Jansons Institute of Technology
 Coimbatore, Tamilnadu, India

S. Murugan
Heat Power Laboratory, Department of
 Mechanical Engineering, National
 Institute of Technology Rourkela
 Odisha, India

Guna Muthuvairavan
Department of Mechanical Engineering
  National Institute of Technology
  Puducherry, Karaikal, Puducherry,
  India

Sendhil Kumar Natarajan
Department of Mechanical Engineering
  National Institute of Technology
  Puducherry Karaikal, Puducherry,
  India

Ram Kumar Pal
Department of Energy Science and
  Engineering, Indian Institute of
  Technology Delhi, New Delhi, India

Amir Hosein Saedi
Department of Energy Systems
  Engineering, School of New
  Technologies, Iran University of
  Science & Technology, Narmak,
  Tehran, Iran

D. Seenivasan
Department of Mechanical Engineering
  Coimbatore Institute of Engineering
  and Technology, Coimbatore,
  Tamilnadu, India

Vijayan Selvaraj
Department of Mechanical
  Engineering, Coimbatore
  Institute of Engineering and
  Technology Coimbatore,
  Tamilnadu, India

C. Shanmugam
Department of Electronics and
  Communication Engineering,
  Jansons Institute of Technology,
  Coimbatore, Tamilnadu, India

Kamaruzzaman Sopian
Solar Energy Research Institute,
  Universiti Kebangsaan Malaysia,
  Bangi, Selangor, Malaysia

Bhuvana Venkatraman
Department of Commerce
  Guru Ghasidas Vishwavidyalaya
  (A Central University)
  Bilaspur, Chhattisgarh, India

R. Venkatramanan
Department of Mechanical Engineering
  Coimbatore Institute of Engineering
  and Technology, Coimbatore,
  Tamilnadu, India

Preface

Solar energy is the largest and most widely distributed renewable energy resource on the planet and can be utilized for a wide range of applications, such as solar water heating, photovoltaic electricity generation, solar thermal energy generation, and all manner of passive and active solar architectures. Besides environmental consciousness, the dwindling of traditional energy sources also marks solar energy as the appropriate energy source to meet the increasing energy demand worldwide. Solar thermal collectors can be used efficiently to meet the heating needs of industrial processes because many industrial processes in various industries operate at temperatures ranging from 80 °C to 240 °C. Industrial energy analysis shows that solar thermal energy has enormous applications at low (i.e. 20–200 °C), medium, and medium-high (i.e. 80–240 °C) temperature levels. Almost all industrial processes require heat in some parts of their processes.

Despite all its merits, the implementation of solar thermal technology in industries is challenging due to various factors such as awareness, high investment cost, lower conversion efficiency, etc. This book attempts to explore all these challenges and the use of solar thermal energy in various industrial applications along with successful projects implemented in different categories of industries around the world as case studies.

The chapters of this book are compiled into four major parts, and the first part consists of two chapters. Chapter 1 provides an overview of the global energy scenario and highlights the significance of utilizing solar energy in the aspects of environmental impact and energy resource availability. Chapter 2 discusses in detail the fundamental concepts involved in the conversion of solar radiation into useful thermal energy for low and moderate-temperature heating processes in different industries. It also discusses the basic mathematical modeling of non-concentrating and concentrating collectors.

The second part of the book deals with the potential for solar industrial process heating applications in a few sectors along with the case studies. Chapter 3 elaborates on the potential heating processes involved in various industries and the necessary assessments to be implemented for adopting solar thermal technologies in an industrial location and the hurdles to be considered, such as technological, economic, and social barriers, etc. Chapter 4 discusses the processes in a typical pharmaceutical industry and the possible solar thermal integrations, along with highlighting a few case studies in pharmaceutical industries located in various countries. Further, the use of solar energy for small-scale applications in Ayurveda medicinal herb drying is included. Chapter 5 provides the energy consumption profile and the present energy sources used in numerous processes in the automobile manufacturing industry. It discusses the potential processes along with the desired temperatures and the integration methods adopted in conventional heating systems. It also includes the successful implementation of solar thermal systems situated in different parts of the world. Chapter 6 presents an overview of energy usage in the various food industry sectors, and the integration of solar technology in the food processing industries is introduced to overcome issues of the energy crisis and environmental problems and for sustainable economic development. Chapter 7 covers the fundamental techniques and contemporary advancements in thermal-based high-capacity desalination systems, such as multi-stage flash (MSF) evaporation, multi-effect distillation (MED), vapor compression distillation (VCD), low-temperature thermal desalination (LTTD), and others. In addition, the viability of a solar energy integrated desalination system as a long-term solution to the problem of water scarcity and also a case study of existing solar energy-based large-scale desalination has been presented. Chapter 8 deals with the implementation of solar heat for various processes in the oil and gas industries. In addition, the selection of appropriate solar thermal technologies to meet the process heat demand is discussed in detail, along with their economic and environmental effects. Further, the integration of solar thermal technologies with the existing setup of the oil and gas industries has been discussed.

The third part provides insight into the methods that can be adopted to assess the effective implementation of solar thermal systems for industrial heating applications from technical and economic perspectives. Chapter 9 discusses the technique for integrating solar heating systems into industrial settings using pinch analysis. Integration at the supply level and process level is discussed, and pertinent case examples are presented in depth. Chapter 10 deals with the sustainability of solar thermal systems by using life cycle assessment. It discusses various solar thermal technologies, lifetime assessment, and the environmental impact of solar thermal energy on two different-sized plants.

The final part of the book has two chapters, and they are devoted to elaborating the concepts of modern tools and an overview of several solar energy promotion policies adopted by various countries in the world. Chapter 11 discusses the role of contemporary tools, including artificial neural networks, convolution neural networks, fuzzy logic, machine learning, and genetic algorithms, in the prediction of solar radiation, the design, effective integration, and control of solar thermal systems used in industrial heating processes. Chapter 12 explains the global energy model and international solar energy strategies for chosen countries. Worldwide energy usage and regional climate scenarios are provided. International energy policies to promote renewable energy systems are described in depth.

We would like to express our sincere thanks to all the contributing authors, reviewers, book commissioning editor Ms. Kyra Lindholm, and CRC Press for their continuous support and guidance to bring out this book.

T.V. Arjunan
Vijayan Selvaraj
M.M. Matheswaran

Foreword

The scientific odyssey of Guru Ghasidas Vishwavidyalaya indicates that greatness is achievable. Our successes are the outcome of self-disciplined thinking and conduct. We recognize that perseverance with enthusiasm is the only way to achieve outcomes that surpass expectations. One of our core beliefs is unquestionably being enthusiastic about our work.

When it comes to solar energy technologies, the more you discover, the more you realize how little you know about its potential in different sectors. Therefore, there is a need for a book that explores the difficulties associated with switching from fossil fuels to clean technologies, treating these difficulties not as obstacles to change but rather as problems deserving careful consideration, investigation, and contemplation. We need a framework for clarifying the real-world difficulties, advantages, chances, and potential dangers of switching to a renewable energy future.

Professor T.V. Arjunan has been an important part of our odyssey. I have often found Professor Arjunan's way of executing work commendable and feel that one can learn from him that self-control at all levels—mental, emotional, and behavioral—is essential for reaching one's full potential.

The monograph edited by Prof. T.V. Arjunan, Dr. Vijayan Selvaraj, and Dr. M.M. Matheswaran is a compilation of extremely high-quality chapters authored by eminent researchers in the field of solar thermal conversion technologies. The contributors to this book are well-known experts in their respective fields, and the book describes the possible uses of solar energy for addressing the requirements of low-, medium-, and high-temperature industrial process heating applications. The chapters cover a wide range of topics, from simple solar thermal conversion techniques to large-scale industrial applications and various processes for heating purposes in different industries. The book also presents well-researched knowledge about solar thermal conversion technologies, effective integration techniques with industries, modern tools, and government policies particularly useful to researchers, practicing engineers, and policy formulators working in this field.

My hearty congratulations to all the readers of this book. The contents of this book will give you essential knowledge that can be utilized practically and the knowledge acquired can be employed beneficially in solar thermal energy projects. I trust that you will continue to make significant advances in your chosen field if you maintain your commitment to your fundamental convictions and keep building momentum.

Prof. Alok Kumar Chakrawal
Vice-Chancellor
Guru Ghasidas Vishwavidyalaya (A Central University)
Bilaspur, Chhattisgarh, India

Part I

Solar Thermal Conversion Technologies

Overview

1 Global Energy Scenario with a Special Reference to Solar Systems for Sustainable Environment

N. Kannan

CONTENTS

DOI: 10.1201/9781003263326-2

1.1 INTRODUCTION

Renewable energy technology is becoming highly important to meet the growing energy demand of the world as it causes minimum impact to the environment and its processes [1]. Renewable energy contributes significantly to the total world's energy demand and usage is reported to be 14% [2]. There are various renewable energy sources such as hydropower, geothermal, wind and biomass commonly used to meet increasing energy demand [3, 4]. It has been anticipated that the contribution of renewable energy sources will increase significantly in 2100 by 30–80% [5]. More-over, industry, population growth and technological advancements will drastically increase the use of fossil fuel resources to pollute the environment. Because of this consequence, dramatic climate change has been happening all over the world and is causing detrimental impacts to the survival of living organisms [6].

It is interesting to address that most of the countries have now realized the need for renewable energy sources and their contribution to waste minimization, pollution mit-igation, environmental sustainability and balancing of an ecosystem [7]. Solar energy and wind energy are two main sources freely available in nature. These resources can be utilized effectively to mitigate environmental problems caused by the overuse of fossil fuel resources. Solar energy is most popular among renewable energy sources because of its nature and availability in the environment. Furthermore, advance tech-nologies such as photovoltaic (PV) systems are used to generate electricity from the harvested solar energy. It has been stated that the proper use of solar energy technol-ogies to harvest available solar energy will be sufficient to meet the world's energy demand [8]. Furthermore, around four million EJ (exajoules) of solar energy are received by the earth's surface annually [9]. It is important to note that around 5×10^4 EJ solar energy can be utilized effectively out of the total solar energy available at the earth's surface [9]. However, the use of solar energy is still limited especially due to technical barriers in spite of its availability and environmental sustainability [9].

Moreover, intensive use of freely available solar energy to meet growing energy demand can lead to a significant reduction in global CO_2 emissions from the burning of fossil fuel resources [9]. It has been reported that around 696544 Mt of CO_2 has been reduced in the state of California, the United States, by implementing 113533 household solar systems [10]. Therefore, an effective solar energy system supported by appropriate policies and subsidies to harvest and utilize solar energy in an effi-cient manner must be practiced for energy sustainability in future. In addition to solar energy, wind energy is also used significantly to meet the energy demand of the world. Wind energy systems are actively practiced in 82 countries worldwide [10]. The growth of wind energy has happened remarkably over the last 20 years [10]. It has now reached its maturity and has been used productively to contribute to the energy demand of the dynamic world [10]. A significant number of wind turbines have been installed in the world. The total availability of the wind energy in the world is around 26000 TWh/yr. However, the utilization is limited to 9000 TWh/yr due to various constraints and economic implications.

As discussed previously, environmental pollution by excessive use of fossil fuel burning is an emerging concept all over the world to be addressed systematically to implement appropriate solutions. As a consequence of this, the concept of utilizing renewable energy sources is growing rapidly as it provides reasonable solution for problems caused by the excessive use of fossil fuel resources in the energy sector. Solar energy is used by many countries with room for future expansions to meet their annual energy demand. There are many reports and research works published to address the significance of this popular renewable energy source. However, there is very limited information available on the review of solar energy systems and environmental sustainability. This short critical review therefore addresses a comprehensive view of solar energy systems in terms of fundamentals, applications, potentials and future prospects to develop a clean and sustainable environment.

Moreover, this chapter is presented with an intention of filling the knowledge gaps identified by providing critical discussions on the global energy scenario, need of renewable energy sources (RES) and climate change scenarios, solar energy technologies, potentials and barriers of solar energy sources with respect to environmental sustainability. Therefore, this structured chapter will help researchers, policy makers and stakeholders to critically understand and justify real-world scenarios related to solar and wind energy systems in order to better select and expand these systems for viable applications.

1.2 GLOBAL ENERGY SCENARIO

World energy demand is steadily growing because of population growth and industrial revolutions. Figure 1.1 shows the past, present and future energy demand in the world. The energy demand was 1.81×10^5 TWh in 2020 as shown in Figure 1.1. However, it is expected to increase to 2.25×10^5 TWh in 2035. The main cause for

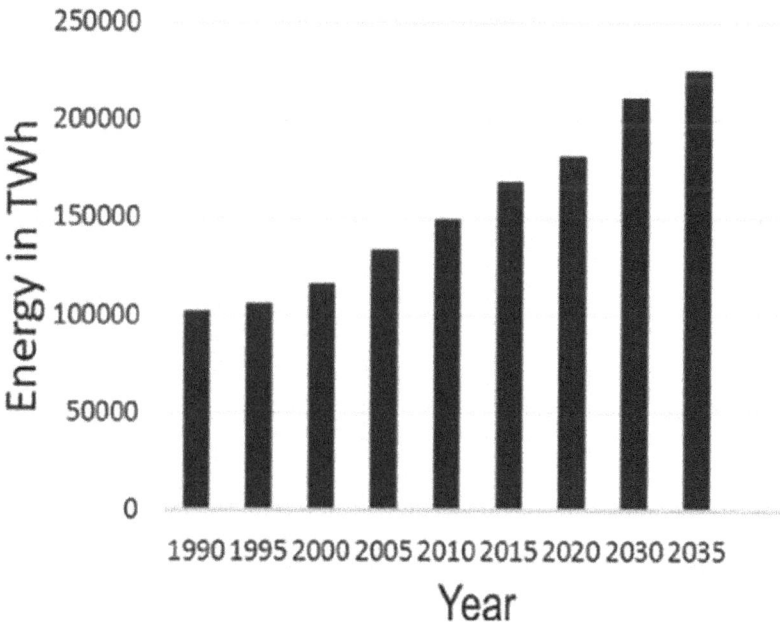

FIGURE 1.1   Past, present and future energy demand in the world [1, 11].

this is population growth. Many energy sources are used to meet such energy demand worldwide. However, around 77.9% of electricity has been produced by fossil fuel and nuclear sources which are expensive and cause environmental pollution particularly through CO_2 generation, and around 0.7% has been met by solar PV [1].

Therefore, there is an enormous gap for the expansion of the PV industry to use freely available solar energy for a better future. The world gets polluted and causes damage to living things mainly due to CO_2 emissions. This generation has caused several environmental issues such as climate change and global warming. Because of environmental pollution and exhaustion of fossil fuel resources, the world has to stimulate the growth of renewable energy sources for better production. Figure 1.2 shows the world fossil fuel energy status.

It is clear from Figure 1.2 that the energy sources will not be obtainable after the year 2300. Hence, many steps are now being taken to increase the use of renewable sources. A significant increase of around 39%, which has been recorded for solar PV in the year 2013, clearly shows its promising trends towards fulfilling the world's energy need. Photovoltaic technology is developing very fast to generate more electricity for satisfying the needs of people [1].

The significant increase has taken place after the year 2007 because of the combination of new technologies such as tracking, focusing and the development of evacuated tube collectors. In addition, the solar thermal industry is also growing significantly. Various collectors are used to capture sunlight to generate heat with minimum loss. Around 326 GW thermal energy has been formed by solar collectors in the year 2013. The largest installations of solar heating systems occurred in China and Europe. Brazil, by 2015, also devised to install roughly 1000 MW of solar heating systems to reduce the use of fossil fuel sources for energy recovery [1].

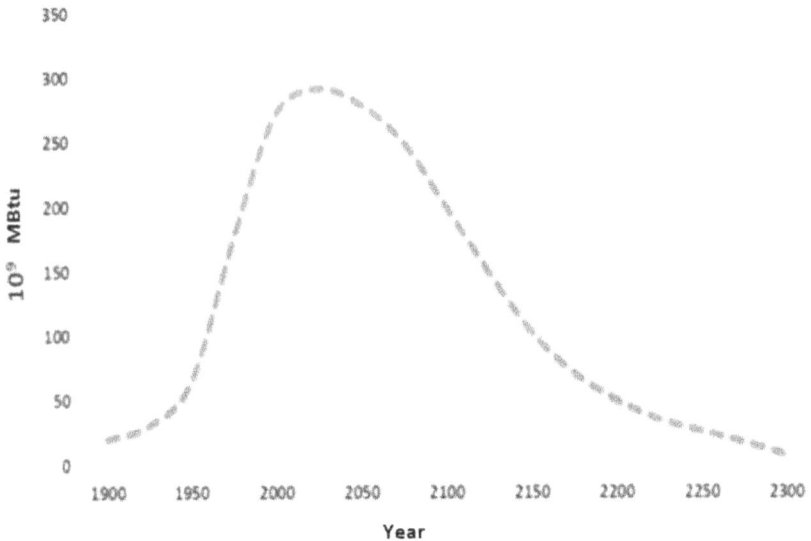

FIGURE 1.2   World fossil fuel energy status [1].

Moreover, the development of renewable energy sources was very significant in the year 2020. Many countries have developed new policies to promote solar industries to generate energy for distinct industrial applications. However, much attention is still being paid to fossil fuel resources for energy recovery as many industrial applications have been set to adapt fossil fuel reserves for energy recovery [12]. The attention to the renewable energy sources and the steps towards the promotion of renewable energy sources are basically due to an increase in the global total energy consumption. The attention to traditional biomass resources are also considered important due to handling ease.

As indicated in Figure 1.3, around 11.2% of the total final energy consumption was contributed by energy sources of so-called modern renewables in the year 2019. However, it was 8.7% in 2009. Moreover, the contributed values of fossil fuel sources to the total final energy consumption in years 2009 and 2019 are 80.3% and 80.2% respectively. It clearly shows the significant contribution of fossil fuel resources to the total final energy consumption in the world. The total final energy consumption of 11.2% in the year 2019 was shared by the electricity generated from renewable sources (6%), heat energy generated from renewable sources (4.2%) and the biofuels for transport activities (1%) [12].

Anyhow, it is highly essential to promote renewable energy sources for various industrial applications as they are freely available and pollute the environment at a minimum level compared to fossil fuel resources. Furthermore, the growth of renewable energy sources to share the total final energy consumption during the last

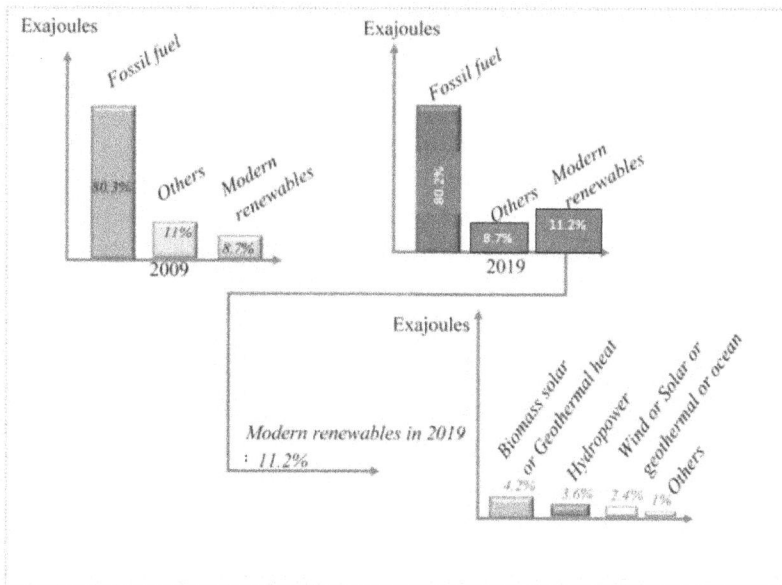

FIGURE 1.3 The estimated renewable energy share of the total final energy consumption between years 2009 and 2019.

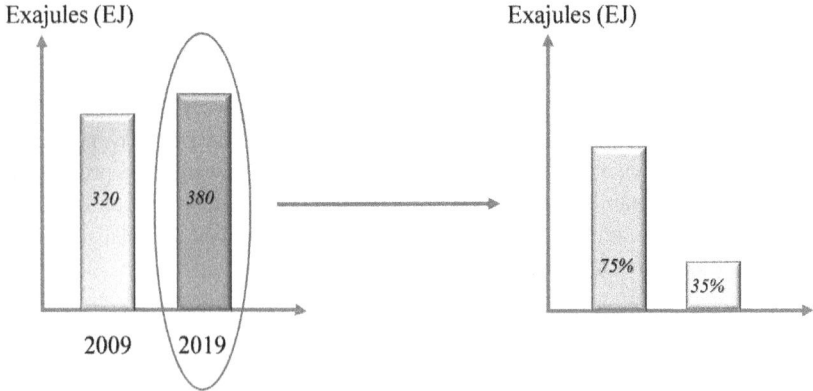

FIGURE 1.4   The estimated growth of modern renewables as a share of total final energy
consumption between years 2009 and 2019.

10 years from 2009 to 2019 has been moderate despite some drastic growth patterns
in some specific renewable energy sources. During this period, modern renewables
grew significantly with the contributed value of 15.1 EJ at an annual growth rate of
4.4% [12].

However, the increase in the total final energy consumption during the period mentioned earlier was 60.9 EJ at the annual growth rate of 1.8%. Hence, it is visible that
the renewable energy sources accounted for 25% of the total growth of the energy
demand as shown in Figure 1.4. It is obvious from the aforementioned fact that the
growth of the modern renewable energy sources are considerably prominent compared to the growth of total final energy consumption during the same period. Moreover, as shown in Figure 1.4 and the facts discussed previously, it is obvious that
fossil fuel resources grew at a slow rate of 1.7% annually during such period, and
the contribution of other energy sources was 75% of the total growth of the energy
demand [12]. This fact clearly shows the increasing share of the modern renewables
to the total final energy consumption, mainly because of the need to reduce emissions
so as to maintain environmental sustainability for future generations. The contribution
of modern renewables can further be improved by developing novel strategies that can
increase the efficiency of renewable sources by minimizing unwanted energy losses
in the harvesting process.

During the past years, the highest share (17%) of the renewable energy was for
electrical applications, mainly lighting and electrical instruments. The lowest share
of 3.4% of modern renewables was for transport activities, whereas around 10.2% of
the share of the modern renewables was used for industrial process heating, heating
and cooling of water and the space [12]. Therefore, it is clear that industrial process
heating operations can be improved by modern renewables with proper design configurations and integrations. These operations will ultimately help conserve environmental processes and their balances.

Moreover, the cost associated with the generation of energy from various solar
systems is still problematic all around the world. Different solar thermal systems

have different costs associated with their manufacture and operations. This obstacle of the solar thermal system makes it complex to deal with. The cost associated with concentrated solar power systems, parabolic troughs and solar towers is specific and isolated. A study conducted in the USA clearly shows the differences in costs associated with different solar systems as mentioned. The cost values are not uniform as far as manufacturing of solar thermal systems is concerned.

Researchers reported that the construction costs associated with concentrated solar power parabolic troughs and concentrated solar power solar towers were 5213–6672 $/kW and 6084 $/kW respectively without thermal storage. However, the addition of thermal storage systems can increase the construction costs significantly [13]. The actual costs associated with solar power parabolic troughs and concentrated solar power solar towers were 8258 $/kW and 9227 $/kW respectively with thermal storage arrangements. Moreover, the construction costs associated with solar photovoltaic systems are lower than solar thermal systems. Solar photovoltaic systems can be constructed with 4739 $ per kW capacity [13]. This is because of less complexity of accessories used in solar photovoltaic systems compared to solar thermal systems. However, there is a need to perform in-depth research activities to reduce the costs associated with solar thermal systems for effective industrial applications in future.

1.3 GREENHOUSE EFFECT AS A MAJOR DRIVING FORCE TO IMPLEMENT RENEWABLE ENERGY SOURCES

Environmental pollution due to the emission of greenhouse gases is a highly significant reason to propose solutions for the development of a sustainable future. Moreover, prolonged environmental pollution by anthropogenic activities including various industrial operations leads to dramatic climate change, which is one of the most important concerns in the 21st century [14]. Drastic change in climate due to such complex industrial processes can lead to many detrimental impacts such as sea level rise, acid rain, flooding, drought, escalation of heat waves, spread of contagious diseases and food scarcity. Developing countries, compared to developed countries, are more prone to the impact of adverse climate change, as the technologies adopted by developing countries to deal with climate change impacts are weak compared to developed countries. However, a significant effect of heat waves was recorded in Europe in 2003 [15]. It clearly shows the sudden impact of unexpected climatic events that influence environmental balance significantly.

Hence, it is obvious that dramatic climate change due to human-made activities, especially industrial operations producing greenhouse gas emissions, is a big threat to both developed and developing countries. The use of fossil fuels to meet the increasing energy demands of the world due to industrial revolution, population growth and technological advancements increased exponentially over the last century. Moreover, industries such as paper production, metal plating, textile and paint production lead to a significant emission of greenhouse gases into the environment. This significant increase in fossil fuel use ultimately released a considerable addition of greenhouse gases (CO_2, NOx, CH_4, CFC and ozone) into the environment. These greenhouse gases are responsible for the greenhouse effect as shown in Figure 1.5.

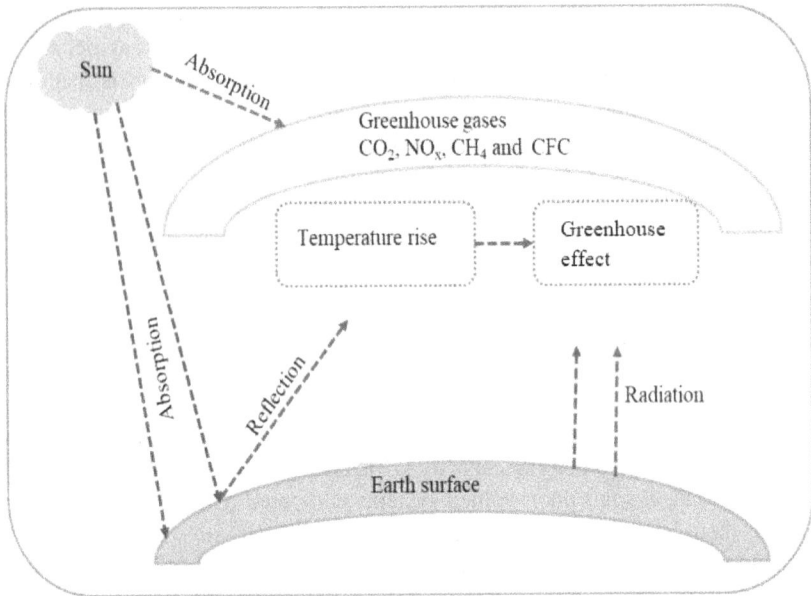

FIGURE 1.5   Schematic of the greenhouse effect.

Greenhouse gases form a layer over the earth's surface. Part of the radiation from the sun falls directly on the surface of the earth and is absorbed. A part is reflected back to the atmosphere through the radiating waves as shown in Figure 1.5. Moreover, radiation from the absorbed solar rays at the earth's surface also occurs into the atmosphere. However, the layer of the greenhouse gases acts as a shield to prevent the escape of radiating waves coming from the earth. They further absorb a fraction of the radiation of the sun directly and emit back to the atmosphere. The greenhouse gas layer is transparent to visible light (incoming radiation) of the sun and not transparent to the backward radiation from the earth surface as it is in the infrared (IR) form [16].

It has been noticed that the increase of human activities to meet escalating energy demand contributes significantly to greenhouse gas accumulation. Greenhouse gases, CO_2, CH_4, N_2O, R-11 (trichlorofluoromethane) and R-12 (dichlorofluoromethane) are likely potential greenhouse gases produced by human activities as shown in Figure 1.6 (a). Moreover, these gases pose an ability to retain IR radiation which is highly responsible for the heating effect. It is obvious from Figure 1.6 (a) that the annual growth rate of CO_2, CH_4, N_2O, R-11 and R-12 is 0.4, 1, 0.2, 5 and 5% respectively.

The annual growth rate of R-11 and R-12 is comparatively high compared to other gases as shown in Figure 1.6 (a) due to the increased use of refrigerators. Moreover, the growth rate of CH_4 and CO_2 is basically due to the burning of fossil fuel for meeting the increasing energy demand. Furthermore, the highest share of 71% of the greenhouse effect is held by CO_2 along with its highest contribution of 50% increase with 5% standard deviation due to human activities. As shown in Figure 1.6

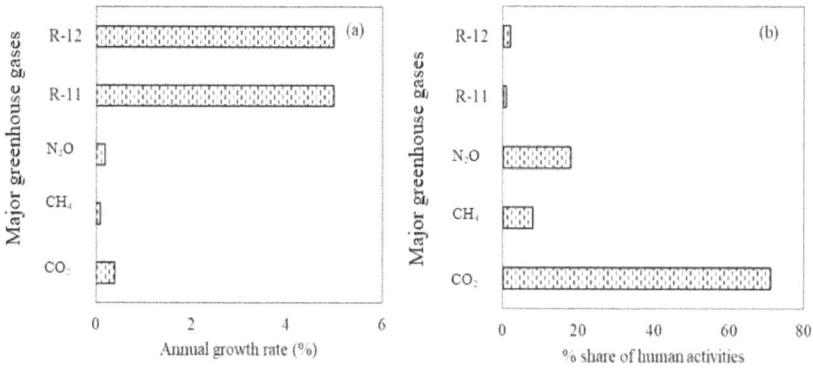

FIGURE 1.6   Annual growth rate of major greenhouse gases (a); percent share of human activities to major greenhouse gases (b).

(b) the most significant greenhouse gas is CO_2 which is produced via human activities. However, the growth rate of R-11 and R-12 is also critically important compared to other gases. Therefore, there is a prompt action needed to mitigate such harmful gases into the atmosphere to keep the environmental system sustainable for future generations.

In addition to the effect of various greenhouse gases on climate change, consideration of the countries that generate such greenhouse gases is also very crucial to propose viable mitigation strategies effectively. As shown in Figure 1.6, CO_2 emissions have grown by an average of 1.7% during the period of 24 years from 1971 to 1995. Moreover, this growth rate has been modeled to be at 2.2% for the year 2020. Furthermore, developing countries will account for half of this growth as shown in Figure 1.7. China and Organization for Economic Development and Cooperation (OECD) countries will contribute significantly to the emission of CO_2 into the atmosphere as shown in Figure 1.7 due to increased industrial activities.

Moreover, it has been reported that around 31% of CO_2 level increase occurred over the last 200 years. In addition, around 20 Gt of carbon has been added to the environment due to the cutting of trees [17]. Moreover, the ozone layer depletion has been severe due to an increase in CH_4 levels. All these greenhouse effects ultimately have caused a rise in global surface temperature by 0.4–0.8 °C during the last century above the reference temperature of 14 °C [17]. This rise in surface temperature in the end caused a sea level rise at an annual rate of 1–2 mm during the last century. However, around 37% of the contribution to the global greenhouse gas emission has been made by industries. The significant fraction of around 80% of the total greenhouse gas emissions is contributed by fossil fuel consumption for recovering energy [17].

According to the discussion made in this section, it is obvious that the emission of CO_2 into the environment is increasing at a faster rate due to human-made activities, technological advancements and the significant use of fossil fuel resources for meeting the energy demand. As evidenced in this section of the discussion, appropriate

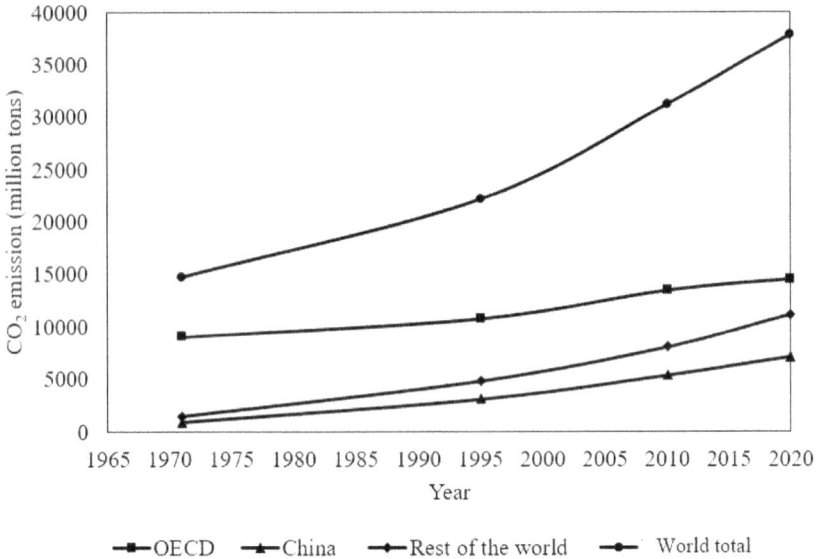

FIGURE 1.7 CO_2 emission rate by OECD countries, China, the rest of the world and the world total.

strategies need to be developed systematically in order to flatten this growing issue of greenhouse gas emissions into the environment. The most reliable option for this problem is the comprehensive, effective and economic use of renewable resources available in the world. The following section is therefore to describe how renewable energy sources have grown significantly to mitigate climate change scenarios with a special emphasis on the solar and wind energy sections as there is limited research presented so far.

1.3.1 HIGHLIGHTS OF THE DEVELOPMENT OF THE SOLAR ENERGY SECTOR

As discussed in Section 1.3, structured development of the renewable energy sector is extremely important for the betterment of the sustainable future. This part therefore gives an overview on the growth, development and importance of the renewable energy sector with a special emphasis on solar energy sources for environmental sustainability. The growth of the renewable energy sector has been occurring steadily over the last few decades. It has been reported that the contribution of renewable energy sources would reach 47.7% from 13.6% in 2001 as shown in Figure 1.8 [17]. It is a remarkable growth of the renewable energy sector which significantly tends to mitigate adverse climate change scenarios.

Moreover, solar thermal contribution would be 480 million tons equivalent in 2040 from 4.1 million tons equivalent in 2001. Photovoltaic and solar thermal energy will reach their contribution of 784 million tons equivalent and 68 million tons equivalent, respectively, in 2040, as shown in Figure 1.9. As shown in Figure 1.9, solar sectors have a potential to meet the energy demand by 2040. The growth of the solar

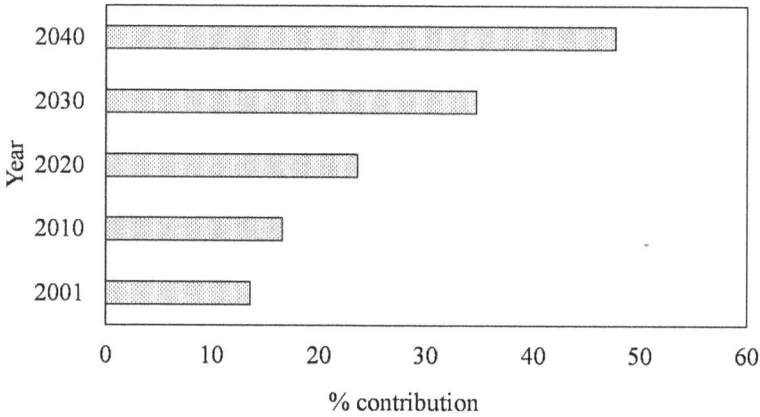

FIGURE 1.8   Percentage contribution of RES.

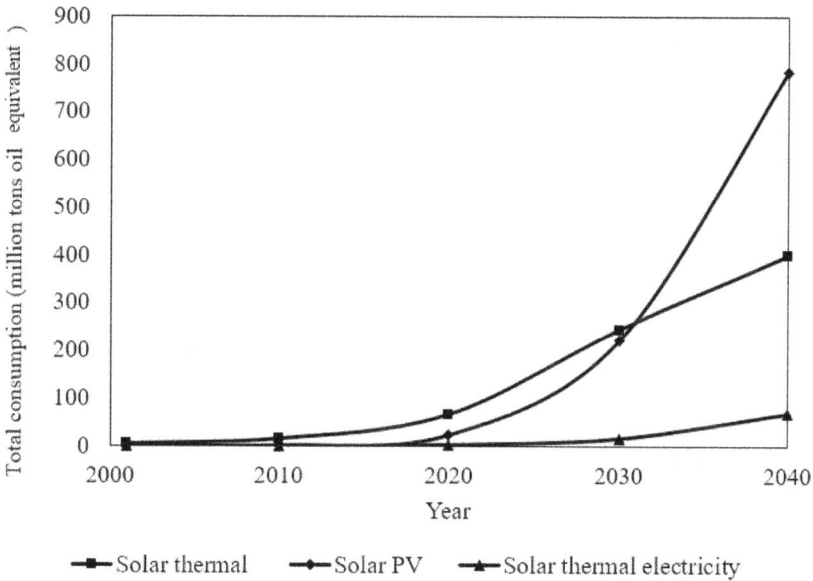

FIGURE 1.9   Total consumption of wind energy, solar thermal power, solar PV and solar electricity.

energy sector is highly significant because of considerable research and investment initiatives made to harvest as much solar energy as possible.

This difference is basically due to various policies and activities of various countries related to energy sectors and their functions. The use of these renewable energy sources will lead to zero or nearly zero emissions of pollutants such as aerosols and greenhouse gases. Moreover, development of the renewable energy sector will make the world sustainable energy wise with minimal impact to the environment compared to the fossil fuel burning energy recovery [18]. The contribution of solar energy sources to environmental sustainability is incredible in terms of distribution and reliability.

1.3.2 SOLAR THERMAL ENERGY USE FOR INDUSTRIAL PROCESS HEATING SYSTEMS

Solar thermal technology is generally obtained from solar radiation upon its heat conversion. Moreover, solar thermal energy systems can replace the use of fossil fuel systems significantly. For the application of solar energy for industrial heat applications, various solar thermal structures along with concentrators are being used. Nowadays, it is important to consider optimum working parameters of solar collectors, concentrators and heat exchangers for having high industrial outcomes through process heating.

The major barriers of the solar heating industrial systems are associated with solar radiation distribution over a day. However, it is vital to obtain some hybrid systems for better process heating applications. Moreover, around 17%, 44% and 10% of global energy demand are occupied by electricity, low temperature heat applications and high temperature process heat respectively [19]. As solar radiation is freely available in the world, its application to industrial heating process is broad and effective. Industrial processes, solar drying, cleaning, washing, water heating, pasteurization and sterilization, are highly prominent examples of industrial applications of solar thermal heat. A simple water solar thermal system used for industrial process heating is represented by Figure 1.10. All these aforementioned applications require a temperature value less than 250 °C.

Furthermore, for increasing applications of industrial solar heating systems, various solar collectors such as flat-plate solar collectors, evaluated tube solar collectors, unglazed collectors, vacuum tube collectors and fresh collectors are used. The flat-plate collector has plates to transmit the absorbed energy to the working fluid. This collector generally has a high absorption efficiency. The design system at the flat-plate collectors provides thermal stability and handling easiness. However, the evacuated tube collectors are very effective compared to the flat-plate collectors [19]. The heat loss from the absorber plate in this system is prevented by a well-organized glass envelope. Furthermore, the collector system is isolated from the industrial process heating systems.

The use of an unglazed system is basically only for low temperature applications. Additionally, in a vacuum tube collector, water is circulated inside pipes placed in vacuum glass tubes. The heat loss in this system is very low, and it can develop high temperature profiles. It is important to note that temperature values above 400 °C can be obtained for industrial process heat applications using the Fresnel collector. The efficiency at this system can be improved by the incorporation of a tracking system. However, there is a need to develop appropriate protocols for the better development of solar collectors for effective industrial heating applications. The research activities on this concept are still at an embryonic stage, and it requires further in-depth research.

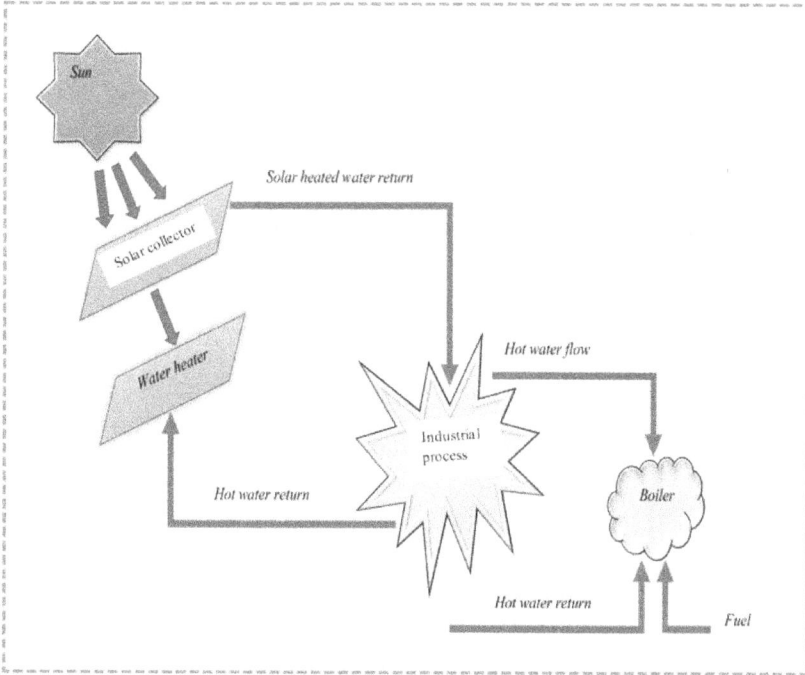

FIGURE 1.10   Simple representation of solar thermal heat for industrial process applications.

More importantly, it is highly vital to make a critical note on the current status of solar thermal technologies all around the world so as to provide readers with effective pieces of information. Europe, Asia and North America have been successfully using solar thermal systems for industrial process heating applications. In addition to this, solar thermal energy for electricity generation is also becoming popular in the world. However, the effectiveness of solar thermal systems are mainly influenced by the distribution of the solar radiation. Most of the Asian countries, some parts of Africa and Australia have a potential to have stabilized solar radiation throughout the year, and therefore, it is possible for them to implement solar thermal systems for the effective process heating applications [19].

Globally, solar thermal energy is used for various industrial operations based on the processing conditions. Automobile industries of South Africa heavily use solar thermal energy for painting purposes. It is also used heavily in various agricultural operations such as drying and water heating by countries like Spain, China and Germany. Drying of agricultural produce is mainly used by solar thermal energy in countries such as the USA, India, China and Germany [19]. The global importance of solar thermal energy is based on the nature and the stability of various industrial processes. Overall, it is used for water heating, drying, preheating, steam heating, pasteurization, sterilization and washing at different temperatures according to the need.

1.3.3 HEAT ENERGY DEMAND BY A RANGE OF INDUSTRIAL PROCESS TEMPERATURES

Different industrial processes are set with various temperature ranges. The actual temperature required by a preferred industrial process should be addressed properly for accurate heat requirement calculations. Moreover, it is also highly important to note that most industrial processes are set with a medium heat range. However, the scientific findings of heat demand by temperature range of various industrial processes are still very limited [19]. Hence, the need for further systematic scientific research in this regard is highly needed. Table 1.1 shows the temperature range of different industrial processes commonly in use.

TABLE 1.1
The Temperature Range of Different Industrial Processes [19]

Division	Process Type	Industrial Temperature Range (°C)
	Biochemical reaction	20–60
	Distillation	100–200
Chemicals	Compression	105–165
	Cooking	80–100
	Thickening	110–150
	Blanching	60–100
	Scaling	45–90
	Evaporation	40–130
	Cooking	70–120
	Pasteurization	60–145
Food and Beverages	Smoking	20–85
	Cleaning	60–90
	Sterilization	100–140
	Tempering	40–80
	Drying	40–100
	Washing	30–80
	Bleaching	40–150
	De-inking	50–70
Paper	Cooking	110–180
	Drying	95–200
	Picking	40–150
	Chroming	20–75
	Degreasing	20–100
Fabricated Metal	Electroplating	30–95
	Phosphating	35–45
	Purging	40–70
	Drying	60–200

Division	Process Type	Industrial Temperature Range (°C)
Rubber and Plastic	Dying	50–150
	Preheating	50–70
Machinery and Equipment	Surface heating	20–120
	Bleaching	40–100
	Coloring	40–130
Textile	Dying	60–90
	Washing	50–100
	Fixing	160–180
	Pressing	80–100
	Steaming	70–900
Wood	Picking	40–70
	Compression	120–170
	Cooking	60–90
	Drying	40–150
	Pasteurization	60–80
Diary	Sterilization	100–120
	Drying	120–80
	Concentrates	60–80
	Boiler feed water	60–90
	Sterilization	110–120
Tinned Food	Pasteurization	60–80
	Blanching	60–90
	Washing	60–90
Heat	Sterilization	60–90
	Cooking	90–120
	Thermo diffusion beam	80–100
Flour and Byproducts	Drying	60–100
	Preheating water	60–90
	Preheating pump	100–120
Bricks and Blocks	Curing	60–140
	Preparation	120–140
	Distillation	140–150
Plantation	Separation	200–220
	Extension	140–160
	Drying	180–200
	Blanching	120–140
Automobile	Water heating	~90
	Cleaning	~120
Pharmacy	Different processes	7–180

(Coninued)

TABLE 1.1
Coninued

Division	Process Type	Industrial Temperature Range (°C)
	Cleaning	~60
Mine	Electric-mining	~50
	Other processes	~80
Agriculture	Drying	~80
	Water heating	~90
Leather	Retaining	~80
	Other processes	~90
Metal	Heating	~180
	Washing	~160

Moreover, the solar thermal heating systems can be incorporated into the industrial process heating systems in two ways: either at the process level or at the supply level. Integration of solar thermal systems in industrial process is carried out based on the temperature desired. The supply level temperature requirement is greater than the process level temperature. The capacity of solar heating systems is different from different industrial processes. Around 36–51% of the installed solar thermal systems in industries around the world are for water heating or washing process, 14% systems for heating baths, 6% of the systems used for drying applications, and approximately 29% of the water heating systems utilized for other purposes like car washing, etc. The temperature profile of the industrial process is very important as it determines the yield of the industry [19]. Many industrial process are associated with temperature values less than 40 °C or between 40 °C and 60 °C.

Moreover, there are there major types of solar collectors such as flat-plate, evacuated tube and dish, mainly used for various industrial applications. The temperature ranges of the flat-plate collector, evacuated tube collector and dish collector are 40–80 °C, 80–150 °C and >150 °C respectively as shown in Table 1.2.

Process series of an industry can take either one of these collectors or the combination of two or more based on their need, mainly the process temperature. For

TABLE 1.2

Operational Temperatures of Various Solar Collectors Commonly in Use [20]

Operational Temperature (°C)	Solar Collector Type
40–80	Flat plate
80–150	Evacuated tube
>150	Dish

TABLE 1.3
Suitable Solar Collectors for Different Industrial Applications [20]

Industry	Suitable Solar Collector	Temperature (°C)
Textile	Evacuate tube collector	80–150
	Flat plate	40–80
Pulp and Paper Industry	Flat plate	40–80
	Evacuate tube collector	80–150
Leather Industry	Flat plate	40–80
	Evacuate tube collector	80–150
Automobile Industry	Evacuate tube collector	80–150
	Dish	>200

industries such as textile, pulp and paper, dairy and leather, flat-plate and evacuated tube collectors can be used based on the temperature range of the sub-processes. However, for the automobile industries, dish collectors are used in addition to the use of evacuated tube collectors so as to meet temperatures above 200 °C for some drying purposes as shown in Table 1.3. The pieces of information in Tables 1.1, 1.2 and 1.3 can be used effectively for the selection of solar collectors for various industrial applications. Moreover, there is a need to induce structured research activities in order to develop a common protocol for the confirmation of solar collectors for industrial operations based on the mass and energy balance concept of various industrial processes.

1.3.4 Factors That Are to Be Considered for Thermal Collectors in Industrial Process Heating Systems

The degree of a solar thermal collection is important for effective industrial solar heating processes. The space effectiveness, temperature control and the integration process are important for the effective design of solar collectors. A simplified flowchart is displayed in Figure 1.11 to help readers understand the important factors that influence the selection of solar collectors for industrial applications. Most importantly, climatic factors, design factors and operational factors must be considered for the effective use of solar collectors in industrial applications [21]. Solar collectors should be designed in such a way that the absorption of solar energy is maximum for reasonable industrial process heating. The space required for an industry must be considered as it is very limited for many industrial processes. Hence, rooftop solar collector installations are mostly prepared. As temperature profiles of industrial processes are complex, the collection system should be set to meet such requirements precisely. The use of heat carrier fluid is important in industrial solar heating applications. Solar heating systems of industrial processes are mostly associated with conventional heating systems.

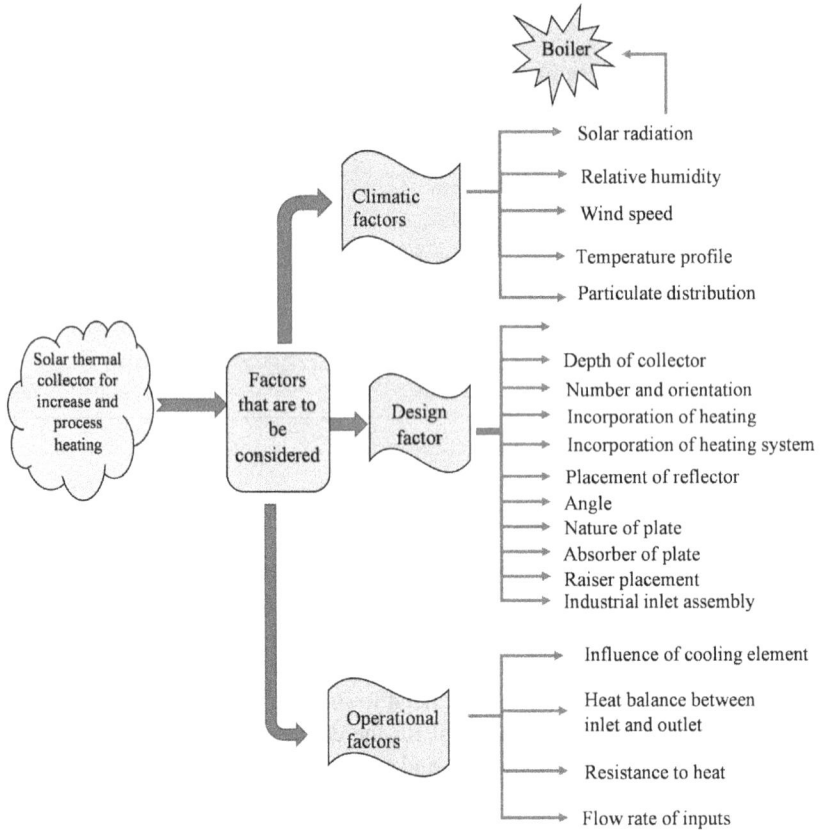

FIGURE 1.11   Simple representation factors that influence solar thermal collectors for industrial process heating applications [21].

1.3.4.1 Appropriate Industries and Processes for Solar Thermal Applications

Moreover, it is not possible to use solar thermal energy for all industrial applications. Hence the identification of suitable industries whose processes can be dealt effectively with an incorporation of solar thermal heat. The industries can be selected with the heat demand value below 300 °C. Based on the literature, machinery, automobile and electrical equipment are not considered suitable for this process since their major fraction of the heat energy is used for space heating. Most importantly, the sectors that deal with chemicals and the products related to chemicals have a significant potential to use solar thermal heat for process applications since there is considerable heat demand at low (<100 °C) and medium (<300 °C) temperatures [20].

This sector is mostly associated with sub-processes that require temperature values below 100 °C most of the time. Furthermore, the food and beverages sector, like chemicals and chemical products sector as discussed earlier, has a potential heat demand of its processes at low and medium temperatures. Hence, it is also highly suitable for the incorporation of solar thermal energy systems. The sub-processes of food industries such as drying, evaporation, blanching, hot water cleaning and pasteurization are set with low heat demand, below 300 °C [20]. Therefore, this sector can be successfully developed with the use of solar thermal heating inputs.

Furthermore, the paper and paper product industries have the potential to utilize solar heat for their operations. Around two-thirds of the processes of the paper and paper product industries is associated with temperature values between 100 °C and 500 °C. This temperature range is basically for drying purposes. However, one-third of the heat demand is still associated with temperature values below 100 °C. Moreover, the efficiency of the dyeing systems and de-inking systems of paper industries must be improved by appropriate strategies to save considerable energy cost. In addition to paper industries, metal products industries have an ability to utilize low temperature heat energy for their operations. It is basically associated with process heat, space heat and hot water production at temperature values below 100 °C. The processes such as degreasing, pickling and electroplating do require slow heat at temperatures below 100 °C. Appropriate insulation techniques can minimize the heat loss significantly, resulting in the saving of energy costs [20]. The processes associated with rubber and plastic product industries have a potential ability to use low temperature heat energy at temperature values below 100 °C. The solar thermal energy in plastic and rubber industries is mainly used for drying of plastic products and space heating activities at low heat energy demand. Besides, a significant number of processes associated with textile and wood industries utilize low temperature heat energy for better outcomes. In textile industries, processes like washing, coloring and drying are set below 150 °C. Around 82% of the total processes in the wood industry is associated with slow drying at temperatures below 100 °C so as to preserve the quality of the wood for better appearance [20].

From the facts discussed, it is possible to identify the most promising industrial sectors that can use solar thermal energy for their various processes effectively. For industries such as motor vehicles, metal and electrical equipment, most of the heat energy is used for space heating, whereas chemicals, food and beverages industries have the strongest potential to use solar thermal heat effectively for their processes because of less process complexity and simplified production steps. However, there is a need to perform systematic research investigations to improve the use of solar thermal energy in these promising, industrial sectors.

It is vital to note that the industries discussed in this section have different industrial processes set with different heat requirements and temperature profiles. As clearly shown in Table 1.1, different industrial processes have an ability to use solar thermal energy effectively based on their processing configurations. Moreover, for the effective utilization of solar thermal energy in such processes, it is necessary to avoid unwanted heat losses in and from industrial process structures. The development of a process-specific solar thermal energy usage protocol is greatly required nowadays for the effective utilization of solar thermal energy in industrial process heating applications.

1.3.5 CHALLENGES AND POLICY RECOMMENDATIONS FOR INDUSTRIAL SOLAR HEATING SYSTEMS

Though solar industrial process heating systems are considered to be effective, there are some common practical barriers. The most important problem is the high initial cost associated with industrial solar process heating systems. The barriers to solar industrial process heating can be grouped into economical barriers, policy barriers, technology barriers, technical and human resource barriers and social barriers. The market fluctuation and limited government subsidies to the solar market are potential risks associated with solar industries. These are critically influencing solar industrial processing heating systems [19].

Furthermore, around 25% of the global final energy consumption is associated with industrial processes. Many countries have a strong willingness to promote renewable energy sources, especially solar energy, to meet industrial heat energy requirements in a sustainable manner. As an evidence to support the aforementioned statement, in 2020, significant policy support was developed via financial incentives. For instance, 189 million USD was allocated to support industries that developed strategies to cut greenhouse gas emissions by the incorporation of renewable energy sources. Moreover, in 2020, the European Union developed a renewable hydrogen policy to increase the use of renewable energy sources for industrial processes. As developed and developing countries realize the importance of renewable energy sources for industrial processes to reduce greenhouse gas emissions, incentives and low-interest credits are promoted for the incorporation of solar thermal energy into various industrial processes [12].

Moreover, the policy frameworks that are related to industrial solar heating systems are still confusing. In addition, the subsidy schemes are mostly directed to fossil fuel resources, and the system practiced in industrial solar heating systems is also physically weak. Furthermore, some technological barriers are also present. The technology produced for industrial solar heating systems is not well organized, and there is no advance technology available to manufacture solar cells locally.

The availability of updated solar maps to check the solar heating performance of a system is still problematic all over the world. Therefore, there is need to overcome all the previously mentioned technological barriers for better industrial solar heating processes in future. In addition to technological barriers, social barriers such as poor understanding of solar systems in rural areas, the mentality of people to stick to traditional elements, the opposition of solar projects by some local communities and the poor technical skills of people are also affecting the performance of solar heating systems. Hence, for effective industrial solar heating systems, there is a need to overcome these barriers systematically.

1.3.5.1 Policy Reasons Related to Industrial Solar Heating Systems

Industrial solar heating systems are highly effective as they are sourced with freely available solar radiation. However, as discussed, they have been experiencing so many challenges. Hence, it is important to develop a highly structured policy framework in order to overcome such barriers [19]. The following points can be considered while making a good policy framework for industrial solar heating systems.

- Local people should be educated about the barriers of industrial solar heating systems.
- Government and stakeholders should be linked strongly for the sources of industrial solar heating systems.

- Some micro-finance industries should be developed in rural areas to facilitate industrial solar heating projects.
- Policies should be developed in such a way that they can attract foreign investors to make an investment on solar heating projects in rural areas.
- The importance of renewable energy sources as opposed to conventional energy sources should be explained in the policy framework.
- Marketing systems should be strengthened to promote industrial solar heating systems.
- The technical skills of professionals should be improved via appropriate training systems.
- Government should allocate enough money for industrial solar heating projects.
- Government should open new windows in order to develop international collaboration for future projects (solar heating).

Industrial solar heating systems will definitely be improved as these technical points are incorporated into policy frameworks developed by various countries in the world.

1.4 FUNDAMENTALS OF SOLAR ENERGY TECHNOLOGIES

The understanding of fundamental concepts of solar power systems is very important to make appropriate arrangements to have as much energy as possible for various utilizations. This part therefore tends to bring such concepts in a comprehensive manner to facilitate researchers and relevant personnel to grasp required solid information in order to make actions that are needed to increase the use of solar and wind energy for better environmental sustainability. Solar energy is one renewable energy source used for various environmental applications. It is broadly categorized into two major groups, active technology and passive technology. These technologies are basically defined to harvest the maximum amount of solar energy as possible [22]. The classification of solar energy technologies is presented in Figure 1.12. Passive technology is usually set to utilize solar energy (light and heat) without any further transformation [23].

The typical example of passive solar technology is a home heating system as shown in Figure 1.13, where solar energy is directly used to heat homes for the better survival of people. The heat penetration is possible via rooftops, walls and windows as heat wave radiation. The penetrated heat should be protected from loss so as to have a long heating effect in homes. Hence, an insulation should be provided to the space heated by solar radiation. However, the actual configuration of home heating systems and their optimization are still at the embryonic stage and need further research investigations for better development of home heating systems.

Active solar technology, which is very different from passive technology, is associated with some technical devices (e.g. mirrors or solar cells) to convert the solar energy into either heat or electricity. The most common classification of active solar technology is based on the outcome of the system. It is classified into solar thermal systems and solar photovoltaic systems as shown in Figure 1.12 [24]. Moreover, the

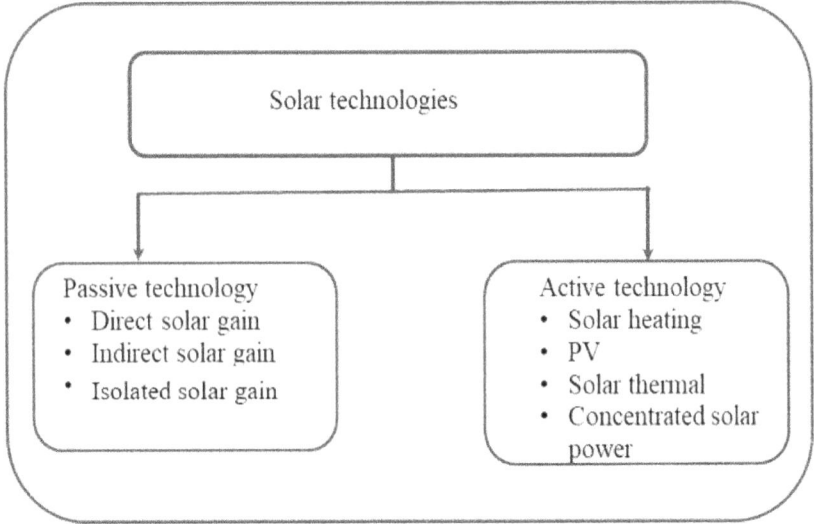

FIGURE 1.12   Classification of solar energy technologies [22].

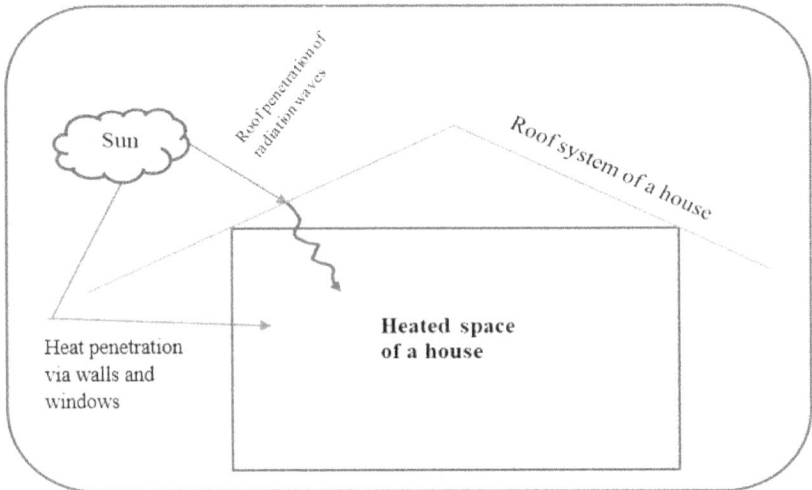

FIGURE 1.13   A simple schematic of a home heating system.

photovoltaic technology (PV) is used to produce electricity from solar energy with the help of semiconductor materials and inverters as shown in Figure 1.14 [25]. Solar PV technology is becoming very popular and a significant amount of well-structured research works have been carried out to increase the efficiency of solar PV technology. However, there is still a need for further research activities to improve the efficiency of solar cells used in PV systems in order to make them commercially viable.

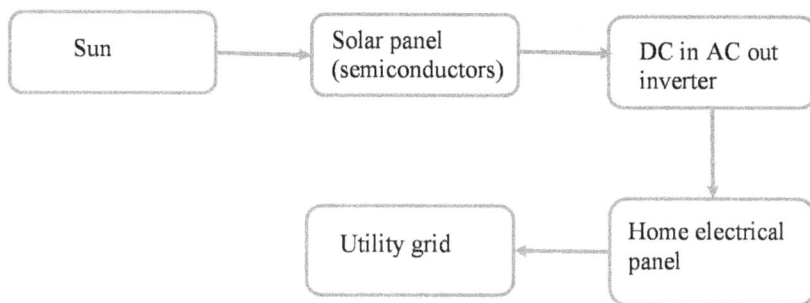

FIGURE 1.14 A simple schematic of a solar PV system.

As a result of intensive research activities, hybrid perovskite solar cells ((CH_3NH_3) PbI_3) have acquired an efficiency of 18% [26]. However, there are many different types of solar cells such as crystalline silicon, commercial thin film cells and perovskite cells available in solar PV systems. The efficiencies of these cells are very different and influenced by manufacturing configurations [27]. Furthermore, solar thermal technology is mainly used for heating purposes such as solar cooking, solar water heating and solar drying as shown in Figure 1.15 (a), (b) and (c). Solar cooking is a simple form of solar energy use [28]. Different manufacturers manufacture different types of solar cookers, out of which the simple box type cooker is highly efficient and user-friendly because of its less technical complexity as shown in Figure 1.15 (b). However, more research works are still needed to improve the efficiency of the solar cookers and increase the usability among users. In addition to this, solar water heaters (Figure 1.15 (a)) play a vital role in mitigating CO_2 emissions into the environment [29].

It has been reported that a solar water heating plate with 100 L capacity was able to mitigate approximately 1237 kg of CO_2 emission per year at 50% efficiency [30]. At the commercial scale, concentrated solar thermal systems (CST) and concentrated solar power systems (CSP) are highly popular [22]. Fresnel mirrors, parabolic troughs, power towers and solar disc collectors are popular types of concentrated solar power systems. These types are distinct in their operations. Parabolic troughs concentrate solar rays to a receiving tube, whereas Fresnel systems use multiple mirrors to concentrate solar rays into a receiver tube. Moreover, thousands of mirrors are used in power towers with a tracking mechanism for concentrating solar radiation to a desired point. That point is situated above the reflector disc to concentrate solar radiation in solar collectors [31]. Furthermore, solar dryers are simple devices to dry different products as shown in Figure 1.15 (c) in an efficient manner. The system must be well insulated and the heat loss from the dryer system should be minimized for better efficiencies. More research activities in the solar dryer systems are still required to improve the efficiency in order to increase their commercial applicability for different agricultural products.

The overall fundamental mechanism of a solar energy conversion system is that it uses solar radiation for heating purposes and electricity generation via concentrated

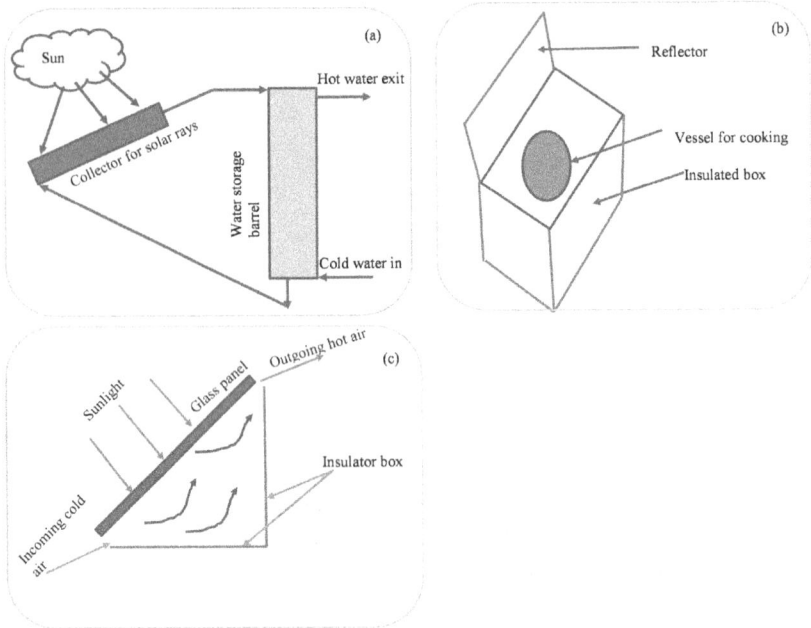

FIGURE 1.15   Solar thermal applications: solar water heating (a); solar cooking (b); solar drying (c).

thermal systems and PV systems, respectively. However, wind energy systems are different from solar energy systems [32, 33, 34, 35, 36], and the research is distinct in this area. Moreover, the research and development related to solar thermal systems are growing at a faster rate to improve the mechanism of harvesting for environmental sustainability. For the continuous supply of effective applications, solar thermal systems are to be associated with appropriate sources of energy so as to make them hybrid in nature. However, more research activities are still required to fabricate economically viable solar energy hybrid systems for energy sustainability and environmental protection.

1.4.1 GLOBAL STATUS OF SOLAR ENERGY SYSTEMS

This part of the comprehensive review provides significant information about the global status of solar energy systems at a glance with necessary comparisons and knowledge gaps identified. The distribution of solar energy is greatly influenced by locational differences among countries. The degree of environmental pollution due to solar systems among countries is different [37]. The solar flux to the earth surface is influenced by seasonal variations, cloud movements, latitude and crop covering [38]. Around $342 \, Wm^{-2}$ of solar energy is available at the surface of the earth. However, only around 70% is available for harvesting as the 30% of the total incoming energy is reflected back into the space [39]. Moreover, worldwide annual active solar radiation gain varies from $60–250 \, Wm^{-2}$ [40]. This variation is

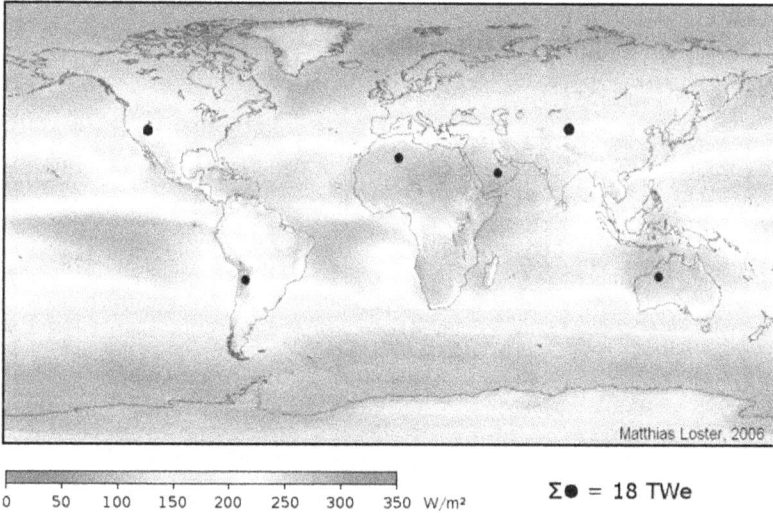

0   50   100 150 200 250 300 350 W/m²   Σ● = 18 TWe

FIGURE 1.16   Distribution of solar radiation on the earth's surface [22].

clearly shown in Figure 1.16 in which the locations marked with black dots could contribute to the total energy demand of the world with at least 90% efficiency [22, 39].

As shown in Figure 1.16, the sunniest locations are found around the continent of Africa. It has been reported that the capacity of CSP and PV systems is 470 and 660 Petawatt hours (PWh) respectively in the sunniest place, Africa [22]. Moreover, Middle East regions; the hottest plains of India, Pakistan and Australia; some parts of the United States, Central and Southern America; and north and south parts of Africa have the limited potential of 125 GWh/km^2 [41]. Furthermore, it has been stated that a significant portion of unused land is available in the Northern and Southern regions of China where electricity generation capacity is 13000 GW per 6300 km^2 [42]. Moreover, reports from the National Renewable Energy Laboratory (NREL) states that the annual solar energy potential in the US is around 400 Zetta-watt hours [22].

Furthermore, some countries which have high solar radiation availability have started improving their solar technologies to increase the harvesting potential. For example, the world's largest CSP and PV projects have been activated in Morocco to have improved power generation of 2000 MW by the year 2020 [22]. This move by the Moroccan government is highly productive because of their locational merit to harvest the maximum available sunlight. In addition to this, the highest solar radiation power per meter square of land is available in Australia. Moreover, the Australian continent has been reported to have the highest solar radiation of 4–6 $kWhm^{-2}$ [22]. It is therefore obvious that the nature and distribution of solar radiation are influenced by spatial differences. In order to gain the highest potential of solar energy for future developments, many countries have started upgrading their solar energy harvesting

Total Capacity (GW)

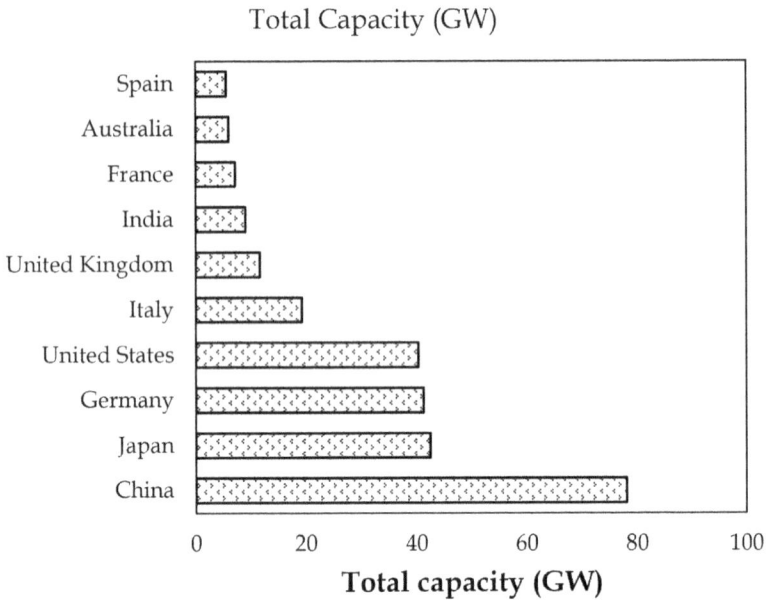

FIGURE 1.17   Solar power capacity of different countries.

systems. For example, the solar power generation capacity of China, Japan and Germany is 78.07, 42.45 and 41.22% respectively [22] as shown in Figure 1.17.

The growth of the solar PV systems is also happening at a faster rate because of advanced research activities. It has grown to produce 2.25 GW in 2015 from 3.7 GW in 2004 [22]. China recorded the most solar power plants installed in the world and its plan was to reach the capacity of 150 GW by the year 2020 [22]. However, it has been a challenging task to meet such expectations. Moreover, countries like America and Australia have also significantly grown in solar PV system development. The USA set the target to reach 45 GW PV capacity in 2017 [43]. Furthermore, the solar PV capacity of 913 MW has been recorded in Australia in 2015.

All these increases contributed significantly to global emission reduction. Apart from these developments, India reached a 8062 MW solar power grid system in 2016 [44] and planned to achieve 100000 MW by 2022 [44]. It is also important to consider wind energy sources so as to develop renewable hybrid power systems that can keep the energy line continuous [45, 46, 47, 48]. Furthermore, a planned structure is being developed by the government of France to construct a 100 km solar roadway system with an intention of providing clean energy for 5000 houses/km [22]. Moreover, most of the countries have recorded drastic growth in solar energy systems. China is leading the world's growth of solar energy system investment. The sudden growth in the solar energy market in the year 2016 has been recorded in the USA due to the anticipation of a federal investment tax credit. It is therefore obvious that the growth of solar systems is very crucial all over the world to achieve environmental sustainability in the near future. However, much more research activities are still required to increase the efficiency of solar systems to harvest more freely available solar energy.

Based on the previous facts discussed, solar energy systems can be researched further to enhance clean energy and environmental sustainability. It is reasonable to produce the following highlights and summary points for the benefit of future readers of this chapter.

- The efficiency and distribution of solar energy systems is influenced by the fluctuation of solar radiation intensity worldwide. Hence, an appropriate selection must be made based on appropriate spatial information before making the decision on which system is suitable for an intended use.
- The African region has huge potential to receive the most solar radiation to be harvested along with the Australian continent. However, most of the usable solar radiation is still wasted because of limited technologies available to increase the harvesting potential significantly.
- Solar PV, CSP and CST systems recorded robust growth to contribute more for future energy demands. This robust growth is basically due to the abundance of solar radiation compared to wind energy systems. However, there is still limited information available on specific efficiency of solar energy systems.
- It is important to have global hybrid solar power systems for better utility and applications.

1.5 MERITS AND DISADVANTAGES OF SOLAR ENERGY SYSTEMS WITH RESPECT TO A CLEAN AND SUSTAINABLE ENVIRONMENT

This section of the review critically summarizes the merits and disadvantages of solar for developing better utilization strategies without damaging the environment.

1.5.1 MERITS OF SOLAR ENERGY SYSTEMS

Solar energy systems are becoming popular worldwide due to the abundant nature of solar radiation on the earth's surface. This solar energy potential, if properly harvested, is enough to meet the energy demand of the world. Moreover, there is nothing to worry about the depletion of solar energy unlike fossil fuel resources since it is renewable in nature [49]. Fossil-fuel-driven power plants are significant sources of greenhouse gas emissions responsible for global warming potentials. They further contribute to 255 of anthropogenic emissions worldwide [50]. The use of solar power systems contributed significantly to minimizing greenhouse gas emissions though the manufacturing, maintenance and installation activities reported to produce 0.03–0.09 kg of CO_2 per kWh power generation, which is significantly low compared to coal and natural gas with 0.64–1.63 kg and 0.27–0.91 kg respectively. This clearly shows its significant potential in the reduction of greenhouse gas emissions [22]. Thus, solar energy systems, if properly structured and installed, can contribute significantly to minimizing global warming potential by reducing greenhouse gas emissions from coal-powered power plants in order to develop a sustainable future.

Another important advantage of solar energy systems is the reduction in the particulate matter and other toxic gases which are generally generated when fossil fuel resources are used for power generation. Fossil fuel burning to meet energy demand produces toxic gases like SO_2, volatile organic compounds, suspended particulate matter and soot. These substances are highly toxic and can cause health problems like dysfunctions of the nervous system, heart failure and respiratory complications [51]. A research study insisted that the use of renewable energy sources could minimize infant mortality and expenditure for treating diseases [52]. Moreover, power plants driven by fossil fuel sources require a significant amount of water for cooling. It then leads to the wastage of good quality water and to water scarcity. Inefficient use of water in power plants can drastically reduce the power generation capacity of plants. However, the use of solar energy systems requires no water and produces no toxic substances as power plants driven by fossil fuel resources do. Therefore, the use of solar energy in this sense can minimize this environmental problem significantly.

Furthermore, fossil fuel systems used for power generation are capital intensive, whereas solar energy systems are labor intensive. As solar energy systems are considered to be labor intensive, they can help improve employment opportunities of people worldwide. This is a wise outcome of the solar energy sector and its expansion, as, at present, most of the rural people are suffering from unemployment problems. It has been stated by the Solar Foundation of the USA that the solar industry provided employment opportunities for 208859 people in the USA as part-time or full-time workers for manufacturing, installation, sales and maintenance. Moreover, around 20.2% growth has been recorded in a year [22]. Furthermore, solar energy systems help circulate money within the country. It also lowers the importation of fossil fuel and currency flow into the foreign countries. Therefore, it can help strengthen the economic structure of the country.

Furthermore, as vast research activities progress, the efficiency of solar systems has been significantly improved. As a result of this improvement, the cost of the PV modules has dropped considerably. It is obvious from reports that the cost of a PV module has been reduced from USD 1.3 per watt in 2011 to USD 0.5 per watt in 2014 with around 60% cost reduction [22]. As this scenario happens steadily and progressively in solar markets, the cost of PV modules and related structures will be highly affordable in the near future.

In addition, the economic cycle of the fossil fuel market is highly turbulent with dramatic price instabilities. However, unlike fossil fuel markets, the solar market is comparatively stable for a longer period of time. It further produces no noise, as solar PV systems are set to have no moving parts which are common in generators used to generate power from fossil fuel resources. Additionally, the installation of solar power systems to buildings and rooftops is simple and reliable. Furthermore, large structures in solar systems are rare as solar PV systems are combinations of individual modules. Moreover, a fault to one module can cause no significant impact as others in the system continue to work in order to maintain the supply. These advantages are highly productive as far as the expansion of solar energy systems are concerned.

1.5.2 BARRIERS TO SOLAR ENERGY SYSTEMS

There are some limitations observed for solar energy systems though it is considered to be an effective source of renewable energy. It can only provide a constant power supply if it is well structured with other components like inverters and high-quality batteries. This section highlights barriers related to solar energy usage. The initial cost associated with the installation of solar systems is high compared to other renewable energy sources. It has been reported that the cost for a watt of solar energy was \$3.7 in 2016 in the USA [22]. Moreover, a solar system of 5kW for a household costs around \$13000 with 30% cost reduction due to federal tax credit. The long payback period diminishes the good value of the system [22]. Another critical bottleneck of the solar systems is that its efficiency is low. The commonly used solar panels with homes currently have an efficiency value from 10 to 20% [22]. However, highly efficient solar cells like perovskite cells are very expensive.

As discussed, for a continuous supply of electricity, PV panels should be associated with high-quality batteries. The cost associated with such batteries and inverters is high. In addition to the cost, the life of batteries is unstable and short. The quality of batteries and other associated components should be improved in order to reduce the cost of production and increase the durability. The batteries which are currently used in solar systems have some metal ions (Pb and Cu). The disposal of such batteries after use is a challenging task in the solar industry. Moreover, commonly available batteries are big in size with significant weight resulting in larger space requirements for safe storage.

Additionally, maintenance of solar systems and installation of components require technically sound human resources, and these are barriers to solar energy systems. In addition to these, people who live in rural areas tend to have poor technical knowledge about solar systems and their proper maintenance, resulting in damage to solar panels, batteries and circuits. This ultimately incurs additional cost for repair and maintenance. Another significant problem associated with solar PV modules is the formation of cracks, which lead to the intrusion of water into the solar cells. This intrusion can ultimately yield algal growth that can reduce the overall efficiency of the solar system. Moreover, solar panels can actively work only during the day time.

It is not a viable option in areas where abrupt weather conditions are generally reported. In addition to bad weather conditions, severe air pollution due to particulate matters and smog can reduce the efficiency of solar panels [53]. It has been reported that the exposure of solar panels to extreme smokes and aerosols resulted in 10 and 7% current reduction respectively in silicon solar cells [53]. Furthermore, the land area requirement to install larger solar panels at commercial scale is high. For example, solar panels made up of crystalline cells with 1 MW capacity require four acres of land, whereas panels made with thin-film technology need six acres of land for effective operation [54]. However, concentrated attention needs to be given to preserve the environment while implementing different renewable energy projects: solar, wind, etc. [55, 56, 57, 58].

As discussed in this section, major bottlenecks associated with solar systems are high cost of manufacturing, installation, the requirement of a skilled workforce, poor battery life, technical complexity of inverters, significant influence of weather and air pollution, and larger land area requirements for commercial-scale applications. These bottlenecks do prevent the expansion of solar energy systems for environmental sustainability.

1.6 CONCLUDING REMARKS WITH IDENTIFIED KNOWLEDGE GAPS FOR FUTURE RESEARCH

The aim of this chapter is to bring all basic aspects of solar energy sources under one umbrella to inform readers about relevant information for research activities, policy frameworks and educational purposes in order to achieve a clean and sustainable environment. Therefore, the information about the need of renewable energy sources, contribution of solar sources to climate change mitigation, fundamental aspects of solar energy technologies, global status of solar energy technologies and highlights of the merits and disadvantages of solar energy systems were analyzed and critically presented. The following points are major conclusions made, along with knowledge gaps identified for future improvements of solar and wind energy systems.

- Environmental pollution due to increasing fossil fuel burning is a critical problem because of industrial revolution, population growth and technological advancements. The emission of CO_2 into the atmosphere creates a lot of environmental problems including a greenhouse effect. Around 31% of the CO_2 level increase has been recorded over the last 200 years, with significant contribution from the burning of fossil fuels to meet increasing energy demand. There are various sources contributing to the increase in CO_2 levels in the atmosphere. Though there are reports available on CO_2 emissions due to various processes, process mapping of contributors of global CO_2 emissions is still a limited task to be completed. This map is highly required in future to identify potential sources of CO_2 emission and to propose viable mitigation strategies for a clean and sustainable future.
- The growth of renewable sources is happening at a significant rate due to the reasons indicated in the previous point. Renewable energy sources are expected to meet 47.7% of the total energy demand of the world in 2040. As sources are abundant in the world, appropriate strategies need to be developed to harvest the maximum amount of energy as possible from these resources. It is evidenced from the literature that most of these resources are still wasted and there is a need to develop and expedite research innovations to fill this knowledge gap.
- Solar energy is used for electricity generation through PV technology using semiconductors. In addition to this, it is used for concentrated thermal systems and direct uses like cooking and heating of spaces. However, wind energy is mainly used for generating electricity using wind turbines. It is obvious that solar and wind energy systems have distinct utility phases. However, availability of these two resources are highly affected by environmental factors and locational differences. In order to maintain continuity in the supply system from these two resources, an appropriate integration (solar-wind hybrid systems) should be developed to maintain the uniformity in the supply line for better consumer preferences. There are many research activities reporting on the performance reliability of solar energy systems and wind energy systems in isolation. However, further studies are required to make a comprehensive evaluation of solar-wind

hybrid systems for better applications since there is a lack of scientific understating in this sector.

- The costs for manufacturing solar systems are extremely high. Solar cells are made to harvest as much solar energy as possible. However, its efficiency has not gone beyond 30% so far. It is interesting that researchers are now working on hybrid perovskite solar cells and quantum dots to increase the efficiency further. However, the efficiency of solar cells has not been increased significantly. Therefore, there is an opening in the manufacturing of novel solar cells to fill this reach gap. Solar energy systems have their own cons and pros. Materials used in the manufacturing of solar systems are very toxic in nature. There is still a limited facility available to manage these toxic waste materials after use. Moreover, little has been known so far about specific impacts of such toxic materials on environmental processes. For example, heavy metals and radioactive substances used in the manufacturing of PV cells can significantly damage the health of living organisms. It is therefore required to investigate in-depth the general and specific effects of commonly used toxic manufacturing materials on living organisms. Moreover, cost-effective systems are needed for managing such toxic waste materials generated by the solar energy sector in order to maintain environmental sustainability.
- Industrial processes use solar thermal energy effectively. However, proper research activities are needed to develop appropriate protocols for the promotion of solar thermal energy for various other industrial applications.
- Though solar energy systems are reliable sources of renewable energy, it is highly necessary to identify their risks associated with environmental processes. For example, the effect of heat waves on the population dynamics of flora and fauna in areas where highly active concentrated solar power systems are used has not been studied well yet. Hence, highly structured research activities are needed to be conducted to fill this knowledge gap for the better application of solar and wind energy systems without damaging natural environmental processes.

As discussed, solar energy systems are highly reliable sources of renewable energy in the world. This short critical review made a comprehensive overview of such resources in terms of global status, fundamentals, advantages and disadvantages and highlights of knowledge as identified for future researchers in order to achieve a clean and sustainable environment for future generations. This short critical review will definitely be useful for researchers from various disciplines, policy makers and stakeholders to make necessary steps to gain as many benefits as possible from these renewable energy sources of nature's gifts.

ACKNOWLEDGMENTS

The author would like to acknowledge the Department of Agricultural Engineering, Faculty of Agriculture, University of Jaffna, Sri Lanka, for providing necessary resources for completing this work successfully.

REFERENCES

1. Kannan, N. and D. Vakeesan, *Solar energy for future world:—A review*. Renewable and Sustainable Energy Reviews, 2016. **62**: p. 1092–1105.
2. Demirbas, A., *Effects of moisture and hydrogen content on the heating value of fuels*. Energy Sources, Part A: Recovery, Utilization, and Environmental Effects, 2007. **29**(7): p. 649–655.
3. Dincer, I., *Environmental issues: II-potential solutions*. Energy Sources, 2001. **23**(1): p. 83–92.
4. Bilgen, S., K. Kaygusuz, and A. Sari, *Renewable energy for a clean and sustainable future*. Energy Sources, 2004. **26**(12): p. 1119–1129.
5. Fridleifsson, I.B., *Geothermal energy for the benefit of the people*. Renewable and Sustainable energy Reviews, 2001. **5**(3): p. 299–312.
6. Farhad, S., M. Saffar-Avval, and M. Younessi-Sinaki, *Efficient design of feedwater heaters network in steam power plants using pinch technology and exergy analysis*. International Journal of Energy Research, 2008. **32**(1): p. 1–11.
7. Sims, R.E.H., *Bioenergy to mitigate for climate change and meet the needs of society, the economy and the environment*. Mitigation and Adaptation Strategies for Global Change, 2003. **8**(4): p. 349–370.
8. Blaschke, T., M. Biberacher, S. Gadocha, and I. Schardinger, *'Energy landscapes': Meeting energy demands and human aspirations*. Biomass and Bioenergy, 2013. **55**: p. 3–16.
9. Rutovitz, J., E. Dominish, and J. Downes, *Calculating global energy sector jobs: 2015 methodology*. 2015.
10. Arif, M.S., *Residential solar panels and their impact on the reduction of carbon emissions*. University of California, Berkeley. Retrieved from https://nature.berkeley.edu/classes/es196/projects/2013final/ArifM_2013.pdf, 2013.
11. Reddy, K.G., T.G. Deepak, G.S. Anjusree, S. Thomas, S. Vadukumpully, K.R.V. Subramanian, S.V. Nair, and A.S. Nair, *On global energy scenario, dye-sensitized solar cells and the promise of nanotechnology*. Physical Chemistry Chemical Physics, 2014. **16**(15): p. 6838–6858.
12. Murdock, H.E., D. Gibb, T. Andre, J.L. Sawin, A. Brown, L. Ranalder, U. Collier, C. Dent, B. Epp, and C. Hareesh Kumar, *Renewables 2021—Global status report*. 2021.
13. Boretti, A., *Cost and production of solar thermal and solar photovoltaics power plants in the United States*. Renewable Energy Focus, 2018. **26**: p. 93–99.
14. Tingem, M. and M. Rivington, *Adaptation for crop agriculture to climate change in Cameroon: turning on the heat*. Mitigation and Adaptation Strategies for Global Change, 2009. **14**(2): p. 153–168.
15. Kobayashi, T., K. Ishiguro, T. Nakajima, H.Y. Kim, M. Okada, and K. Kobayashi, *Haines A, Kovats RS, Campbell-Lendrum D, Corvalan C. 2006. Climate*. Health, 2007. **120**: p. 585–596.
16. Dincer, I., *Energy and environmental impacts: present and future perspectives*. Energy Sources, 1998. **20**(4–5): p. 427–453.
17. Panwar, N.L., S.C. Kaushik, and S. Kothari, *Role of renewable energy sources in environmental protection: A review*. Renewable and Sustainable Energy Reviews, 2011. **15**(3): p. 1513–1524.
18. Zakhidov, R.A., *Central Asian countries energy system and role of renewable energy sources*. Applied Solar Energy, 2008. **44**(3): p. 218–223.
19. Farjana, S.H., N. Huda, M.A.P. Mahmud, and R. Saidur, *Solar process heat in industrial systems—A global review*. Renewable and Sustainable Energy Reviews, 2018. **82**: p. 2270–2286.
20. Suresh, N.S. and B.S. Rao, *Solar energy for process heating: a case study of select Indian industries*. Journal of Cleaner Production, 2017. **151**: p. 439–451.

21. Elbreki, A.M., M.A. Alghoul, A.N. Al-Shamani, A.A. Ammar, B. Yegani, A.M. Aboghrara, M.H. Rusaln, and K. Sopian, *The role of climatic-design-operational parameters on combined PV/T collector performance: A critical review.* Renewable and Sustainable Energy Reviews, 2016. **57**: p. 602–647.

22. Kabir, E., P. Kumar, S. Kumar, A.A. Adelodun, and K.-H. Kim, *Solar energy: Potential and future prospects.* Renewable and Sustainable Energy Reviews, 2018. **82**: p. 894–900.

23. Sun, D. and L. Wang, *Research on heat transfer performance of passive solar collector-storage wall system with phase change materials.* Energy and Buildings, 2016. **119**: p. 183–188.

24. Herrando, M. and C.N. Markides, *Hybrid PV and solar-thermal systems for domestic heat and power provision in the UK: Techno-economic considerations.* Applied Energy, 2016. **161**: p. 512–532.

25. Mohanty, P., T. Muneer, E.J. Gago, and Y. Kotak, *Solar radiation fundamentals and PV system components*, in *Solar Photovoltaic System Applications.* 2016: Springer. p. 7–47.

26. Jeon, N.J., J. Lee, J.H. Noh, M.K. Nazeeruddin, M. Grätzel, and S.I. Seok, *Efficient inorganic-organic hybrid perovskite solar cells based on pyrene arylamine derivatives as hole-transporting materials.* Journal of the American Chemical Society, 2013. **135**(51): p. 19087–19090.

27. Alharbi, F.H. and S. Kais, *Theoretical limits of photovoltaics efficiency and possible improvements by intuitive approaches learned from photosynthesis and quantum coherence.* Renewable and Sustainable Energy Reviews, 2015. **43**: p. 1073–1089.

28. Biermann, E., M.Y. Grupp, and R. Palmer, *Solar cooker acceptance in South Africa: results of a comparative field-test.* Solar Energy, 1999. **66**(6): p. 401–407.

29. Kalogirou, S., *Thermal performance, economic and environmental life cycle analysis of thermosiphon solar water heaters.* Solar Energy, 2009. **83**(1): p. 39–48.

30. Kumar, A. and T.C. Kandpal, *CO2 emissions mitigation potential of some renewable energy technologies in India.* Energy Sources, Part A, 2007. **29**(13): p. 1203–1214.

31. Romero, M. and J. González-Aguilar, *Solar thermal CSP technology.* Wiley Interdisciplinary Reviews: Energy and Environment, 2014. **3**(1): p. 42–59.

32. Singh, S., T.S. Bhatti, and D.P. Kothari, *Indian scenario of wind energy: problems and solutions.* Energy Sources, 2004. **26**(9): p. 811–819.

33. Pryor, S.C. and R.J. Barthelmie, *Climate change impacts on wind energy: A review.* Renewable and Sustainable Energy Reviews, 2010. **14**(1): p. 430–437.

34. Nagrath, I.J., *Power system engineering.* 2007: Tata McGraw Hill India.

35. Ozgener, O., K. Ulgen, and A. Hepbasli, *Wind and wave power potential.* Energy Sources, 2004. **26**(9): p. 891–901.

36. Balat, M., *A review of modern wind turbine technology.* Energy Sources, Part A, 2009. **31**(17): p. 1561–1572.

37. Balat, M., *Usage of energy sources and environmental problems.* Energy Exploration & Exploitation, 2005. **23**(2): p. 141–167.

38. Holm-Nielsen, J. and E.A. Ehimen, *Biomass supply chains for bioenergy and biorefining.* 2016: Woodhead Publishing.

39. Hart, M., *Hubris: The troubling science, economics, and politics of climate change.* 2015: Lulu. com.

40. Luqman, M., S.R. Ahmad, S. Khan, U. Ahmad, A. Raza, and F. Akmal, *Estimation of solar energy potential from rooftop of Punjab government servants cooperative housing society Lahore using GIS.* Smart Grid and Renewable Energy, 2015. **6**(05): p. 128.

41. Adaramola, M., *Solar energy: application, economics, and public perception.* 2014: CRC Press.

42. Hang, Q., Z. Jun, Y. Xiao, and C. Junkui, *Prospect of concentrating solar power in China—the sustainable future.* Renewable and Sustainable Energy Reviews, 2008. **12**(9): p. 2505–2514.

43. Hemmeline, C., *Overview of solar energy in Texas-Texas solar market update*. CATEE conference, 2017.

44. Indora, S. and T.C. Kandpal, *Institutional cooking with solar energy: A review*. Renewable and Sustainable Energy Reviews, 2018. **84**: p. 131–154.

45. Kalvig, P. and E. Machacek, *Examining the rare-earth elements (REE) supply-demand balance for future global wind power scenarios*. GEUS Bulletin, 2018. **41**: p. 87–90.

46. Kar, S.K., A. Sharma, and B. Roy, *Solar energy market developments in India*. Renewable and Sustainable Energy Reviews, 2016. **62**: p. 121–133.

47. Marugán, A.P., F.P.G. Márquez, J.M.P. Perez, and D. Ruiz-Hernández, *A survey of artificial neural network in wind energy systems*. Applied Energy, 2018. **228**: p. 1822–1836.

48. Dincer, F., *The analysis on wind energy electricity generation status, potential and policies in the world*. Renewable and Sustainable Energy Reviews, 2011. **15**(9): p. 5135–5142.

49. Görig, M. and C. Breyer, *Energy learning curves of PV systems*. Environmental Progress & Sustainable Energy, 2016. **35**(3): p. 914–923.

50. Jerez, S., I. Tobin, R. Vautard, J.P. Montávez, J.M. López-Romero, F. Thais, B. Bartok, O.B. Christensen, A. Colette, and M. Déqué, *The impact of climate change on photovoltaic power generation in Europe*. Nature Communications, 2015. **6**(1): p. 1–8.

51. Burt, E., P. Orris, and S. Buchanan, *Scientific evidence of health effects from coal use in energy generation*. Chicago and Washington: School of Public Health, University of Illinois and Health Care Without Harm, 2013.

52. Machol, B. and S. Rizk, *Economic value of US fossil fuel electricity health impacts*. Environment International, 2013. **52**: p. 75–80.

53. Radivojević, A.R., T.M. Pavlović, D.D. Milosavljević, A.V. Đorđević, M.A. Pavlović, I.M. Filipović, L.S. Pantić, and M.R. Punišić, *Influence of climate and air pollution on solar energy development in Serbia*. Thermal Science, 2015. **19**(suppl. 2): p. 311–322.

54. Castillo, C.P., F.B. e Silva, and C. Lavalle, *An assessment of the regional potential for solar power generation in EU-28*. Energy policy, 2016. **88**: p. 86–99.

55. Owusu, P.A. and S. Asumadu-Sarkodie, *A review of renewable energy sources, sustainability issues and climate change mitigation*. Cogent Engineering, 2016. **3**(1): p. 1167990.

56. Ahmed, S., A. Mahmood, A. Hasan, G.A.S. Sidhu, and M.F.U. Butt, *A comparative review of China, India and Pakistan renewable energy sectors and sharing opportunities*. Renewable and sustainable Energy reviews, 2016. **57**: p. 216–225.

57. Miller, L.M. and D.W. Keith, *Climatic Impacts of Wind Power*, Joule, *2018*. **2**: p. 2618–2632. 2018.

58. Nazir, M.S., A.J. Mahdi, M. Bilal, H.M. Sohail, N. Ali, and H.M.N. Iqbal, *Environmental impact and pollution-related challenges of renewable wind energy paradigm—a review*. Science of the Total Environment, 2019. **683**: p. 436–444.

2 Low and Medium Temperature Solar Thermal Collectors

V.P. Chandramohan and Pritam Das

CONTENTS

DOI: 10.1201/9781003263326-3

2.1 INTRODUCTION

Solar thermal applications are drawing more attention among researchers and industrialists over the last few decades due to their enormous advantages. Utilization of solar energy is being considered as a potential and viable option for global warming related issues and towards achieving the goal of sustainable development. It is a well-known fact that solar radiation is not constant all day. Hence, the aim of solar energy systems is to capture maximum energy during peak hours and store it efficiently for later use. Thus, solar thermal collectors are the main component in any solar energy system. Collectors are used to convert solar energy into heat for thermal applications or directly into electricity for power generation applications (Sharma *et al.* 2017).

Solar thermal collectors are mainly used in various fields of applications such as residential applications for water heating, space heating, drying of agricultural food products and the desalination process. Also, it is used in low temperature industrial applications such as food, beverage, textile, automobile and pharmaceutical, etc. (Kumar *et al.* 2019). These solar collectors are the alternatives to fossil fuel based conventional resources. In these applications, the outlet fluid temperatures must be higher, so picking a convenient solar collector plays an important role. These collectors are mainly used for warming heat transfer fluids (HTF) such as air, water or oil (Tian and Zhao 2013) in different applications.

The solar collectors are mainly classified into two types, non-concentrating and concentrating (Tian and Zhao 2013, Iparraguirre *et al.* 2016). This classification is done based on concentration ratio (CR) which is the ratio between aperture area or intercepting area to the area of collector.

$$CR = \frac{\text{aperture area}}{\text{area of collector}} \qquad (1)$$

Aperture area is the area where solar radiation enters into the collector and is mentioned in Figure 2.1.

The operating temperature ranges of solar collectors are identified as either low (below 100 °C), medium (up to 400 °C) and high (more than 400 °C) (Kalogirou 2003, 2004). Different industries choose different temperature ranges as per their applications. The lower and medium temperature solar thermal collectors are mainly applicable for various residential and industrial process heating applications whereas high temperature solar thermal collectors are mainly applicable for power generation applications. Most of the industrial process heating applications require a temperature range between 30 to 250 °C (reportedly 60% of thermal energy requirement) (Fernández-García *et al.* 2010, Jradi and Riffat 2014).

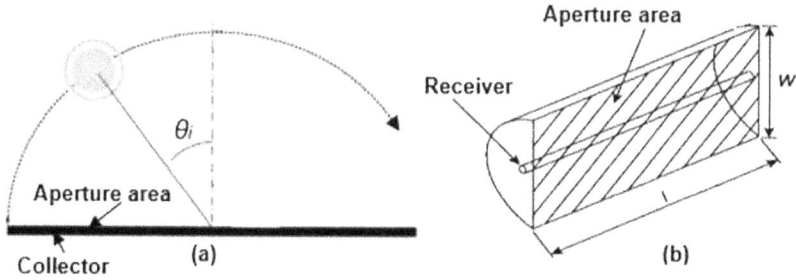

FIGURE 2.1 Aperture area (a) non-concentrating and (b) concentrating solar collectors.

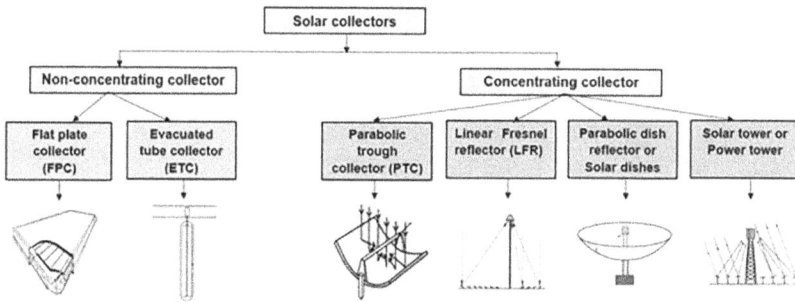

FIGURE 2.2 Classifications of solar collectors.

In this chapter, an effort is made to discuss different kinds of solar collectors and their applications. Later, the solar industrial process heating system and its application to various industries have been discussed in detail. The discussion has been done based on the types of industries, processes, collectors used and temperature ranges, etc. In the end, global status and future outlook are also addressed.

2.2 SOLAR COLLECTORS

The solar collector is one of the energy transfer units which convert solar radiation into heat and transfer heat to flowing fluid, which may be water, air or oil. This conversion process is for solar thermal applications. The heat received by the fluid can be used either for domestic or industrial applications or stored in the thermal energy storage system for later use. In other instances, solar radiation can be directly converted to electricity in photovoltaic applications.

Among the two categories (non-concentrating and concentrating), the flat plate collector (FPC) and evacuated tube collector (ETC) belong to the family of non-concentrating collectors whereas the parabolic trough collector (PTC), linear Fresnel reflector (LFR), solar dish and solar tower belong to the family of concentrating solar collectors (Iparraguirre *et al.* 2016). FPCs and ETCs are mostly used for lower to medium temperature applications. This is due to the same intercepting and absorbing area which results in a CR of 1. The classification of solar collectors is shown in Figure 2.1. The concentrators such as PTC, LFR and solar dishes are used for medium to high temperature range applications because the CR value varies up to 1500. It is worth mentioning here that a higher value of CR, the higher outlet temperature of the fluid (Kalogirou 2003).

2.2.1 NON-CONCENTRATING SOLAR COLLECTOR

Non-concentrating collectors are mostly economic and widely used collectors for various low temperature based residential and industrial applications. These collectors capture both direct beam radiation (straight line rays reach the earth) and diffuse beam radiation (scattered rays), hence a solar tracking system is not necessary. In these types of collectors, the area of the collector is equal to the receiver area resulting in the concentration ratio (CR) being equal to 1 (Kalogirou 2008).

The thermal analysis of non-concentrating solar collectors can be derived under a steady-state condition. The energy balance equation for solar collector yields,

$$q_u = A_a S - q_l \tag{1}$$

Where, q_u is the rate of heat transfer to the working fluid, A_a is the aperture area, S is solar flux per unit effective area of aperture and q_l is the rate of heat loss from the absorber.

The flux absorbed in the absorber plate is,

$$S = I_b r_b (\tau\alpha)_b + \left[I_d r_d + (I_b + I_d) r_r \right] (\tau\alpha)_d \tag{2}$$

Where, I_b and I_d are the beam and diffuse radiations, respectively. r_b, r_d and r_r are the tilt factor for beam, diffuse and reflected radiations, respectively. τ and α are transmissivity of the collector cover and absorptivity of the absorber plate, respectively. The $(\tau\alpha)_b$ and $(\tau\alpha)_d$ are the transmissivity-absorptivity product of solar beam radiation and solar diffused radiation which are falling on the collector cover and absorber plate, respectively.

The collector efficiency (η_i) can be estimated as

$$\eta_i = \frac{useful\ heat\ gain}{radiation\ incident\ on\ the\ collector} = \frac{q_u}{A_{cc} I_t} \tag{3}$$

Where, A_{cc} is the collector cover area (usually 15 to 20% more than the absorber plate surface area) and I_t is the solar flux on the tilted surface. This is also can be estimated as

$$I_t = I_b r_b + I_d r_d + (I_b + I_d) r_r \tag{4}$$

Due to the absence of a tracking system, these category solar collectors are cheaper compared to concentrating types. Its two types such as FPC and ETC are discussed in the following sections.

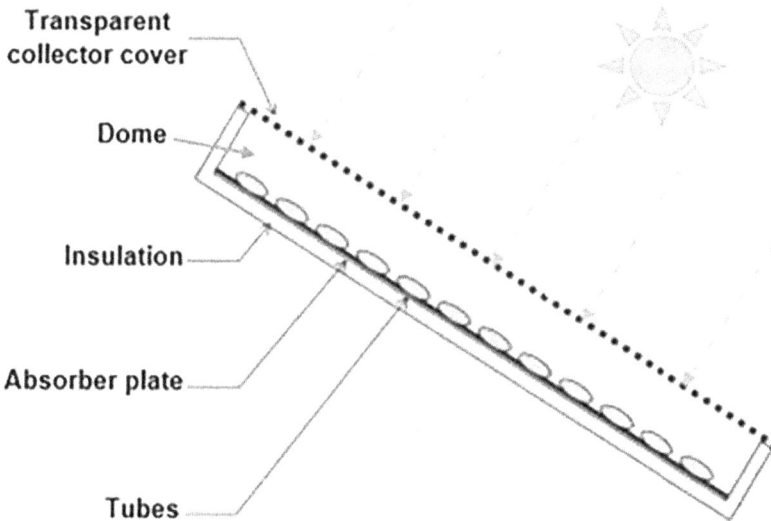

FIGURE 2.3 Schematic layout of flat plate collector (FPC).

Source: Orosz and Dickes (2017), with permission from Elsevier.

2.2.1.1 Flat Plate Collector (FPC)

Flat plate collectors (FPC) consist of an absorber plate, transparent cover and a bunch of tubes welded into the absorber plate to carry fluid as shown in Figure 2.3. Both sides of the FPC are perfectly insulated to avoid conduction heat losses. The solar radiation falls on the transparent collector plate, is transmitted through the collector plate and absorbed by the absorber sheet. A large quantity of absorbed energy is transferred to the tubes that are carrying fluid for various applications or storage. The transparent glass cover also minimizes the energy losses from the absorber plate due to radiation. The transmitted shorter wavelength solar radiation is being trapped inside the dome (as shown in Figure 2.3) as the collector cover acts as an opaque body for longer wavelength radiation radiated from the absorber plate.

FPCs can be classified based on fluid flow and structural design. FPCs are mainly liquid heaters or air heaters (Sakhaei and Valipour 2019). The liquid heater is again classified based on the fluid type (water, nanofluid and heat transfer fluids, HTFs), fluid channel type (pipe and fin, liquid sandwich and semi-liquid sandwich) and fluid flow type (parallel, serpentine and coil). The air heater is classified based on absorber plate design (corrugated, finned, perforated, matrix, V-grooved, porous media, etc.), collector cover design (unglazed, single glazed and double glazed) and fluid flow type (single pass, double pass and cross flow). Apart from that, transpired solar collectors come in the family of FPCs which are the unglazed type. Unglazed solar collectors are used for low temperature applications, even sometimes less than 30 °C. It has no glass cover and is made up of black plastic and hence it can absorb a large part of solar radiation.

The FPCs are placed in a fixed position with an angle of inclination while the direction faces either south for the northern hemisphere or north for the southern hemisphere. The optimum inclination angle should be equal to the latitude of the region or 10 to 20° more or less than the latitude (Kalogirou 2003, Das and Chandramohan 2019). The various types of absorber plates are flat, corrugated plates with tubes/fins attached (Sakhaei and Valipour 2019). Glazing glass is used as collector cover due to its higher potential for transmission of solar radiation. A few other materials such as plastic films, commercially available window glasses are also used as collector covers (Das and Chandramohan 2021). An anti-reflexive coating layer may be used to improve the transmittance of the collector cover further (Kalogirou and Lloyd 1992). To reduce the radiation and convection losses, low cost and higher temperature resistant insulating materials are used in the FPC (Sakhaei and Valipour 2019, Vengadesan and Senthil 2020).

2.2.1.1.1 Mathematical Modeling Involved in FPC Collector

The mathematical modeling of FPC can be comprehensively discussed with the proposed model of Hottel-Whillier-Bliss (Barone *et al.* 2019). In general, the model is discussed with a set of equations for calculation of optical losses, thermal losses and estimation of useful heat gain.

The solar thermal performance of the collector is estimated based on the quantity of solar radiation effectively absorbed by the absorber plate. It depends upon the optical parameters of the considered materials. As already discussed in Section 2.2.1, the absorbed solar radiation is a function of the product of transmissivity and absorptivity ($\tau\alpha$). The absorber plate is covered by the collector cover which is generally considered a glass material, ensuring the transmittance of solar radiation. The transmitted radiation is absorbed by the absorber plate. However, in a practical scenario, multiple reflections of solar beams inside the dome (space between collector cover and absorber plate) improve the actual transmissivity-absorptivity product to 1% higher than its value. The actual transmissivity-absorptivity factor can be estimated as $(\tau\alpha) = 1.01\tau\alpha$.

The rate of heat loss from the absorber is mentioned as

$$q_1 = U_1 A_{ap} \left(T_{ap} - T_{amb} \right) \tag{5}$$

Where, U_1 is the overall loss coefficient, A_{ap} is the area of the absorber plate and T_{ap} and T_{amb} are the average temperature of the absorber plate and ambient air, respectively. In general, the U_1 varies in the range of 2 to 10 W/m²K (Barone *et al.* 2019).

$$U_1 = U_s + U_b + U_t \tag{6}$$

The U_s, U_b and U_t are side, bottom and top loss coefficients, respectively. The side and bottom loss coefficients can be estimated using,

$$U_s = \frac{(L + W)tK_1}{LW\delta_s} \tag{7}$$

$$U_b = \frac{K_i}{\delta_b} \tag{8}$$

Where, K_i is the conductivity of the insulation, δ_s and δ_b are side and back insulation thickness, respectively, and t is the total height of the collector casing. In general, the typical values of side loss and bottom loss coefficients vary in the range of 1 to 2.5 W/m²K (Barone et al. 2019).

To determine the U_t, it is assumed that the possible heat losses from the absorber plate to the collector and from the collector to ambient are the same.

Hence the amount of heat loss from the top surface would be

$$q_t = U_t A_{ap} \left(T_{ap} - T_{amb} \right) \tag{9}$$

The amount of heat loss from the absorber plate to the collector cover can be estimated by,

$$q_t = h_{ap-cc} A_{ap} \left(T_{ap} - T_{cc} \right) + \frac{A_{ap}\sigma \left(T_{ap}^4 - T_{cc}^4 \right)}{\left(\dfrac{1}{\varepsilon_{ap}} + \dfrac{1}{\varepsilon_{cc}} - 1 \right)} \tag{10}$$

Where, h_{ap-cc} is the convective heat transfer coefficient between the absorber plate and collector cover. T_{ap} and T_{cc} are the temperatures attained by the absorber plate and collector cover, respectively. ε_{ap} and ε_{cc} are the emissivity of the absorber plate and collector cover, respectively. σ is the Stefan-Boltzmann constant.

Similarly, the amount of heat loss from the collector cover to ambient air can be estimated by,

$$q_t = h_{cc-amb} A_{ap} \left(T_{cc} - T_{amb} \right) + A_{ap}\sigma\varepsilon_{cc} \left(T_{cc}^4 - T_{sky}^4 \right) \tag{11}$$

Where, T_{sky} is the effective temperature of the sky with which radiative heat exchange takes place and h_{cc-amb} is the heat transfer coefficient in between the collector and ambient.

Equating the three Eqs. (9–11), U_t can be estimated, and substituting the same value in Eq. (6), the value of U_l can be found.

Once the total heat loss is estimated using Eq. (5), the useful heat gain can be estimated as already mentioned in Eq. (1).

The useful heat gain accumulated after all losses can be expressed through

$$q_u = A_a \left[S.(\tau\alpha) - U_l \left(T_{ap} - T_{amb} \right) \right] \tag{12}$$

This useful heat gain is utilized to heat the working fluid, thus,

$$q_u = A_a \left[S.(\tau\alpha) - U_1 \left(T_{ap} - T_{amb} \right) \right] = \dot{m}c_p \left(T_{in} - T_{out} \right) \tag{13}$$

Where, T_{in} and T_{out} are the inlet and outlet temperatures of the working fluid, respectively, [Insert Equation Here] is mass flow rate and c_p is the specific heat of the fluid.

To link the useful heat gain with the fluid inlet temperature, another term is introduced, a collector heat removal factor (F_R). It is the ratio of q_u to the heat gain which would occur if the absorber plate temperature is at the temperature of the fluid inlet.

$$F_R = \frac{q_u}{A_a \left[S(\tau\alpha) - U_1 \left(T_{in} - T_{amb} \right) \right]} \tag{14}$$

$$\Rightarrow q_u = F_R A_\sigma \left[S(\tau\alpha) - U_1 \left(T_{in} - T_{amb} \right) \right] \tag{15}$$

The thermal efficiency of the collector (η) is estimated using,

$$\eta = F_R \left[(\tau\alpha) - \frac{U_1 \left(T_{in} - T_{amb} \right)}{S} \right] \tag{16}$$

2.2.1.1.2 Applications
The potential applications of FPC are drying of agricultural food products, residential applications and low temperature industrial process heating. These are mostly used specifically for lower temperature applications of below 100 °C (Kalogirou 2004, Sakhaei and Valipour 2019). The FPCs show enormous advantage such as less maintenance cost, being easy to manufacture and install, collecting both direct and diffuse beams, and no tracking system is required. Water or air is generally used as a working fluid; during the winter season or in a colder climate, antifreeze fluid is mixed with water to avoid freezing.

2.2.1.2 Evacuated Tube Collector (ETC)

An evacuated tube collector (ETC) consists of a set of modular tubes made up of glass. Each glass tube contains a heat pipe attached with absorber fins as shown in Figure 2.4. The air between each glass tube is removed or evacuated (Kalogirou 2004); therefore, the conduction and convection heat losses are minimized. There are mainly three types of ETC, such as (i) water-in glass ETC, (ii) evacuated tube heat pipe collector and (iii) U-type ETC (Sabiha *et al.* 2015). The important design parameter of the ETC is the design of the absorber tube. Various designs of absorber tubes are available such as finned tube (Case-1), U-tube welded inside the circular fin (Case-2), U-tube welded in the copper plate (Case-3) and a U-tube placed inside the rectangular tunnel (Case-4). These designs are illustrated in Figure 2.5. U-tube welded inside the circular fin is the best performing system compared to all other existing designs. However, the double glass tubular design is one of the most efficient absorbers as it can absorb solar radiation from different directions (Sabiha *et al.* 2015).

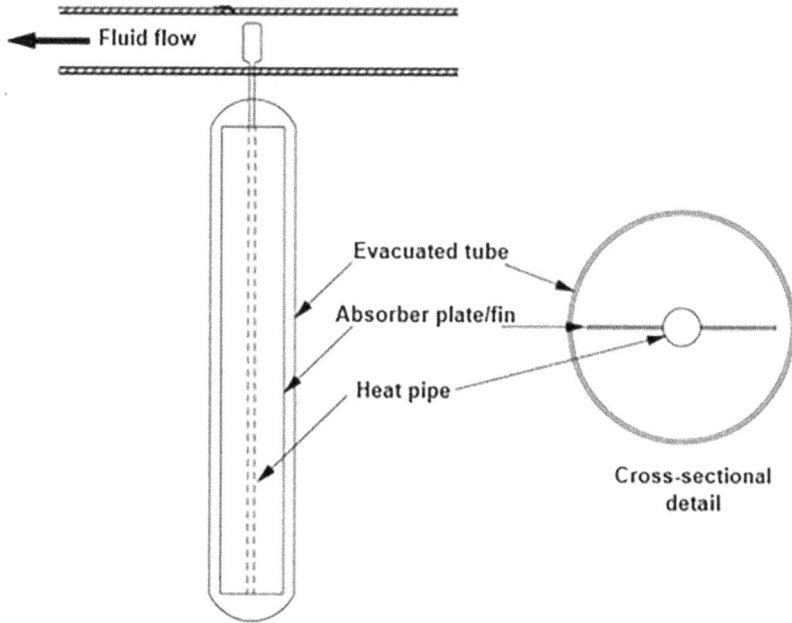

FIGURE 2.4   Schematic layout of evacuated tube collector (ETC).

Source: Kalogirou (2004), rights received from Elsevier.

The lifetime of 'vacuum maintenance' varies from collector to collector which is in the range of 5 to 15 years (Bhatia 2014). ETC is more favorable for use in colder climates or cloudy weather because the energy output in ETC is more compared to flat plate collectors, but the flat plate collectors are more efficient during sunshine hours. The fluid (methanol) inside the heat pipe undergoes an evaporating-condensing cycle to transfer heat at higher efficiency (Kumar *et al.* 2019). However, the heat pipe protects from overheating and freezing. This self-temperature control of heat pipes is a unique inherent feature of ETCs.

2.2.1.2.1 Mathematical Modeling Involved in ETC

Similar to FPC, the useful heat gain in ETC can also be estimated using Eq. (15).

$$q_u = F_R A_a \left[S.(\tau\alpha) - U_l \left(T_m - T_{amb} \right) \right] \tag{17}$$

Where, T_m is the mean temperature of working fluid which is generally considered as the average of T_{in} and T_{out}.

The instantaneous efficiency of the collector can be expressed using,

$$\eta = \eta_0 - a \frac{\left(T_m - T_{amb} \right)}{S} - b \left(\frac{T_m - T_{amb}}{S} \right)^2 \tag{18}$$

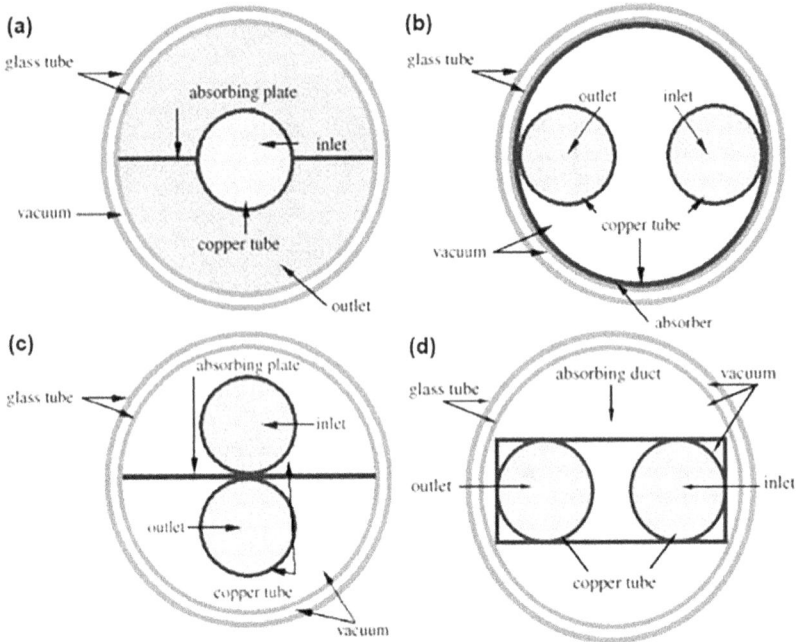

FIGURE 2.5 Absorber tube of ETC collector; cut-section of (a) Case-1, (b) Case-2, (c) Case-3 and (d) Case-4.

Source: Sabiha *et al.* (2015), with permission from Elsevier.

The efficiency of the collector is termed as instantaneous efficiency as it accounts only the instantaneous parameters (climatic conditions such as solar insolation, ambient temperature, wind speed, etc.) for the estimation of the efficiency. Where, η_o, a and b are constants, which can be evaluated either experimentally or analytically and in general, these values specified by the manufacturer are 0.8, 1.2 and 0.007, respectively (Hayek *et al.* 2011).

2.2.1.2.2 Applications

The range of temperature can be achieved up to 200 °C in ETC (Kalogirou 2008, Sharma *et al.* 2017). The performance of ETCs can be enhanced (for achieving the higher temperature range) by providing selective surface coating (Kalogirou and Lloyd 1992). At present, ETCs are very much useful for residential applications such as domestic water heating, space heating, space cooling, as well as medium temperature range industrial process heating (Jradi and Riffat 2014, Kumar *et al.* 2019). Heat pipe ETCs and U-tube glass ETCs are the most suitable collectors for domestic applications due to their easy design and flexibility for high pressure conditions. Compared to standard flat plate collectors (FPC), ETCs offer significant advantages such as being less prone to corrosion, occupying less space, performing excellently under extreme cold conditions, having greater durability and being easy to maintain

(Jradi and Riffat 2014). ETCs also can capture more sunlight as the area of surface exposed to the sun is higher (Bhatia 2014).

2.2.2 Concentrating Solar Collector

The outlet fluid temperature of solar systems can be increased with the application of concentrating solar collectors. These types of collectors utilize the small concentrating area to capture a greater quantum of solar radiation (Kalogirou 2008, Jradi and Riffat 2014). These types are used for various mid and high temperature ranges of applications. PTCs, LFRs, parabolic dish reflectors and solar towers come under the category of concentrating types of solar collectors. These solar collectors offer an enormous advantage over non-concentrating types of solar collectors such as,

1. The working fluid can achieve higher temperature and offers higher thermal efficiency compared to typical FPCs.
2. Less surface area is required for the receiver of concentrating type collectors.

(Kalogirou and Lloyd 1992)

The thermal analysis of concentrating solar collectors can also be derived using similar equations as already discussed for non-concentrating collectors (Section 2.2.1).

The rate of heat loss from the absorber can be estimated using Eq. (1).

If both Eqs. (1) and (5) are combined,

$$q_u = A_a \left[S - \frac{U_1}{CR} \left(T_{ap} - T_{amb} \right) \right] \tag{19}$$

Where, CR is the concentration ratio $= A_d/A_{ap}$.

Selective coating and vacuum insulations are required to avoid heat loss from the system. Concentrators or absorbers can be any shape such as cylindrical, parabolic, continuous or segmented, whereas receivers can be convex, concave, flat, cylindrical or covered with glazing, etc. The concentration ratio (CR) can be more than 1 to 1500 (Kalogirou 2008). The higher CR means the exit temperature from the collector is higher. However, the positioning, optical properties and precision play a major role in the CR of the solar collectors. Various kinds of solar collectors are explained in the following sections.

2.2.2.1 Parabolic Trough Collector (PTC)

A parabolic trough collector (PTC) is the most progressive solar concept for various process heating applications. It consists of a collector (also called a trough or reflector) which focuses the sun rays on a cylindrical/tubular absorber by tracking the sun (single axis tracking) as shown in Figure 2.6. The tracking mechanism should be a reliable one and it should follow the sun with a certain precision over the day and return to its original position in the post sunset hours. Additionally, a tracking

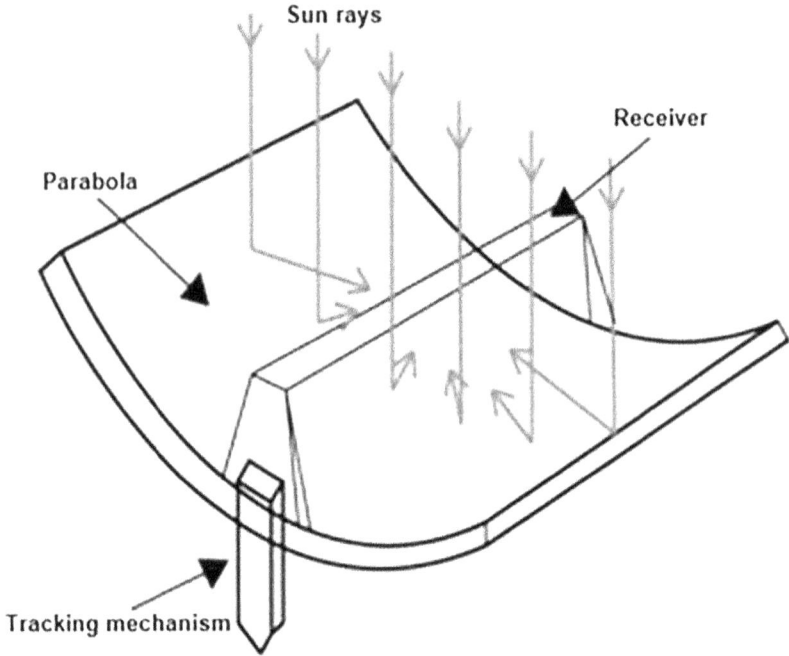

FIGURE 2.6 Schematic layout of a parabolic trough collector (PTC).

Source: Kalogirou (2004), with permission from Elsevier.

mechanism is also used to protect the collector from windy and overheating conditions (Kalogirou 2008, Tian and Zhao 2013). The absorber tube has a glass covering.
The space (annular) between the absorber tube and glass covering has been evacuated to reduce the heat loss due to convection. Mostly, copper is used as the absorber
tube metal because of its higher storage capacity (Kalogirou 2004).

The glass covering must be kept clean and a selective coating or anti-reflective
coating must be done to improve the transmissivity. A heat transfer fluid (HTF) is
pumped through the absorber tube. The HTF gets heated up by absorbing heat from
the walls of the absorber tube through convection; the walls in turn are heated by
concentrated solar flux. The HTF transfers the heat to the water so that steam is
produced. Then the steam is allowed to expand in a turbine for producing electricity.
The HTF, after exchanging heat with steam, returns to the PTC module to be heated
again in a closed loop.

Several HTFs are used in the PTC system, namely water or steam, synthetic oil,
molten salt, nanofluids, etc. PTCs can be adopted in different fields of applications. In
general, they can be utilized for industrial process heating (60–400 °C) applications
such as desalination, water heating, space heating, etc. Apart from that, PTC can also
be used for power generation applications as it has enormous potential for carbon-free
power generation (Tian and Zhao 2013). In some instances, the fossil fuel plants can
be coupled with PTC so that the CO_2 emission is reduced (Kalogirou *et al.* 1994).

The integration of a thermal energy storage system also improves the system run time because it can run even in post sunshine hours. For this purpose, nano-phase change materials (PCM with nano powders), water and molten salt, etc. can be used as energy storage materials (Nawsud *et al.* 2022).

2.2.2.1.1 Mathematical Modeling on PTC

For the PTC, the useful heat gain can be estimated using,

$$q_{u} = S_{b}\eta_{o}A_{a} - A_{r}U_{l}\left(T_{r} - T_{amb}\right) \tag{20}$$

Where, S_{b} is beam radiation, A_{a} and A_{r} are the aperture and receiver areas, respectively. T_{r} is the receiver temperature.

The possible useful thermal power and efficiency are estimated using,

$$q_{u} = F_{R}\left[S_{b}\eta_{o}A_{a} - A_{r}U_{l}\left(T_{r} - T_{amb}\right)\right] \tag{21}$$

$$\eta = F_{R}\left[\eta_{o} - U_{l}\left(\frac{T_{r} - T_{amb}}{S \times CR}\right)\right] \tag{22}$$

2.2.2.1.2 Advantages of PTCs

The major advantage of PTCs is that they are one of the commercially proven solar energy technologies for various applications. Furthermore, they have a longer life span as the working temperature range is low to moderate. The PTC system also lowers emissions and significantly contributes towards carbon mitigation (the effort of emission reduction). Whereas, the reduction in the concentration of CO_2 emission is termed CO_2 mitigation. However, the sun tracking system is necessary for PTCs to have an enhanced potential of harnessing solar energy hence the cost associated with maintenance is also increased compared to other collectors (Kalogirou 2004).

2.2.2.2 Linear Fresnel Reflector (LFR)

A LFR system includes a series of long, narrow and curvature/flat mirrors that focus light to a linearly placed absorber as shown in Figure 2.7. The concentration ratio varies between 10 to 40 and sometimes can be achieved up to 100 using stationary secondary collectors (Kalogirou 2008, Hess 2016). The operating temperature range of the LFR is up to 450 °C (Mills 2012). The design of LFR is based on the principle of power tower and PTC. The LFR is similar to PTC but contains a fixed receiver with more mirrors for tracking the sunlight. Due to the similar concentration ratio and operating temperature range, similar HTFs for PTCs can be applicable for LFRs (Kalogirou 2008). Possible applications of a few other HTFs are liquid sodium, supercritical carbon dioxide, etc. as they create a higher working temperature range (Tian and Zhao 2013). Molten salt can be also used as HTF due to its properties as a storage medium. Water and all PCMs also can be used as an energy storage medium (Tian and Zhao 2013). The receiver acts a major impact in the performance of LFR.

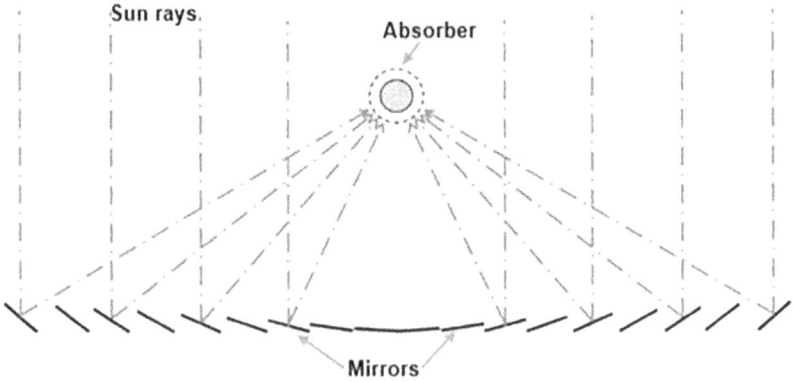

FIGURE 2.7   Layout of linear Fresnel reflector (LFR).

Source: Orosz and Dickes (2017), with permission from Elsevier.

Many studies reported increasing the collected heat by increasing optical efficiency (Kalogirou 2004, Singh *et al.* 2010, Barone *et al.* 2019). It can be achieved by using an array of cylinders (Kalogirou 2004, Barone *et al.* 2019) or making the shape of the receiver as trapezoidal (Singh *et al.* 2010).

2.2.2.2.1 Mathematical Modeling Involved in LFR

For LFR, the useful heat gain would be estimated using the same Eq. (1).

The rate of energy absorbed by the absorber tube can be estimated using,

$$q_{ap} = \eta_o \sum_{i=1}^{N} \left[\left(WL - A_s - A_b \right) S_b cos\theta_i \right] \tag{23}$$

Where, η_o is optical efficiency, θ is the angle between solar beam flux and normal to the plane. In the previous equation, the term "$(WL - A_s - A_b)$" refers to the effective area from which reflected radiation reaches the receiver. A_s and A_b are termed as shaded and blocked areas, respectively. L and W are the length and width of the collector, respectively.

A part of the energy absorbed is lost from the absorber tube to the environment, hence the heat loss rate can be expressed using,

$$q_l = U_1 A_{ap} \left(T_m - T_{amb} \right)$$
$$\Rightarrow q_l = \frac{U_1}{CR} \left(T_m - T_{amb} \right) nWL \tag{24}$$

The instantaneous efficiency of the collector on beam radiation is given by

$$\eta_i = \frac{q_u}{WLI_b \sum_{i=1}^{n} cos\theta_i} \tag{25}$$

2.2.2.2.2 Advantages

The system offers various advantages such as higher reliability, low capital cost, flexibility in design and higher robustness. The mirrors are also attached close to the ground hence less requirement of the structures (Kalogirou 2004). One major difficulty in the LFR system is the shading and blocking between successive reflectors (Kalogirou 2008). It can be avoided by enhancing the altitude of absorber towers but it translates into higher investment costs.

2.2.2.3 Parabolic Dish or Solar Dish

A parabolic dish reflector is otherwise known as a parabolic dish or solar dish and is mentioned in Figure 2.8. The main elements are parabolic reflector, parabolic solar receiver and solar tracking system. The shape of the dish is paraboloid where the solar radiation is focused into one focal point. The dish is supported by a structure and piping is done to carry HTFs to the receiver and the storage unit. To maintain the focus, the dish has to track the sun using a two axis or double axis tracking mechanism (Hess 2016). These collectors operate with a higher concentration ratio (100–1000) and the outlet temperature range of the fluid up to 1000 °C or more (Hess 2016).

Point focus systems (such as solar dishes and towers) have lower heat loss than line focus systems (such as LFR and PTC) and hence the concentration ratio is higher. In this system, the solar radiation received by the receiver is converted into thermal energy. This heat energy is transferred to the circulating fluids or HTFs. Sometimes the heat can be directly converted to electrical energy by coupling a generator to the

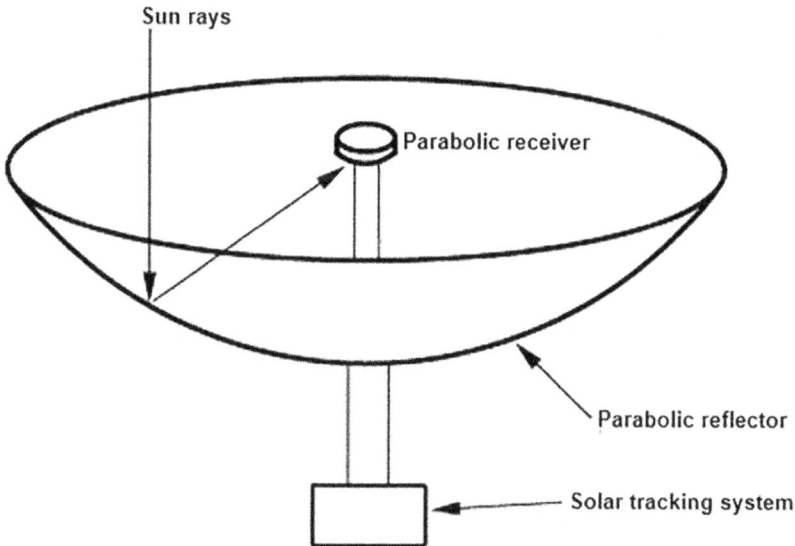

FIGURE 2.8 Schematic layout of parabolic dish or solar dish.

Source: Kalogirou (2004), with permission from Elsevier.

receiver. Or in some cases, the heat is directly transferred through the pipes to the central power conversion system.

Special attention must be taken to the piping and pumping requirements of this system as it involves heat transfer in HTFs and heat losses associated with it. Various parameters must be taken care of while discussing the design aspect of the parabolic dish reflector. They are the material of the reflector, solar radiation data, shape and size of the concentrator and receiver, focal point, focal length of the parabolic dish, etc. (Hafez *et al*. 2016).

Due to the higher concentration ratio and temperature range, these collectors can be used for commercial-scale power generation applications. However, due to the absence of a storage system, the system can be integrated into fossil fuel-based plants for a partial reduction in CO_2 emission (Kalogirou 2004).

2.2.2.4 Solar Tower

The solar tower is surrounded by a large number of heliostats as depicted in Figure 2.9. Heliostats or flat mirror plates are the tracking mirrors used to intensify the solar radiation into the receiver of the tower (Tian and Zhao 2013). The concentrated heat energy is transferred to the circulating HTFs for direct use or can be stored for future trials. The other names of the solar tower are power tower, heliostat field collector or central receiver collector. The heliostats are individual elements built in the ground resulting in increasing reflective area. Due to the higher concentration ratio, the solar tower offers higher efficiency and is thus suitable for high temperature applications. There is some loss in optical efficiency due to the presence of the number of heliostats instead of a continuous collector but it produces better overall collector efficiency and cost-effectiveness.

These collectors operate with a higher concentration ratio (100–1500) and the outlet temperature range of the fluid is up to 2000 °C or more (Kalogirou 2008). The HTFs used are air, water or steam. Liquid sodium, molten salt may also be used

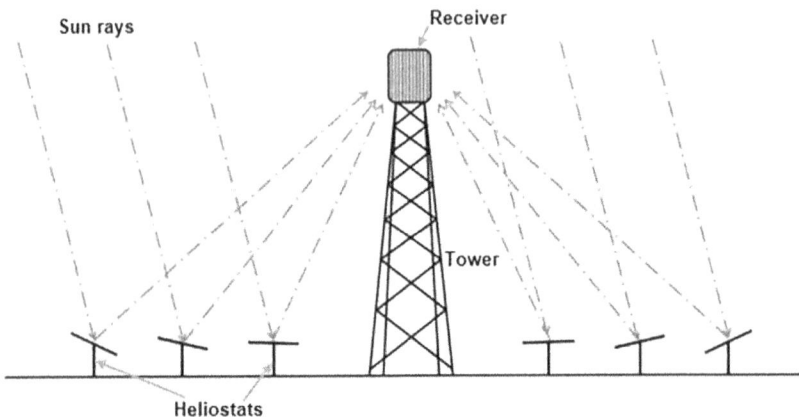

FIGURE 2.9 Schematic layout of the solar tower.

Source: Orosz and Dickes (2017), with permission from Elsevier.

TABLE 2.1
Comparative Assessment of Different Types of Solar Thermal Collectors

Solar Concentrating Technology	FPC	ETC	PTC	LFR	Solar Dishes	Solar Towers
Type of Absorber	Flat	Flat	Tubular	Tubular	Point	Point
Motion	Stationary	Stationary	Single-axis tracking		Two-axis tracking	
Working Fluid or HTFs	Water, air, oil		Water, steam, oil	Water, steam	Air	Water, steam, air
Output Temperature (°C)	30–80	50–200	60–400	100–450	Up to 1000	Up to 2000
Optical Efficiency (%)	30–45	50–65	70–80	Up to 70	75–85	Up to 70
Thermal Energy Storage	PCM, water, molten salt			PCM, Water	–	Concrete, PCM, molten salt
Concentration Ratio	1	1	15–45	10–40	100–1000	100–1500
Potential Applications	Drying of agricultural food products, low temperature Industrial process heating	Space heating or cooling, medium range industrial process heating	Medium range industrial process heating	Medium range industrial process	High temperature industrial process heating and power generation applications	

Note: FPC—flat plate collector, ETC—evacuated tube collector, PTC—parabolic trough collector, LFR—linear Fresnel reflector, HTF—heat transfer fluid.

Source: Kalogirou (2003. 2004), Hess (2016), Solar thermal I Ipieca (2022).

as HTFs due to their storage capability. Concrete or oil mixed with crushed rock; ceramic bricks also can be used as energy storage materials for solar towers (Tian and Zhao 2013). For higher power applications (in MW scale), solar towers can be used (Kalogirou 2004). The central receiver system can also be integrated with other fossil fuel-based plants so that the system works with hybrid operations and can be used in different industrial process heating applications.

A comparative assessment of different types of solar thermal collectors is made by considering their working fluids, temperature range, efficiency, concentration ratio, potential applications, etc. and tabulated in Table 2.1. The absorber plate used in FPC and ETC is flat, whereas for PTC and LFR, it is tubular. Water, air or oil is mostly used as HTFs for major types of collectors. PCM or water is mostly used as thermal energy storage material, whereas molten salt is also used for higher temperature applications. Among various types, as already mentioned, FPC and ETC are for low temperature (up to 200 °C) industrial process heating applications, whereas PTC and LFR are meant for medium temperature (up to 400 °C) industrial process heating applications. The detailed applications have been discussed in a later section.

2.3 GLOBAL POTENTIAL, WORLDWIDE STATUS AND COST ANALYSIS

The global potential, worldwide status and cost analysis of various solar industrial process heating applications can be discussed in this section. The first sub-section mentions various industrial process heating applications that are carried out across the globe. Additionally, various solar thermal collectors involved in different processes are also discussed. In the second sub-section, the research and development work carried out across the globe for improving the process parameters are explained. Finally, the cost analysis of the solar industrial process heating applications is discussed.

2.3.1 GLOBAL POTENTIAL OF SOLAR INDUSTRIAL PROCESS HEATING

Solar industrial process heating is widely used in several fields of applications. Almost all countries across the globe are utilizing solar energy for their present energy demands. It is used in various fields of applications such as food processing, automobile, pulp and paper, chemical and pharma, textile, cement, steel, leather and rubber industries (Jradi and Riffat 2014, Sharma *et al.* 2017, Ramaiah and Shekar 2018). Most of the industrial sectors use solar energy for water heating, room heating and generating steam for various operations. The operating temperature range can be easily attained with the help of various low and medium temperature solar collectors such as FPCs, ETCs, PTCs and LFRs. Different countries use solar energy for industrial heating applications as mentioned in Table 2.2.

The capacity needed for large solar thermal systems for commercial and residential buildings is identified as 299 MW by the end of 2020. The larger shares of projects are from countries like China, followed by France, Greece, Turkey, etc. Similarly, the total established plants' capacity is 1410 MW. The projects are from Denmark,

TABLE 2.2
Solar Industrial Process Heating Reported in Various Plants and Countries

Country	Potential Applications	Operating Temperature (°C)	Type of Collector Used	Working Medium or HTFs	Collector Area (m²)
Spain, 2001	Food processing	20–300	PTC	Steam, water	8,000,000–10,s,000
Portugal, 2001	Food processing	40–188	PTC/non-selective flat plate	Steam, water	1,900,000–2,500,000
USA, 2001	Textile industry	55–105	PTC	Steam, water	725–10,000
Italy, Switzerland France, 2001	Food processing	80–120	PTC/ETC	Steam, water	400–1728
Egypt, 2004	Pharmaceutical industry for steam generation	173	PTC	Pressurized water	1900
Greece, 2001	Food and beverages, textile, pharma, leather products	Up to 100	ETC	–	4522
Austria, 2004	Food and beverages, beer and malt, textile, automobile machineries	Up to 250	ETC, PTC	Hot water, steam	4,300,000
TVS Group, Chennai, India	Degreasing and phosphating process in automobile industry	55–65	ETC	Hot water	1200
Sunil Health Care, India	Capsule shell manufacturing industry	Up to 75	FPC	Hot water	11,760
New York City Subway Washing Plant, USA, 2010	Transportation	120–220	ETC	–	164
Germany, 2012	Chemicals, food and beverages, textile, pulp, paper, etc.	88% of total heat demand is below 300 °C	FPC/ETC/PTC	Steam	–

Continued

TABLE 2.2
Continued

Country	Potential Applications	Operating Temperature (°C)	Type of Collector Used	Working Medium or HTFs	Collector Area (m²)
St. Pauls, North Carolina, USA, 2012	Food and dairy	80–120	FPC	-	7804
Mexico, 2013	Agriculture and food processing	60–180	PTC	Steam	496
Morocco, 2015	Food, chemical, textile and leather	Below 100	FPC, ETC	Steam, water	2,300,000
Egypt, 2015					4,600,000
Pakistan, 2015					7,100,000
Milma Dairy Kozhikode, India	Hot water for steam boiler	60–80	FPC	Hot water	276
India, 2017	Dairy and paper industry	50–250	ETC, PTC	Steam, water	2,940,000
Salt Lake City, Utah, USA	Power plant	Up to 400	PTC	Therminol	656
Latur, India	Steam heating for milk pasteurization	180	LFR	Pressurized hot water	160
Ludhiana, India	For bleaching and washing garments in textile industry	60–80	FPC	Hot water	360
Chennai, India	Cleaning in automobile industry	55–70	ETC	Hot water	1365
Anul dairy, India	Milk pasteurization	140	PTC	Steam	560
Spain	Power production	293–393	PTC	Oil	

Note: FPC—flat plate collector, ETC—evacuated tube collector, PTC—parabolic trough collector, LFR—linear Fresnel reflector.
Source: Taibi et al. (2012), Ramos et al. (2014), Sharma et al. (2017), Ramaiah and Shekar (2018), Ramaiah and Shekar (2018). Solar Heat for Industrial Processes Technology Brief (2022), Solar Thermal Plants Database l Solar Heat for Industrial Processes (SHIP) Plants Database (2022).

followed by Germany, Sweden, Austria, China, etc. (Solar thermal capacity globally 2018 | Statista 2022). Despite the global pandemic in 2020, the solar heating market share increased significantly, such as in Germany (+26%), Brazil (+7.3%) and Netherlands (+6.5%). The global installed capacity of solar heating for industrial processes reaches up to 791 MW (Solar Heat for Industrial Processes Technology Brief 2022, Solar Heat Worldwide 2021—a rich source of global, national and sector-specific data—Solarthermalworld 2022).

Apart from the application of industrial process heating, it is worth mentioning that there is a significant decrement in energy demand from conventional sources as a considerable part is attained through solar energy. Also, it contributes towards the reduction of environmental emissions. More efforts should be devoted to further evaluation of the potential application of industrial process heating in other sectors.

2.3.2 WORLDWIDE STATUS OF INDUSTRIAL PROCESS HEATING

The solar thermal collectors are applicable for various industrial process applications that are either low or medium temperature based. The main research and development carried out across the globe are to enrich the thermal performance of the plants and improve their cost-competitiveness. The thermal efficiency of the collector can be improved by different methods (Bellas and Lidorikis 2017, Zhu *et al.* 2017, Verma, Gupta, *et al.* 2020) such as incorporating design/structural changes, picking advanced materials and selective coating for higher absorption, optimizing process parameters and modification of working medium or HTFs. Whereas, for higher temperature process heating or power generation applications, solar dish or solar towers are mainly used.

There are many studies (Dović and Andrassy 2012, Goudarzi *et al.* 2014, Nikolić and Lukić 2015, Verma, Sharma, *et al.* 2020) that reported the structural changes on the collector to improve the thermal performance. In particular to FPCs, design modifications are performed on corrugated collectors, spiral collector tubes, cylindrical solar collectors. In double exposure collector, the absorber is placed between glazing and reflector. Few studies (Javaniyan Jouybari *et al.* 2017, Zhu *et al.* 2017) proposed thermal performance enhancement by introducing porous metallic media and increasing the surface area contact by introducing arrays of micro heat pipes. The thermal efficiency is improved from 7 to 69% because of the aforementioned design modifications (Zhu *et al.* 2017).

Furthermore, few studies mentioned the incorporation of turbulators (Figure 2.10) which allows improving the heat transfer by generating turbulence and swirling flow (Goudarzi *et al.* 2014, Sakhaei and Valipour 2019). Few of these can be achieved by introducing a concentric coil, twisted tapes, wire mesh, coil inserts, etc. (Sakhaei and Valipour 2019). A few reported design modifications in ETCs are ETCs with hot air dryers (Lamnatou *et al.* 2012), thermo-syphon heat pipes with ETCs (Redpath 2012), selective coating design (Bellas and Lidorikis 2017), etc.

Introducing nanoparticles widens the solar absorption spectrum (Verma, Gupta, *et al.* 2020) and therefore lower wavelengths of solar radiations are absorbed. It produces improvement in operating temperature and thermal efficiency of the system by 112% and 30%, respectively (Bellas and Lidorikis 2017). PCM integrated design collectors enhance the heat absorption characteristics (Chopra *et al.* 2018, Verma,

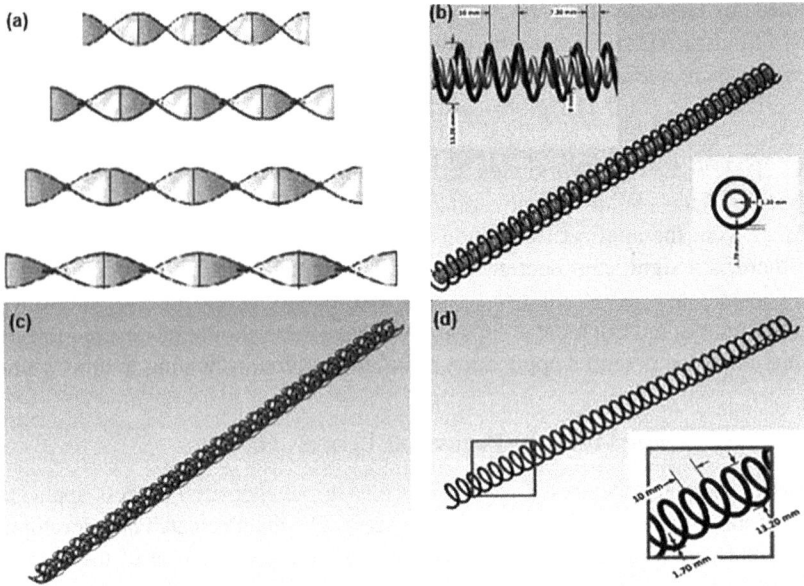

FIGURE 2.10   Different turbulators incorporated in FPCs: (a) twisted tapes, (b) concentric
coils, (c) wire mesh and (d) coil inserts.

Source: Sakhaei and Valipour (2019), with permission from Elsevier.

Gupta, *et al.* 2020). Similarly, cylindrical parabolic reflectors integrated with PCMs
further reduce radiation losses from the outer glass of the absorber tube of ETCs
(Chopra *et al.* 2018).

Most of the studies on concentrating types of collectors (PTCs and LFRs) are
based on the modification of the receiver part of the collector (Bellos and Tzivanidis 2018, Bellos, Daniil, *et al.* 2018, Bellos, Tzivanidis, *et al.* 2018). These studies
include a change in absorber tube shape (absorber tube with toroidal rings, Z-Shape,
spiral, helical ribs, etc.) (Bellos, Tzivanidis, *et al.* 2018) to increase the absorber surface area and introduce a variety of inserts (star shaped, cylindrical, multiple cylindrical, pentagonal, etc.) (Bellos and Tzivanidis 2018). However, various insertions
improve heat transfer rate significantly but the enhancement in thermal efficiency
is comparatively low (less than 2%). This is because of more energy losses while
increasing pumping work (Verma, Gupta, *et al.* 2020).

Water, air, oil, glycol, etc. are used as a working fluid or HTF in various applications. But the thermal conductivity of these HTFs is much less. To improve the
heat transfer rate, many studies suggested adding nanoparticles (Faizal *et al.* 2013,
Bellas and Lidorikis 2017, Alawi *et al.* 2021). In general, nanofluids such as Al_2O_3,
CuO, SiO_2, TiO_2, MgO, etc. are dispersed in water and used as HTFs to enrich the
thermal efficiency. Most of the studies concluded that the thermal performance can
be improved from 4 to 60% while the nanofluid is used as HTF (Faizal *et al.* 2013,
Alawi *et al.* 2021). Another advantage is that the addition of nanoparticles reduces
the surface area requirement of the collector by 21.5–25.6% (Faizal *et al.* 2013).

Some studies reported the influence of different collector materials and coating on thermal performance (Kalogirou and Lloyd 1992, Bellas and Lidorikis 2017). High transmissivity, chemical resistance, low emissivity and cost-effectiveness are the criteria for the selection of collector covers (Kalogirou 2004). The coating can be applied on the collector covers to improve the glazing quality of a surface and hence its overall performance. Various metals are used as coating materials such as silver, copper, and metal oxides such as zinc oxide, tin oxide, aluminum doped zinc oxide, etc. (Kalogirou and Lloyd 1992, Kalogirou 2004). Sol-gel coating is used as a coating material which improves the transmittance significantly (Bellas and Lidorikis 2017).

Conventionally, the absorber plates are made up of either aluminum or copper sheets. The thermal conductivity and cost of the material play an important role while selecting the material for the absorber plate. Selective coatings are used to improve the absorptance and emittance. Black chrome is generally used as a coating material due to its higher thermal performance and aging effect (Kalogirou 2004).

Insulation plays an important role while reducing the losses in the collectors. Rock wool, glass wool, polyurethane foam, etc. are generally used as insulation materials (Balijepalli *et al.* 2017). These insulation materials should be weather resistant and with good chemical stability during higher temperature conditions.

Apart from these modifications, integration of an energy storage system can also improve the system run time during off-sunshine hours and thus, the performance of the system significantly. There are three types of storage materials, sensible, latent and chemical. However, chemical storage materials are yet to be explored for different types of applications. Due to the requirement of large chemical reactors to attain chemical stability and durability of the reaction, further efforts should be devoted towards this. Sensible and latent heat storage materials are used for low and medium collectors but the system size increases significantly. Sensible storage materials have very low storage capacity. Although the storage capacity of latent heat storage materials is higher, they have poor heat transfer. Therefore, more studies are required on the compactness of the storage system and various techniques to improve the storage capacity.

For the commercial aspect of collectors in industrial process heating and power generation applications, economic analysis is needed. The economic feasibility analysis has three different aspects, capital investment, operational and maintenance cost, and equipment cost. The fabrication cost, auxiliary equipment and their cost required for solar tracking, land and labor cost, material cost, integration of thermal energy storage system, etc. have a significant contribution towards the cost addition of the system. For commercial acceptance, various deciding factors such as cost of the project, cost of electricity, internal rate of return, payback period, etc. must be considered and then only the economic feasibility analysis will be a complete one.

2.3.3 Cost Analysis

The cost of a solar industrial process heating system largely depends on many factors such as the size of the plant, type of application, temperature range required, site for the installation of the plant, etc. In general, for solar thermal systems, 50 to 70% share is held for capital investment, whereas the remaining is used for installation and integration costs. Various components of solar thermal systems hold 50% of the total investment cost, 11% for storage and heat exchanger expenses and 5% for control strategy. The piping adds an additional 20% to the cost burden. For FPC and ETC, the total investment cost in Europe varies in the range of 275 to 1098 $/kW, whereas it is 220 to 330 $/kW for countries like India, Turkey, Mexico, South Africa, etc. The installation cost for other concentrated systems such as PTC, LFR and solar dishes varies in the range of 600 to 2000 $/kW, 1200 to 1800 $/kW and 400 to 1800 $/kW, respectively (Solar Heat for Industrial Processes Technology Brief 2022).

The integration of a hybrid solar district heating plant also is one of the cost-effective methods for commercial scale solar heating applications. The hybrid solar district heating system consists of FPC and PTC of 5960 and 4039 m² of the collector area resulting in a 5 to 9% reduction in levelized cost of heat (Tian *et al.* 2018). Therefore, the choice of solar collectors can be made as a hybrid mode to reduce the levelized cost of heat (LOCH).

2.4 SOLAR THERMAL TECHNOLOGIES FOR PROCESS HEATING

Growing energy demand in the industrial sector results in overdependency on various energy sources. Therefore, apart from fossil fuels, researchers and industrials are looking for various other potentials of renewable energy sources, mainly on the utilization of solar energy. It is one of the clean energies with the potential for solar thermal and power generation applications in various residential and industrial sectors.

Industrial process heating can be classified mainly into two categories, combustion-based and electricity-based. The process heating system has a heating device and a distribution system. The heating device generates heat, whereas the distribution system transfers heat from the source to the destination. There are two modes of process heating, direct and indirect, as mentioned in Figure 2.11. In the direct mode of heating, heat is generated within the material itself through different modes of heat transfer mechanisms (such as conduction, convection and radiation). Examples of various processes are hardening, rolling, forging, etc.

In the indirect mode, heat is generated from another source and transferred to the material. Steam or pressurized water is used as the medium of transfer. Various applications in the systems include boilers, furnaces, heat pumps and solar collectors. Solar thermal applications are comparatively more energy-efficient than power

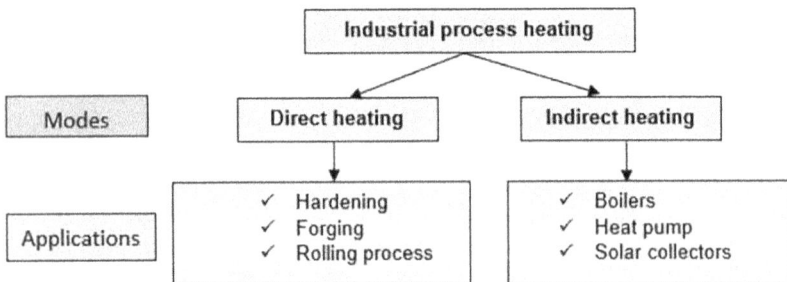

FIGURE 2.11 Classification of industrial process heating.

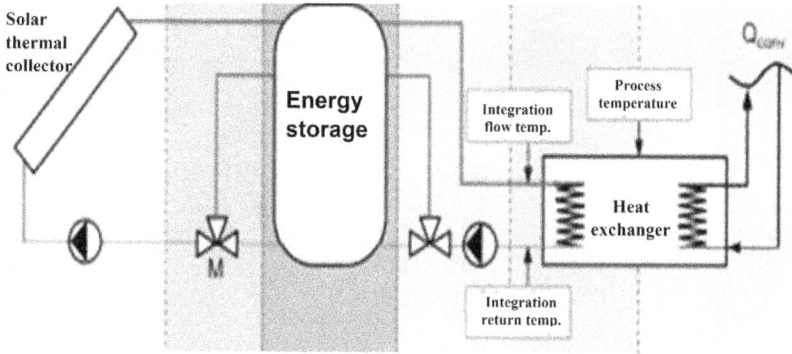

FIGURE 2.12 Schematic layout of a solar thermal system.

Source: Hess (2016), with permission from Elsevier.

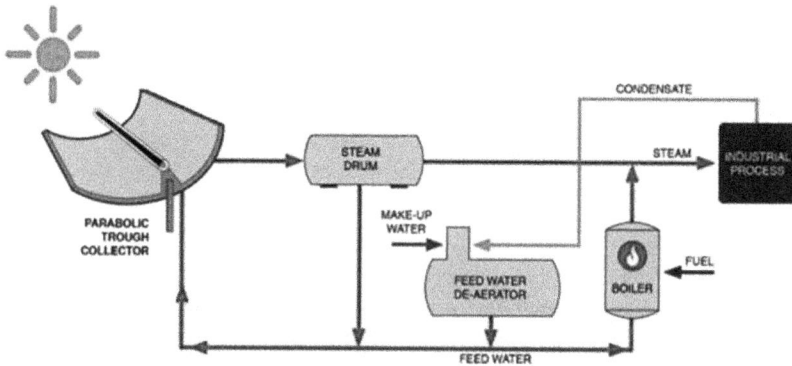

FIGURE 2.13 Schematic layout of direct steam generation.

Source: Solar Process Heat Technologies & Solar Collectors (2022).

generation applications. In the conversion of solar energy into heat energy, the efficiency is achieved up to 70%, whereas for power generation applications, it is 15 to 25% (Sharma *et al.* 2017).

Solar thermal systems (Figure 2.12) mainly consist of a heating unit (solar thermal collectors), distribution system (pump and interconnecting pipes), heat exchanger, HTF and energy storage system (Hess 2016). There are many deciding factors for the selection of solar thermal collectors based on their application, such as operating temperature range, efficiency, cost and annual energy potential. However, some of the other factors also need to be focused on, such as availability of space and foundation support for the installation of solar collectors. Various parameters such as solar insulation data, climatic condition, the efficiency of the collector and the required outlet temperature are also required for the estimation of the area of the solar collector.

Hot water and steam generation through industrial process heating can be described in two methods: direct steam generation and indirect steam generation. In the direct

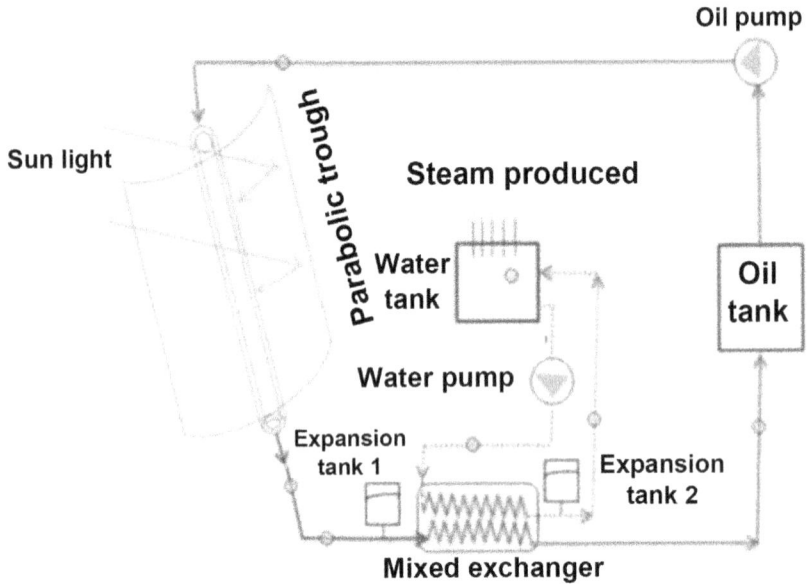

FIGURE 2.14 Schematic layout of indirect steam generation.

Source: Ghazouani *et al.* (2019), with permission from Elsevier.

system, water is boiled in the solar collector loop as depicted in Figure 2.13. Then the hot water comes in contact with a steam drum, where steam is retrieved from the water and feed water is added and recirculated. In this system, there is no provision of the thermal energy storage system.

However, in the indirect steam generation system (Figure 2.14), HTF is used to transfer the heat from the solar collectors with a heat exchanger, and steam is generated indirectly. The unused heat energy is usually stored in the thermal energy storage system. Most of the medium temperature range (up to 400 °C) process heating applications use concentrators for direct steam and pressurized hot water generation. The detailed applications of process heating systems in various fields are discussed in detail in the next section.

2.4.1 Solar Thermal Network (STN)

A number of collectors, either connected in series or parallel in the network to generate the thermal load requirement at the desired temperature, is called a solar thermal network (STN) or a solar collector network. It is used as a heating device and it is one of the integral parts of the solar system for industrial process heating. The collectors are connected in series to get the target temperature, whereas these must be connected in parallel to get the required thermal load (Martínez-Rodríguez *et al.* 2018a, 2018b). The STN is represented in Figure 2.15 where collectors are connected in series (1 to *n*) to get the higher temperature, whereas the collectors are connected in parallel (1 to *m*) to get a higher thermal load.

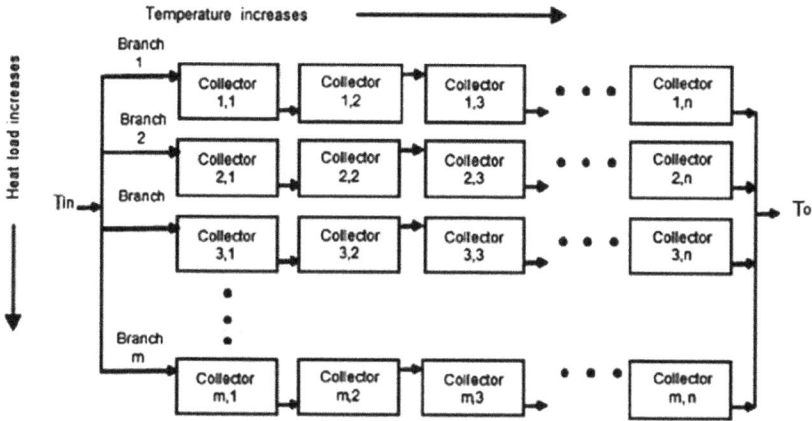

FIGURE 2.15 Layout of solar collector networks.

Source: Martínez-Rodríguez *et al.* (2018b), with permission from Elsevier.

2.4.2 MATHEMATICAL MODELING OF STN USED IN INDUSTRIAL PROCESS HEATING

The number of total collectors used in the STN and outlet temperature of the fluid can be estimated using mathematical expressions as discussed next.

The surface area of the solar collector network (A_{scn}) can be estimated as,

$$A_{scn} = LW\, N_c \tag{26}$$

Where, L and W are the length and width of the collector, and N_c is the number of collectors in the STN.

The N_c of STN can be estimated using,

$$N_c = N_p N_s \tag{27}$$

Where, N_p and N_s are the numbers of collectors connected in parallel and series, respectively.

The solar energy gain for an STN where all the collectors are identical in nature, considering all the collectors connected are FPC type (Zelzouli *et al.* 2012), is,

$$q_u = \sum_{n=1}^{N_c} F_{R,j} A_a \left[S(\tau\alpha) - U_{l,n} \left(T_{in,n} - T_{amb} \right) \right] \tag{28}$$

The outlet fluid temperature can be estimated using,

$$T_{o,n} = \frac{F_{R,n} A_a \left[S(\tau\alpha) - U_{l,n} \left(T_{in,n} - T_{amb} \right) \right]}{\dot{m}c_p} + T_{in,n} \tag{29}$$

Where, n is nth row of collectors connected in series.

2.5 APPLICATIONS OF SOLAR THERMAL COLLECTORS IN VARIOUS INDUSTRIAL PROCESS HEATING

Solar thermal collectors are generally applicable in various fields of applications. Most of the energy consumed in the industry is either in the form of electrical energy or heat energy. Electricity is used for electric bulbs, air conditioners or operating motors, whereas thermal or heat energy is used for various industrial processes such as drying, heat treatment, desalination, refrigeration, dyeing, sterilization, pasteurization, etc. The most important applications and the types of solar collector used are tabulated in Table 2.3.

The major part of the heat energy for different industrial processes reportedly comes in the temperature range of 40 to 225 °C (Sharma *et al.* 2017, Ramaiah and Shekar 2018). Most of this process and its operating temperature range are reportedly achievable using low and medium temperature solar collectors such as FPC, ETC, PTC and LFR. The major industrial processes utilizing heat at mean temperature are food processing, automobile industries, textile industries and pharmaceutical industries, etc. Various industries use solar thermal energy and their details such as working medium and temperature range are tabulated in Table 2.4.

TABLE 2.3
Solar Energy Application and Type of Collectors Used

Application	Type of Solar Collector
Solar water heating	FPC, ETC
Solar air heater (for space heating, drying applications)	FPC
Solar updraft tower (for power generation, mitigation of air pollution, drying, desalination for producing fresh water)	FPC and for hybrid applications can be integrated with biomass, PV/T, industrial waste heat
Solar fuel hybrid power plant (indoor space heating applications)	FPC integrated with biomass heater
Space heating and cooling	FPC, ETC, parabolic dish reflector
Solar refrigeration system (absorption, adsorption units)	FPC, ETC
Industrial process heating for low and medium temperature applications (water heating, steam generation, etc.)	FPC, ETC, PTC, LFR
Solar desalination for production of fresh water: solar stills, vapor compression, etc.	FPC, ETC
Solar thermochemical energy system (to convert solar energy into storable solar fuels, CO and H_2)	PTC
Solar thermal power systems	PTC, LFR, solar tower, parabolic dish reflector

Note: FPC—flat plate collector, ETC—evacuated tube collector, PTC—parabolic trough collector, LFR—linear Fresnel reflector.

Source: Tian and Zhao (2013), Hess (2016), Sharma *et al.* (2017), Ramaiah and Shekar (2018).

TABLE 2.4

Applications of Various Industrial Processes with Corresponding Temperature Range and Heat Transfer Medium

Applications	Process	Heat Transfer Medium	Temperature Range (°C)
Food processing	Cooking	Steam	70–120
	Heat treatment	Water	60–80
	Cleaning	Air/Steam	40–80
	Pasteurization	Steam	60–150
	Drying	Air	40–60
	Sterilization	Steam	100–140
Automobile industries	Pre-treatment	Water	40–50
	Drying	Air	150–175
	Baking of paints	Steam	175–225
Textile industries	Dyeing	Water	60–90
	Drying	Steam	100–130
	Pressing	Steam	120–140
	Fixing	Steam	160–180
	Printing	Water/Steam	40–130
Pharmaceutical industries	Distillation	Water	100–200
	Evaporation	Steam	110–170
	Drying	Air/Steam	120–170
Miscellaneous applications	Lime calcining in cement industries	Air	600–1200
	Hardening, rolling, forging, etc. in steel industries	Air	700–1500
	Drying in rubber industry	Air	50–130
	Drying in plastic industry	Air	180–200
	Drying in glass fiber industry	Air	150–175
	Extrusion in plastic industry	Steam	140–160
	Pulp preparation in pulp and paper industry	Pressurized water	120–170
	Drying and finishing in leather industry	Air	70–100
	Drying in pulp and paper industry	Air/Steam	90–200

Source: Jradi and Riffat (2014), Sharma *et al.* (2017), Ramaiah and Shekar (2018).

Apart from these applications, solar energy is also used for some residential purposes such as domestic hot water, space heating and cooling and agricultural applications (Hess 2016). The detailed applications of industrial process heating in the various sectors are explained in the following sections.

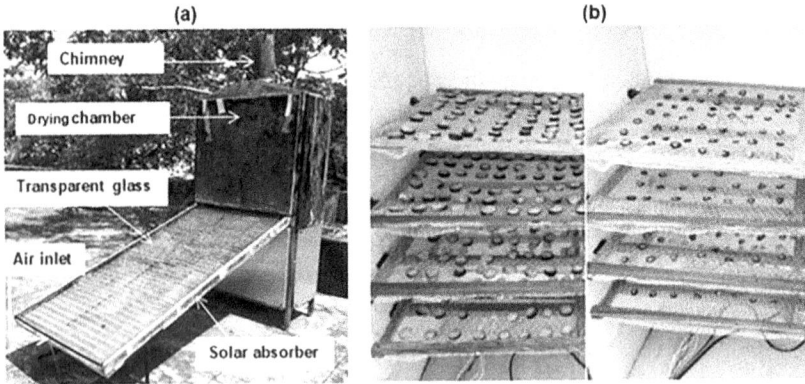

FIGURE 2.16   Indirect type solar dryer used for drying food products.

Source: Gilago and Chandramohan (2022), with permission from Elsevier.

2.5.1 Food Processing

In the food processing industries, the energy needed for various processes such as cooking, cleaning, heat treatment, sterilization and storing or preservation can be met through solar energy. These processes can be carried out in the comfortable temperature range of 40 to 150 °C (Sharma *et al.* 2017). This range can be maintained by low temperature solar thermal collectors such as FPCs, ETCs (Kalogirou 2004, Ramaiah and Shekar 2018). The working fluids or HTFs used are air, water or steam (Sharma *et al.* 2017). The specific applications in the food processing units are drying of various agricultural food products, the dairy industry for processing milk, preparation of milk-derived products, wine and beverages, the bread and bakery industry, etc.

In drying agricultural food products (Figure 2.16), solar air heaters are generally employed. These types of collectors consist of a wooden frame, a glazing glass as a collector to capture and transmit solar radiation and an absorber plate made of high thermal conductivity materials. Solar water heaters are also employed for other industrial processes such as producing hot water and steam. The temperature range used in food processing industries is below 80 °C while drying food products, 40–80 °C for the cleaning process, 60–150 °C for pasteurization, (100–140 °C) for sterilization or evaporation processes (Sharma *et al.* 2017, Ramaiah and Shekar 2018).

2.5.2 Automobile Industries

The automobile sector is one of the major manufacturing sectors as it consists of the production of motorcycles, cars, light and heavy vehicles. Most of the processes in the automobile industry require either heat energy or electricity. A major portion of the total energy requirement (70%) belongs to thermal energy for generating hot water and steam (Ramaiah and Shekar 2018). The operating temperature range in the automobile sector is 40–225 °C (Sharma *et al.* 2017). This temperature range can be easily achieved by employing low and medium temperature solar collectors such as FPCs,

ETCs, PTCs and LFRs. Here also air, water or steam can be used as the heat transferring medium. Other processes involved in the automobile industry are pre-treatment (40–50 °C), drying (150–175 °C) and baking of paints (175–225 °C) (Sharma *et al.* 2017, Ramaiah and Shekar 2018). A few other applications, such as hot water to clean engine components, vehicle steering components before painting, degreasing and phosphating processes, need the heat energy with a temperature range of 55 to 120 °C, which can be done by providing ETC or PTC (Kalogirou 2004, Hess 2016).

2.5.3 Textile Industries

In the textile industry, energy must be used for different processes such as drying, pressing, dyeing, fixing, bleaching and printing, etc.; 60 to 65% of the total energy is used in these processes (Ramaiah and Shekar 2018). Hot water or steam is used as the working fluid for the aforementioned processes. The operating temperature range is 40–180 °C (Kalogirou 2004, Hess 2016). By employing low temperature solar thermal collectors such as FPCs and ETCs, the operating temperature of 40–180 °C can be achieved. The temperature ranges used in the textile industry for various processes are dyeing (60–90 °C), pressing (100–140 °C), printing (40–130 °C), bleaching (90–93 °C) and fixing (160–180 °C) (Sharma *et al.* 2017, Ramaiah and Shekar 2018).

2.5.4 Pharmaceutical Industries

The pharmaceutical industry produces various products such as tablets, ointments, capsules and powders. The manufacturing of these products requires both electrical as well as thermal energy. The various processes involved are chemical synthesis, fermentation, granulation, coating and sterilization. All these processes require thermal or heat energy. More than 50% of total energy is required in these processes, which have an operating temperature of 100 to 200 °C (Ramaiah and Shekar 2018). Solar hot air systems, FPCs, ETCs, PTCs and LFRs can be recommended to achieve the temperature range. Hot water, air or steam can be used as a medium for various pharmaceutical processes such as distillation (100–200 °C), drying (120–170 °C) and evaporation (110–170 °C) (Sharma *et al.* 2017, Ramaiah and Shekar 2018). This industry also needs cold temperatures which can be provided through solar powered refrigeration systems. Therefore, up to 65% of total energy demand for both heating and cooling can be met through solar energy systems (Ramaiah and Shekar 2018).

2.5.5 Miscellaneous Applications

Apart from the aforementioned specific applications, there are a few other potential applications where industrial process heating can be utilized through solar energy. There are several industries, such as the cement, steel, rubber, plastic, glass, fiber and pulp and paper industries, where the major share of energy requirement can be cut down with the application of industrial process heating by solar energy. The operating temperatures in these industries are low to moderate range of 50 to 200 °C (except the steel and cement industry where a high temperature range is used, 600 to 1500 °C) (Sharma *et al.* 2017). Air is used as the working medium for the drying process in various applications. The drying temperature ranges of various industries are rubber (50 to 130 °C), glass fiber (150 to 175 °C), leather (70 to 100 °C), plastic (180 to 200 °C), pulp and paper industry (90 to 200 °C) (Sharma *et al.* 2017, Ramaiah and Shekar 2018). Most of this process can be attained with the application of solar thermal collectors such as FPCs, ETCs, LFRs and PTCs. Lime calcining processes in cement industries; hardening, forging and rolling operations in steel-making industries are based on high temperature applications such as 600 to 1200 °C and 700 to 1500 °C, respectively (Sharma *et al.* 2017, Ramaiah and Shekar 2018). These applications can be attained by employing high temperature solar collectors such as parabolic dish reflectors or solar dishes and solar towers.

2.6 FUTURE OUTLOOK AND RECOMMENDATIONS

The different types of solar collectors, their working conditions, applications and temperature ranges have been discussed for industrial process heating applications. The major limitation of solar collectors is that they perform at low to medium temperature ranges, hence the efficiency is comparatively low.

Improvement of thermal efficiency and cost reduction are the major challenges of solar thermal collectors. The performance of the FPCs can be improved further with the incorporation of structural modifications. Apart from structural modifications, the focus must be given on integration of reflectors, selection of materials, absorber coatings for all-weather conditions and integration of thermal energy storage systems. Therefore, future research needs to answer these issues and salient points.

ETCs are observed to be better and more efficient performers than FPCs, hence they can be recommended for different processes of heating applications up to 200 °C. The performance can be improved by the addition of nanotechnology and the addition of a storage system using PCMs or nano-composite PCMs. However, more studies are required on the compactness of energy storage systems and the stability of nano-composite PCMs. It is worth mentioning that the V-groove configuration is observed to be a cheap and effective method to enrich the performance of FPCs. The application of wire, oil, twisted tape, nanofluids, etc. are also reported to improve the heat transfer. More effort should be devoted to optimizing and improving the overall thermal performance of the system. The collector plate materials with high reflectivity, absorption and low emittance have the potential to improve the thermal behavior of the system. Although structural modification on the receiver of the collector (in PTCs and LFRs) improves the thermal behavior of the system, the studies are limited; therefore, further investigations are required in this regard. More studies are required on the identification of novel and cost-effective materials for reflectors. Reflectors improve the solar incident radiation; therefore, more studies are required to identify proper installation strategies. Additionally new collector designs must be investigated further to integrate into process heating systems for various applications.

2.7 CONCLUSIONS

This chapter discussed the state-of-art of various low and medium temperature solar collectors for industrial process heating applications. Different solar collectors such as concentrating and non-concentrating types have been discussed in detail. The various types of solar collectors include flat plate collectors (FPCs), evacuated tube collectors (ETCs), parabolic trough collectors (PTCs), linear Fresnel reflectors (LFRs), solar dishes and solar towers. Among non-concentrating types, FPCs and ETCs are used for low to medium range of temperature applications (up to 200 °C), whereas concentrating collectors such as PTCs and LFRs are used for an intermediate range (up to 400 °C). The solar dishes and solar towers are mainly for medium to high temperature applications. The mathematical modeling of various low and medium temperature collectors was discussed with mathematical expressions such as calculation of useful heat gain, the thermal efficiency of the collector, etc. Insights on the solar thermal network (STN) and its mathematical modeling were also included in the study. From the STN analysis, it can be concluded that to get the desired temperature, solar thermal collectors must be connected in series, whereas to get the desired thermal load, collectors must be connected in parallel mode. The global potential, worldwide status and cost analysis of various solar industrial process heating applications were also presented. It was observed that the total investment cost of low temperature collectors (FPCs, ETCs) varied in the range of 275 to 1098 $/kW and 220 to 330 $/kW in Europe and other countries like India, Turkey, Mexico, South Africa, respectively. The installation cost for other concentrated systems such as PTCs, LFRs and solar dishes varies in the range of 600 to 2000 $/kW, 1200 to 1800 $/kW and 400 to 1800 $/kW, respectively. Additionally, several applications of solar thermal collectors in industrial process heating such as food processing units including beverages, milk, drying of food products and other industries such as textile, automobile, pharmaceutical, paper and pulp, etc. were discussed and effective results were tabulated. The thermal energy requirement in various processes such as cleaning, heat treatment, drying, sterilization, pasteurization, etc. in food processing units; baking of paints, pre-treatment, etc. in automobile industries; dyeing, pressing and fixing, etc. in textile industries; evaporation, distillation, etc. in pharmaceutical industries could be efficiently attained by the low and medium temperature range solar thermal collectors. It should be noted that apart from these applications, many other applications are not discussed here; they are either in the developing stage or not studied yet. The solar thermal collectors in various industrial process heating applications showed huge potential and could be an economically viable option for a sustainable future. However, to increase further applicability of low and medium temperature collectors, the focus should be given to the compactness of the energy storage system, enhancement of conversion efficiency, materials, etc. Lastly, future scope and recommendations were signposted in this chapter which provides a roadmap to future researchers to choose solar thermal collectors for industrial process heating applications.

NOMENCLATURE

Abbreviations

CR   Concentration ratio
ETC   Evacuated tube collector
FPC   Flat plate collector

HTF   Heat transfer fluids
LFR   Linear Fresnel reflector
LOCH   Levelized cost of heat
PCM   Phase change materials
PTC   Parabolic trough collector
STN   Solar thermal network

Symbols

A Area, m^2
c_p Specific heat of the fluid, J/kgK
F_R Collector heat removal factor
h Convective heat transfer coefficient, W/m^2K
I Solar flux, W/m^2
k_i Thermal conductivity of the insulation, W/mK
L Length of the collector, m
m   Mass flow rate, kg/s
N Number of collectors
q Rate of heat gain, W
S Solar flux per unit effective area of aperture, W/m^2
t Total height of the collector casing, m
T Temperature, K
U_l Overall loss coefficient, W/m^2K
W Width of the collector, m

Greek symbol:

α absorptivity
β Thermal expansion coefficient, $°C^{-1}$
ε Emissivity
ρ Density, kg/m^3
θ Angle between solar beam flux and normal to the plane
σ Stefan-Boltzmann constant
η Efficiency, %
τ transmissivity

Sub-scripts:

a Aperture
amb Ambient
ap Absorber plate
b Beam
cc Collector cover
d Diffused
i Instantaneous
in Inlet
l loss

m	Mean
o	Optical
out	Outlet
p	Parallel
r	Receiver
s	Series
u	Useful

REFERENCES

Alawi, O.A., Kamar, H.M., Mallah, A.R., Mohammed, H.A., Kazi, S.N., Che Sidik, N.A., and Najafi, G., 2021. Nanofluids for flat plate solar collectors: Fundamentals and applications. *Journal of Cleaner Production*, 291, 125725.

Balijepalli, R., Chandramohan, V.P., and Kirankumar, K., 2017. Performance parameter evaluation, materials selection, solar radiation with energy losses, energy storage and turbine design procedure for a pilot scale solar updraft tower. *Energy Conversion and Management*, 150, 451–462.

Barone, G., Buonomano, A., Forzano, C., and Palombo, A., 2019. Solar thermal collectors. In Calise, F., D'Accadia, M.D., Santarelli, M., Lanzini, A., & Ferrero, D. (Eds.) *Solar Hydrogen Production* (pp. 151–178). Academic Press.

Bellas, D.V., and Lidorikis, E., 2017. Design of high-temperature solar-selective coatings for application in solar collectors. *Solar Energy Materials and Solar Cells*, 170, 102–113.

Bellos, E., Daniil, I., and Tzivanidis, C., 2018. Multiple cylindrical inserts for parabolic trough solar collector. *Applied Thermal Engineering*, 143, 80–89.

Bellos, E., and Tzivanidis, C., 2018. Investigation of a star flow insert in a parabolic trough solar collector. *Applied Energy*, 224, 86–102.

Bellos, E., Tzivanidis, C., and Tsimpoukis, D., 2018. Enhancing the performance of parabolic trough collectors using nanofluids and turbulators. *Renewable and Sustainable Energy Reviews*, 91, 358–375.

Bhatia, S.C., 2014. Solar thermal energy. In *Advanced Renewable Energy Systems* (pp. 94–143). Woodhead Publishing India Ltd.

Chopra, K., Tyagi, V.V., Pandey, A.K., and Sari, A., 2018. Global advancement on experimental and thermal analysis of evacuated tube collector with and without heat pipe systems and possible applications. *Applied Energy*, 228, 351–389.

Das, P. and Chandramohan, V.P., 2019. Computational study on the effect of collector cover inclination angle, absorber plate diameter and chimney height on flow and performance parameters of solar updraft tower (SUT) plant. *Energy*, 172, 366–379.

Das, P., and Chandramohan, V.P., 2021. Experimental studies of a laboratory scale inclined collector solar updraft tower plant with thermal energy storage system. *Journal of Building Engineering*, 41, 102394.

Dović, D., and Andrassy, M., 2012. Numerically assisted analysis of flat and corrugated plate solar collectors thermal performances. *Solar Energy*, 86 (9), 2416–2431.

Faizal, M., Saidur, R., Mekhilef, S., and Alim, M.A., 2013. Energy, economic and environmental analysis of metal oxides nanofluid for flat-plate solar collector. *Energy Conversion and Management*, 76, 162–168.

Fernández-García, A., Zarza, E., Valenzuela, L., and Pérez, M., 2010. Parabolic-trough solar collectors and their applications. *Renewable and Sustainable Energy Reviews*, 14 (7), 1695–1721.

Ghazouani, K., Skouri, S., Bouadila, S., and Guizani, A.A., 2019. Thermal analysis of linear solar concentrator for indirect steam generation. *Energy Procedia*, 162, 136–145.

Gilago, M.C., and Chandramohan, V.P., 2022. Performance evaluation of natural and forced convection indirect type solar dryers during drying ivy gourd: An experimental study. *Renewable Energy*, 182, 934–945.

Goudarzi, K., Shojaeizadeh, E., and Nejati, F., 2014. An experimental investigation on the simultaneous effect of CuO—H2O nanofluid and receiver helical pipe on the thermal efficiency of a cylindrical solar collector. *Applied Thermal Engineering*, 73 (1), 1236–1243.

Hafez, A.Z., Soliman, A., El-Metwally, K.A., and Ismail, I.M., 2016. Solar parabolic dish Stirling engine system design, simulation, and thermal analysis. *Energy Conversion and Management*, 126, 60–75.

Hayek, M., Assaf, J., and Lteif, W., 2011. Experimental investigation of the performance of evacuated-tube solar collectors under eastern mediterranean climatic conditions. *Energy Procedia*, 6, 618–626.

Hess, S., 2016. Solar thermal process heat (SPH) generation. In *Renewable Heating and Cooling* (pp. 41–66). Woodhead Publishing.

Iparraguirre, I., Huidobro, A., Fernández-García, A., Valenzuela, L., Horta, P., Sallaberry, F., Osório, T., and Sanz, A., 2016. Solar thermal collectors for medium temperature applications: A comprehensive review and updated database. *Energy Procedia*, 91, 64–71.

Javaniyan Jouybari, H., Saedodin, S., Zamzamian, A., and Nimvari, M.E., 2017. Experimental investigation of thermal performance and entropy generation of a flat-plate solar collector filled with porous media. *Applied Thermal Engineering*, 127, 1506–1517.

Jradi, M., and Riffat, S., 2014. Medium temperature concentrators for solar thermal applications. *International Journal of Low-Carbon Technologies*, 9 (3), 214–224.

Kalogirou, S., 2003. The potential of solar industrial process heat applications. *Applied Energy*, 76 (4), 337–361.

Kalogirou, S., 2008. Recent patents in solar energy collectors and applications. *Recent Patents on Engineering*, 1 (1), 23–33.

Kalogirou, S.A., 2004. Solar thermal collectors and applications. *Progress in Energy and Combustion Science*, 30 (3), 231–295.

Kalogirou, S.A., and Lloyd, S., 1992. Use of solar parabolic trough collectors for hot water production in cyprus. A feasibility study. *Renewable Energy*, 2 (2), 117–124.

Kalogirou, S.A., Lloyd, S., Ward, J., and Eleftheriou, P., 1994. Design and performance characteristics of a parabolic-trough solar-collector system. *Applied Energy*, 47 (4), 341–354.

Kumar, L., Hasanuzzaman, M., and Rahim, N.A., 2019. Global advancement of solar thermal energy technologies for industrial process heat and its future prospects: A review. *Energy Conversion and Management*, 195, 885–908.

Lamnatou, C., Papanicolaou, E., Belessiotis, V., and Kyriakis, N., 2012. Experimental investigation and thermodynamic performance analysis of a solar dryer using an evacuated-tube air collector. *Applied Energy*, 94, 232–243.

Martínez-Rodríguez, G., Fuentes-Silva, A.L., and Picón-Núñez, M., 2018a. Targeting the maximum outlet temperature of solar collectors. *Chemical Engineering Transactions*, 70, 1567–1572.

Martínez-Rodríguez, G., Fuentes-Silva, A.L., and Picón-Núñez, M., 2018b. Solar thermal networks operating with evacuated-tube collectors. *Energy*, 146, 26–33.

Mills, D.R., 2012. Linear Fresnel reflector (LFR) technology. In *Concentrating Solar Power Technology* (pp. 153–196). Woodhead Publishing.

Nawsud, Z.A., Altouni, A., Akhijahani, H.S., and Kargarsharifabad, H., 2022. A comprehensive review on the use of nano-fluids and nano-PCM in parabolic trough solar collectors (PTC). *Sustainable Energy Technologies and Assessments*, 51, 101889.

Nikolić, N., and Lukić, N., 2015. Theoretical and experimental investigation of the thermal performance of a double exposure flat-plate solar collector. *Solar Energy*, 119, 100–113.

Orosz, M., and Dickes, R., 2017. Solar thermal powered organic rankine cycles. In *Organic Rankine Cycle (ORC) Power Systems* (pp. 569–612). Woodhead Publishing.

Ramaiah, R. and Shekar, Kss., 2018. Solar thermal energy utilization for medium temperature industrial process heat applications—a review. *IOP Conference Series: Materials Science and Engineering*, 376 (1), 012035.

Ramos, C., Ramirez, R., and Beltran, J., 2014. Potential assessment in Mexico for solar process heat applications in food and textile industries. *Energy Procedia*, 49, 1879–1884.

Redpath, D.A.G., 2012. Thermosyphon heat-pipe evacuated tube solar water heaters for northern maritime climates. *Solar Energy*, 86 (2), 705–715.

Sabiha, M.A., Saidur, R., Mekhilef, S., and Mahian, O., 2015. Progress and latest developments of evacuated tube solar collectors. *Renewable and Sustainable Energy Reviews*, 51, 1038–1054.

Sakhaei, S.A., and Valipour, M.S., 2019. Performance enhancement analysis of The flat plate collectors: A comprehensive review. *Renewable and Sustainable Energy Reviews*, 102, 186–204.

Sharma, A.K., Sharma, C., Mullick, S.C., and Kandpal, T.C., 2017. Solar industrial process heating: A review. *Renewable and Sustainable Energy Reviews*, 78, 124–137.

Singh, P.L., Sarviya, R.M., and Bhagoria, J.L., 2010. Thermal performance of linear Fresnel reflecting solar concentrator with trapezoidal cavity absorbers. *Applied Energy*, 87 (2), 541–550.

Solar Heat for Industrial Processes Technology Brief [online], 2022. Available from: www.irena.org/-/media/Files/IRENA/Agency/Publication/2015/IRENA_ETSAP_Tech_Brief_E21_Solar_Heat_Industrial_2015.pdf [Accessed 2 Feb 2022].

Solar Heat Worldwide 2021—a rich source of global, national and sector-specific data—Solarthermalworld [online], 2022. Available from: https://solarthermalworld.org/news/solar-heat-worldwide-2021-rich-source-global-national-and-sector-specific-data/ [Accessed 25 Mar 2022].

Solar Process Heat Technologies & Solar Collectors [online], 2022. Available from: www.solar-payback.com/technology/ [Accessed 2 Feb 2022].

Solar thermal | Ipieca [online], 2022. Available from: www.ipieca.org/resources/energy-efficiency-solutions/power-and-heat-generation/solar-thermal/ [Accessed 2 Feb 2022].

Solar thermal capacity globally 2018 | Statista [online], 2022. Available from: www.statista.com/statistics/1064418/solar-thermal-energy-cumulative-capacity-globally/ [Accessed 30 Mar 2022].

Solar Thermal Plants Database | Solar Heat for Industrial Processes (SHIP) Plants Database [online], 2022. Available from: http://ship-plants.info/solar-thermal-plants [Accessed 2 Feb 2022].

Taibi, E., Gielen, D., and Bazilian, M., 2012. The potential for renewable energy in industrial applications. *Renewable and Sustainable Energy Reviews*, 16 (1), 735–744.

Tian, Y., and Zhao, C.Y., 2013. A review of solar collectors and thermal energy storage in solar thermal applications. *Applied Energy*, 104, 538–553.

Tian, Z., Perers, B., Furbo, S., and Fan, J., 2018. Thermo-economic optimization of a hybrid solar district heating plant with flat plate collectors and parabolic trough collectors in series. *Energy Conversion and Management*, 165, 92–101.

Vengadesan, E., and Senthil, R., 2020. A review on recent developments in thermal performance enhancement methods of flat plate solar air collector. *Renewable and Sustainable Energy Reviews*, 134, 110315.

Verma, S.K., Gupta, N.K., and Rakshit, D., 2020. A comprehensive analysis on advances in application of solar collectors considering design, process and working fluid parameters for solar to thermal conversion. *Solar Energy*, 208, 1114–1150.

Verma, S.K., Sharma, K., Gupta, N.K., Soni, P., and Upadhyay, N., 2020. Performance comparison of innovative spiral shaped solar collector design with conventional flat plate solar collector. *Energy*, 194, 116853.

Zelzouli, K., Guizani, A., Sebai, R., Kerkeni, C., Zelzouli, K., Guizani, A., Sebai, R., and Kerkeni, C., 2012. Solar thermal systems performances versus flat plate solar collectors connected in series. *Engineering*, 4 (12), 881–893.

Zhu, T., Diao, Y., Zhao, Y., and Ma, C., 2017. Performance evaluation of a novel flat-plate solar air collector with micro-heat pipe arrays (MHPA). *Applied Thermal Engineering*, 118, 1–16.

Part II

---

*Solar Energy for Industrial
Process Heating*

3 Solar Thermal Energy for Industrial Process Heating
Potentials and Challenges

Prashant Kumar Jangde, Vijayan Selvaraj,
T.V. Arjunan, and P. Kanagavel

CONTENTS

DOI: 10.1201/9781003263326-5

3.1 OVERVIEW

The global energy demand is significantly increasing with each passing day. Fossil fuels are still the major source of energy which accounts for almost 80% of the global energy demand [1] and are also equally responsible for causing global warming and climate change. The industrial sector of any developing country itself makes use of around one-third (37%) of its total energy consumption and releases nearly one-fourth (24%) of CO_2 gas emissions [2]. Figure 3.1 represents the energy consumed by the industrial sectors from total energy consumption in some countries. Moreover, industrial process heating (IPH) applications take a large share of energy in the form of heat. Almost every industrial process requires heat for some portion of its operation. It depends upon a particular process occurring or the product being manufactured in an industry. In the present scenario, almost 90% of industrial heat demands are satisfied through fossil fuels (coal, natural gas, and oils); only a smaller portion (less than 10%) is contributed by renewable energy sources like biomass and solar energy. The temperature requirements for various heating processes may vary from low (<50 °C) heat demanding processes like drying, washing, etc. to very high (>500 °C) heating processes including melting metals in furnaces [3, 4].

The utilization of fossil fuels for energy consumption creates a severe impact on the environment because of greenhouse gas emissions and unstable fuel prices caused by heavy imports as well as the availability of limited reserves.

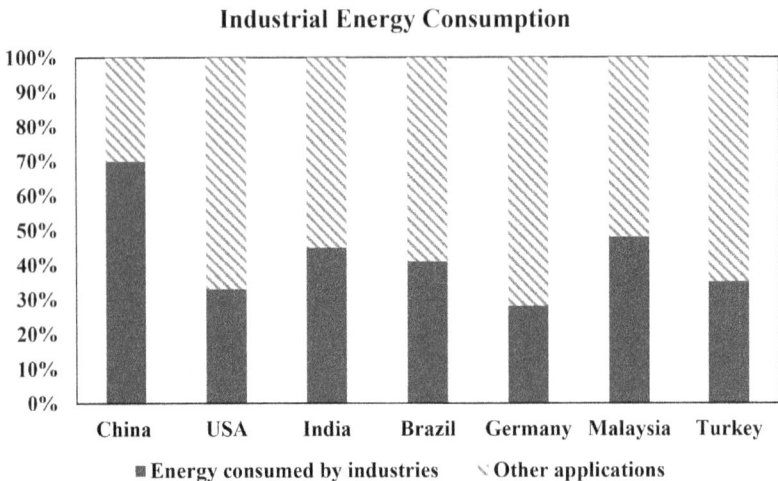

FIGURE 3.1 Industrial Energy Consumption [3].

The increased focus on mitigating carbon emissions to overcome the environmental effects for sustainable development of the countries motivates all the sectors including industries to avail the renewable energy sources for their energy demands.

3.2 INDUSTRIAL PROCESS HEATING DEMAND

Industrial process heating is viewed as the transfer of heat from a heat source to a substance during a manufacturing process for the preparation or conversion of that material from one form to another. Mostly, the heating energy requirement is satisfied either by burning a variety of fossil fuels or by using high-grade electrical energy. The fuel used may be in solid, liquid, or gaseous form, and after combustion the heat energy is exchanged for the process fluid through a well-designed heat exchanger. Electrical energy is also used directly or indirectly to generate the required heat through induction or electromagnetic radiation, etc. The thermal energy requirement is decided by the process and the appropriate source will be selected accordingly. Process heating technologies are categorized into four types according to the energy supply: fuel, steam, electric, and hybrid [5]. A typical industrial process receives thermal energy via different heat transfer mechanisms such as conduction, convection, and thermal radiation. In most cases, the conduction or convection method is employed for low-temperature processes whereas radiation is employed in high-temperature heating load applications, for instance the melting of metals. Process heating systems are as follows.

3.2.1 FUEL-BASED INDUSTRIAL PROCESS HEATING

The desired thermal energy is produced by the combustion of fuels (solid, liquid, or gas) and the generated heat is transferred to the materials directly or indirectly. The contribution of fuel-based process heating is decided according to the availability and cost of the fuel supply. For example, the U.S. is fortunately provided with adequate fossil fuel resources; there, this type of heating accounts for 64% of the total process of heating.

3.2.2 STEAM-BASED INDUSTRIAL PROCESS HEATING

The thermal energy is transported with the help of steam, which, generated using steam generators, is supplied to the process directly or indirectly (with the help of a heat transfer fluid and a heat exchanger) to the material. It can be possible to maintain a constant process heating temperature using the latent heat of the steam, and a considerable amount of energy can be transferred since steam uses latent heat. Industries like food processing industries, brewing, etc. are facilitated with biofuels or byproducts, and processes that require less than 200 °C are mostly preferred for steam-based heating systems. Around 30% of total process heating across U.S. manufacturing industries is accounted as a steam-based system. Among various industries, paper and pulp, chemical, and oil refineries are the top three consumers with energy shares of 82%, 45%, and 18% respectively.

3.2.3 ELECTRICITY-BASED INDUSTRIAL PROCESS HEATING

Electricity-powered process heating systems also use direct and indirect processes to change materials. To accomplish direct resistance heating, for example, electric current is passed directly through the appropriate materials; on the other hand, electrical energy may be inductively linked to the materials that require process heat to create indirect heating.

3.2.4 HYBRID PROCESS HEATING SYSTEMS

Hybrid process heating systems are adopted to achieve high energy performance by combining different energy sources/heating concepts. For example, microwave energy may be combined with convective hot air drying to increase the rate of drying. Optimizing the heat transfer mechanisms in hybrid systems has a huge potential to minimize energy consumption, increase speed/throughput, and enhance product quality.

In industrial process heating, for example, the manufacturing industries deal with the materials like steel, iron, and cement that require very high temperatures (>500 °C) for their processes. On the other hand, food processing, pharmaceutical/chemical, dairy, textile, paper, and pulp industries require relatively lower temperatures for most of the processes (<300 °C).

The total industrial energy demand can be separated into three sections based on the operating temperature range, for instance, low-temperature (below 150 °C), medium-temperature (150–400 °C), and high-temperature (more than 400 °C) applications, which contribute about 30%, 22%, and 48% of energy respectively out of the industrial energy demand [4] (Figure 3.2). As reported by Sharma et al. [6] a significant portion of industrial energy of nearly 60% is required for process heat load in the temperature range of 30–250 °C. Many of the renewable energy sources available are having the potential to fulfill the growing energy demand as well as have negligible pollution emissions. The industrial heating demand is expected to have an annual growth rate of 1.7% by 2030. The potential industrial processes suitable for solar thermal process heating applications are shown in Figure 3.3.

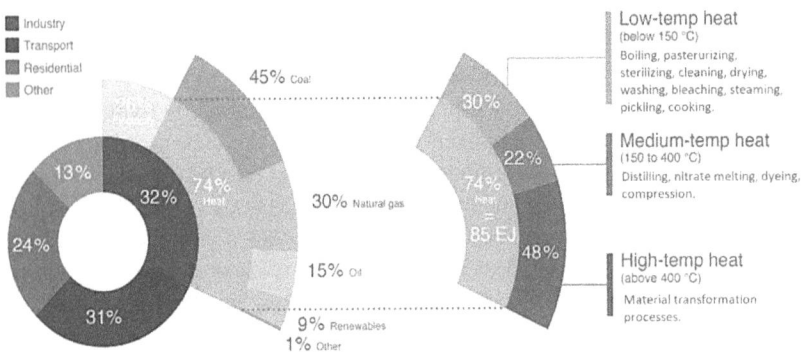

FIGURE 3.2 Share of conventional energy sources with temperature range.

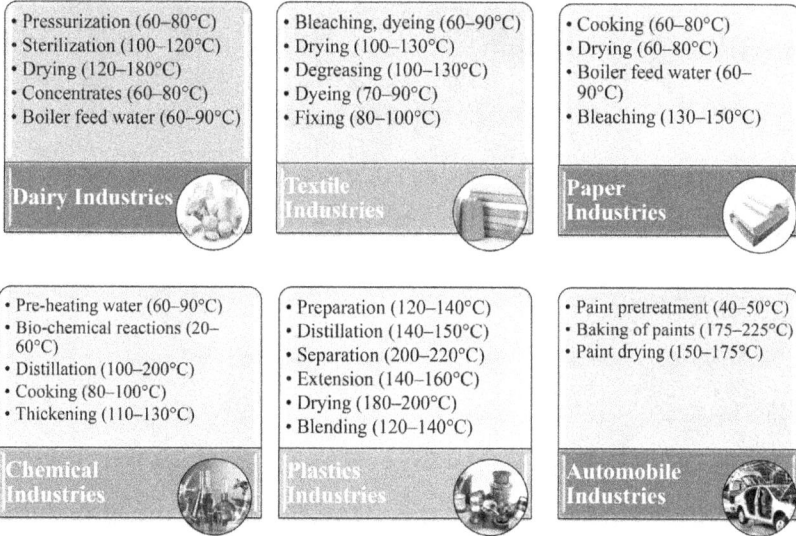

- Pressurization (60–80°C)
- Sterilization (100–120°C)
- Drying (120–180°C)
- Concentrates (60–80°C)
- Boiler feed water (60–90°C)

Dairy Industries

- Bleaching, dyeing (60–90°C)
- Drying (100–130°C)
- Degreasing (100–130°C)
- Dyeing (70–90°C)
- Fixing (80–100°C)

Textile Industries

- Cooking (60–80°C)
- Drying (60–80°C)
- Boiler feed water (60–90°C)
- Bleaching (130–150°C)

Paper Industries

- Pre-heating water (60–90°C)
- Bio-chemical reactions (20–60°C)
- Distillation (100–200°C)
- Cooking (80–100°C)
- Thickening (110–130°C)

Chemical Industries

- Preparation (120–140°C)
- Distillation (140–150°C)
- Separation (200–220°C)
- Extension (140–160°C)
- Drying (180–200°C)
- Blending (120–140°C)

Plastics Industries

- Paint pretreatment (40–50°C)
- Baking of paints (175–225°C)
- Paint drying (150–175°C)

Automobile Industries

FIGURE 3.3 Potential industrial processes with temperature ranges for solar thermal heating [7, 8, 9].

3.3 SOLAR THERMAL ENERGY CONVERSION TECHNOLOGIES

Solar energy can be harnessed in two ways; the first category involves the conversion of solar energy into electricity using photovoltaic (PV) panels and another way is the solar thermal applications converting solar energy into useful heat with the help of solar collectors. The use of solar thermal systems becomes more convenient when solar energy is available with high intensity and the ambient air temperature is high, which enables high collector efficiency [10]. Solar energy has several advantages such as its availability, eco-friendliness and free-of-cost nature, which has become a vital part to supply the energy demand in many applications. These include solar power generation, water heating, drying, desalination, cooking, solar refrigeration, industrial process heating, space heating, etc. Availing solar energy for heating demands can reduce carbon emissions and fossil fuel dependency, which facilitates the attainment of sustainable development. A wide range of temperature needs can be catered using appropriate solar thermal collectors shown in Figure 3.4.

Many solar thermal technologies have been introduced in industries for processes that require low temperatures for their operations. It is found that the non-concentrating collectors, namely flat plate collectors (FPC) and evacuated tube collectors (ETC), can be used to achieve a temperature level of up to 120 °C, which covers around 30% of the industrial process heat demand. For slightly higher temperatures up to 400 °C, concentrating solar collectors, such as parabolic dish reflectors (PDRs), parabolic troughs collectors (PTCs), and Linear Fresnel collectors (LFCs) etc. can be mainly used for steam generation, which accounts for 22% of the energy required in the industrial sector [11].

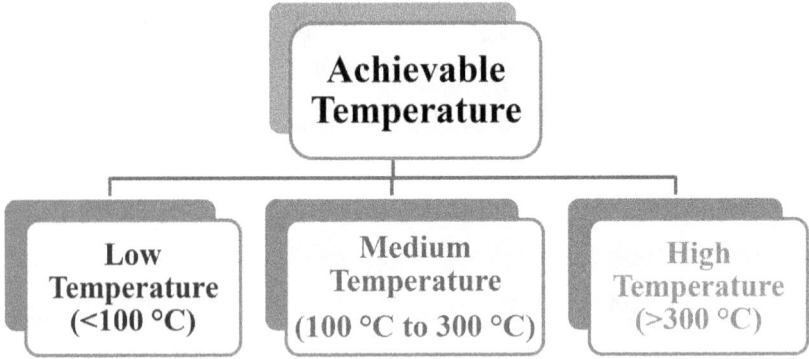

FIGURE 3.4 A

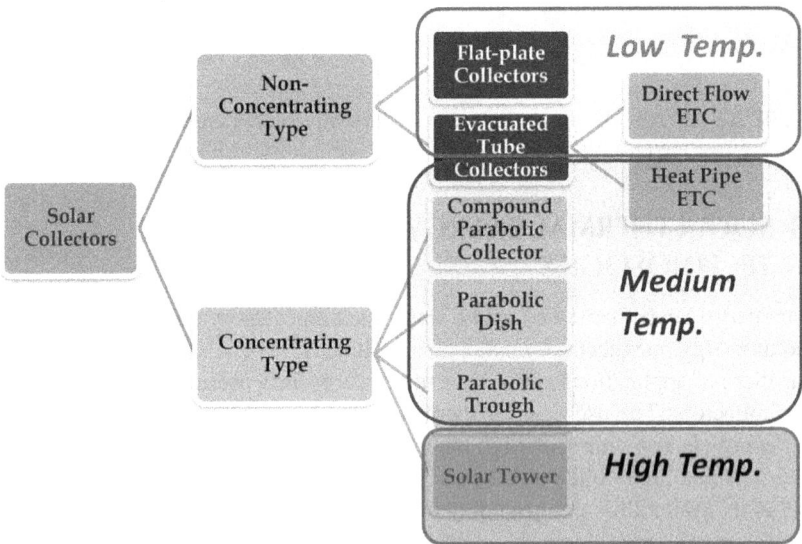

FIGURE 3.4 B   Classification of solar collectors.

As far as the conventional process heating system is considered, there are mainly two types of systems being used in the industries; these are combustion (fuel)-based process heating systems and electricity-based process heating systems [13]. A fuel-based process heating system uses different types of solid, liquid, and gaseous fuels such as wood, coal, pellets, oil, and natural gas, respectively. It also has the advantage of using various types of biomass like vegetable oil, wood waste, etc. The conversion of fossil fuels and electricity into thermal energy for process heating requires many components such as boilers, electric furnaces, burners, etc. [2]. Figure 3.5 represents the contribution of existing energy sources used in industries for process heating applications [14]. Hence, the implementation of solar thermal technologies in industries can significantly reduce the dependency on fossil fuel resources and will also be able to bring down the emission of greenhouse gases which is a matter of great importance nowadays. Although many possibilities exist which encourage the implementation of these technologies in the industrial sectors, many barriers exist in integrating the solar thermal systems into several industrial processes, as solar radiation is periodic and variable. These barriers may include complexity in design considerations as well as the economic feasibility of the system for the long term. Hence this chapter provides some of the possibilities of incorporating solar energy for many processes pertaining to such heating applications and also highlights the various challenges occurring with the use of solar technology with the existing systems in the industry.

3.4 WORKING PRINCIPLE OF SOLAR-ASSISTED INDUSTRIAL PROCESS HEATING SYSTEM

The prime consideration for selecting any solar thermal system is the temperature of the working fluid available for a certain application. It is found that in many applications the use of solar energy has been proved to be convenient and can be used directly, such as crop drying, space heating, hot water for cleaning, etc. Whereas in some applications like steam generation the solar thermal systems must be designed separately along with auxiliary heating systems to compete with the heat load. Various solar collectors are capable of producing these temperature requirements and provide a great opportunity to replace the use of fossil fuels to a certain limit. Some of these are presented in Table 3.1 with their working temperature range.

3.4.1 INTEGRATION

Aspects like availability of space, temperature control, and integration are all important and must be taken into account when determining the most appropriate solar thermal collector technology for a given situation. The use of solar energy is abundant in locations where it is necessary to gather solar energy across a significant amount of land. Energy is used by every company, especially in industrial locations where vacant land is rare. Consequently, for industrial applications using solar energy for process heat, solar-collector technology that is space efficient, such as roof-top installations, should be used. In addition, the temperature ranges of industrial operations are sometimes quite sensitive. Without a doubt, solar-collector systems should be able to provide precise temperatures in response to particular demands. In most sectors, heat carrier fluid is used to distribute heat to various operations. System integration of solar process heating systems with traditional heating systems is possible.

The important step in utilizing the solar thermal energy for industrial process heating applications is integration into the existing system. Determination of the optimal point of integration in the process heating is the crucial step in the integration process. There may be a number of processes which may require different temperature ranges and identifying the suitable collector is very important since selecting different collectors for each process is not a feasible task. The process of integration is influenced by the following criteria.

- Integration temperature level.
- Load profile (daily, weekly, annually).

TABLE 3.1

Operating Temperature Range of Various Collectors [12]

Collector Type		Temperature Range (°C)
Flat-Plate Collector (FPC)		30–80
Evacuated Tube Collector (ETC)		50–200
Compound Parabolic Collector (CPC)		60–240
Linear Fresnel Reflector (LFR)		60–250
Parabolic Trough Collector (PTC)		60–300
Parabolic Dish Reflector (PDR)		100–500
Heliostat Field Reflector (HFR)		150–2000

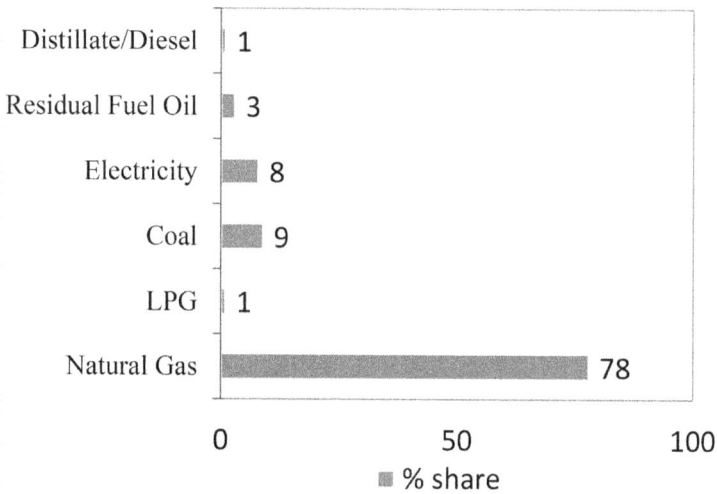

FIGURE 3.5   Share of various sources of energy in conventional process heating systems [14].

- Amount of thermal energy consumed annually.
- Sensitivity to changes.
- Achievable solar fraction.

Analyzing these criteria will ease the process of identification and integration of solar thermal collectors into the existing system. The prime objective is to maximize the amount of solar thermal energy supplied to meet the required amount of energy per unit. Depending on the temperature of the heat demand, the capture temperature may be adjusted so that a greater fraction of the heat demand can be met by solar thermal energy. There is a greater potential for solar thermal energy collection at lower temperatures due to the increased efficiency of the solar capture equipment. But raising the temperature allows for greater heat transfer. Figure 3.6 shows the possible integration points for a solar process heating system into an existing system.

Three basic approaches might be used to integrate these solar-collector technologies into a conventionally operating system.

1. When thermal energy is needed for evaporation, the pre-heating of water is used first. As a result, the condensate might return to the collector at a high temperature, reducing the efficiency of the non-concentrating collector.
2. Installation at the distribution level when high temperatures (mainly steam) are needed and only concentrating solar collectors may be utilized is an example of this. Because it is not tied to a single procedure, it offers the greatest amount of freedom.
3. Direct connection with a particular operation is another key method of integrating solar thermal energy. The utilization of energy, however, may be difficult when the process is modified or interrupted since it is rather rigid.

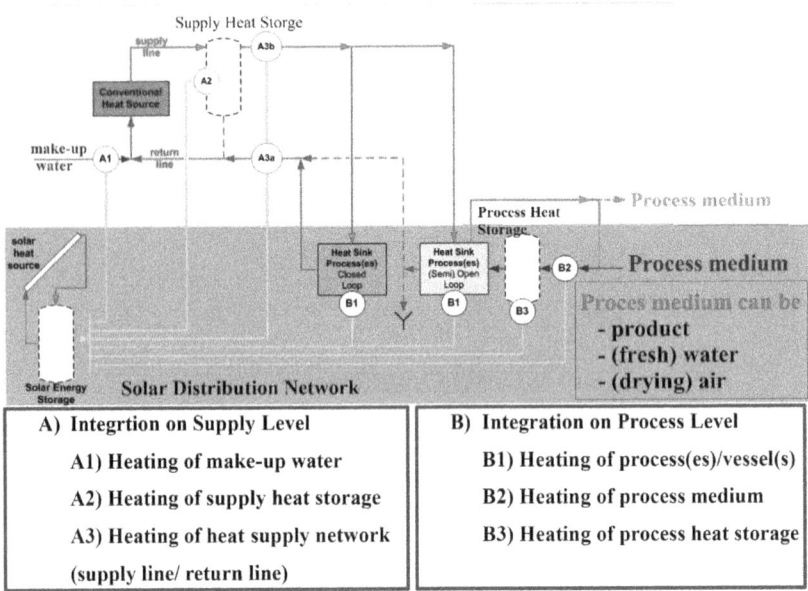

FIGURE 3.6   Possible integration points for solar process heat.

A typical solar-collector system mainly consists of an array of solar collectors to absorb the solar energy, a working fluid (air, water, HTF, etc.) to transmit the absorbed energy to the process, and a heat transfer device such as a heat exchanger. Solar energy is periodic and irregular in nature and varies throughout the year, whereas the load demand of industries remains nearly constant. Hence integrating the solar-collector technology directly with the industrial processes will not be significant without the use of an auxiliary heating device like a boiler, electric heater, etc. On the other hand, the energy produced by such systems in excess of the designed load may be useless without the use of energy storage devices. There are many ways to store thermal energy; these including sensible heat type, latent heat type, and thermo-chemical type. In the sensible heat storage technique, the storage capacity is low and requires a large volume of the working medium which means the system's efficiency is also poor. On the other hand, latent heat type storage systems, since the heat is stored in the form of latent heat, provide better efficiency, constant temperature during phase change, and compact design as it has high heat storage capacity [15]. There are many other techniques to enhance the thermal performance of various collectors. These are, namely, the use of reflectors, nanoparticles, improved coating surface, desiccant wheels, multiple passes of air through the collector, and jet impingement techniques [1].

The performance of any solar thermal energy conversion system is assessed in terms of a parameter known as "solar fraction". It can be defined as the portion of total heat required for any industrial process supplied by the use of solar thermal systems. Solar fraction generally varies according to the availability of solar energy at the location, estimated heat load, type, and area of the collector as well as the size of the energy storage device [16]. A higher value of the solar fraction is an indication of the minimal use of conventional sources of energy hence pollution emissions will be less.

3.5 THE POTENTIAL OF SOLAR INDUSTRIAL PROCESS HEAT (SIPH) IN VARIOUS INDUSTRIES

Solar thermal technology for domestic applications like water heating, pool heating, distillation and building space heating has been successfully implemented and nowadays it can be seen in many places around us. However, in these applications the use of solar thermal technologies has not been explored to a greater extent. It is the industrial sector that provides a wide range of applications depending upon the type of industry, particular process, and operating temperatures [16]. Solar thermal systems are successfully installed for process heating applications in many industries including dairy, food, beverage, pharmaceutical, chemical, textile, automobile, plastics, beverages, paper, etc. (Figure 3.7). The establishment of solar thermal systems in many countries across the world began in the late 1970s. In the 1980s the first industrial-scale SIPH systems based on concentrating solar power (CSP) technology was built in the Mojave Desert of the USA using parabolic trough collectors [18]. Some of the potential studies are mentioned later for different industrial sectors.

3.5.1 TEXTILE INDUSTRY

In cotton-based textile industry, many processes are involved such as the production of yarn, which is obtained from a sequence of operations like blowing, carding, combing, drawing, and dyeing. Yarn is then converted into fabric with the help of a weaving operation. Fabrics are then processed into final finished products. However, around 80% of the total energy is required in the finishing operation which encompasses many operations as mentioned in Table 3.1, which mainly require heat in the temperature range of 40–130 °C [17].

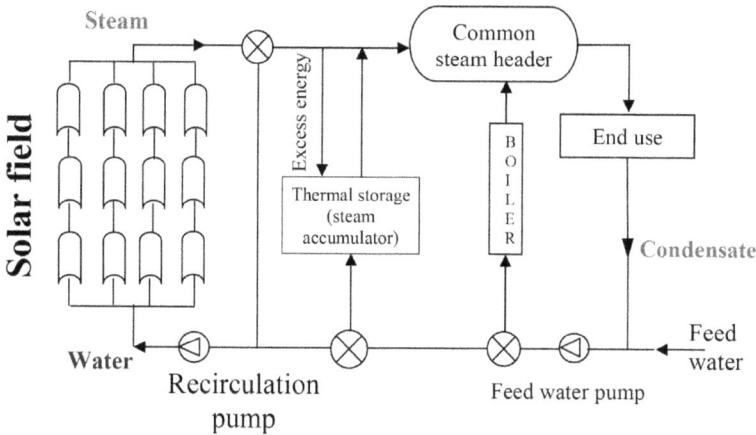

FIGURE 3.7   SIPH system [17].

According to a potential study conducted in the textile industries for Mexico, it was found that around 68% of the total energy requirement is basically in the form of process heat only. This energy is supplied with the help of liquefied gas, natural gas, and diesel. The working fluids being used are water and low to medium pressure steam having a temperature range from 60 to 180 °C. To fulfill this energy demand of mainly concentrating type solar collectors, more precisely the parabolic trough collectors are found to be the appropriate technology. As per their estimation, the system had a total collection area of 75084 m^2 and produced 45051 kW of energy which was equivalent to 131.54 GWh/year. The total capital cost for the installation of the systems was estimated to be 17 million USD [19].

3.5.2 PAPER INDUSTRY

The paper industry is another sector that requires large energy demand for various activities, with 75–85% of the energy required for process heating whereas 15–25% of energy is in the form of electrical power. In general, the production of paper involves many stages such as raw material preparation, pulp production, bleaching, chemical recovery, and paper-making. The raw materials are converted into pulp which is used to produce paper and paper products. Most of the energy is used in form of process heat in pulp production, particularly in digesting, evaporating, and washing.

In the present scenario, the energy is supplied with the help of various fuels like coal, petroleum fuels, furnace oil, agro-residues, biomass, and electricity out of which coal and petroleum fuels themselves contribute more than 68% of the total energy requirement. In paper mills, the basic operations which consume the maximum amount of heat energy are pulping, drying, bleaching, and washing of raw materials. Heat transfer mediums such as water, steam, air, or thermic oils are utilized to transport energy to the specified processes [20].

According to a study conducted to evaluate the SIPH potential in India, the total process heating load for the paper industry with 8 different locations has been estimated to be 25.2 PJ/annum. This demand can be met with the use of a parabolic trough-based process heating system having a collector area of 1.11 million m^2 which can yield an annual solar fraction of 0.25 to 0.30. This system without energy storage is also expected to reduce 0.34 million tons of CO_2 emissions per annum [21].

3.5.3 CHEMICAL INDUSTRY

Today chemical industries are needed for satisfying the highly turbulent routine needs of the customer related to cosmetics, food processing, wall paint applications, wood applications, pharmaceuticals, etc. Hence it is indeed necessary to identify the potential challenges restricting the development of chemical industries globally. This sector has a percentage share of 13.5%, 20.9%, 44.7%, and 11% in the temperature range of below 100 °C, 100–500 °C, 500–1000 °C, and above 1000 °C respectively for process heat applications [22]. In this sector, the largest capacity 1050 kW SIPH plant is in China using PTC for the generation of steam at a temperature of 130 °C and provided with a collector area of 4600 m^2.

A pharmaceutical plant situated in Egypt employs a solar thermal plant-based process heating system in which parabolic trough collectors have a surface of 1,900 m² and produce 1.3 t/h of saturated steam. The plant makes use of a saturated steam system operating at 7.5 bar; however, in the collector loop first the pressure of water is reduced with the help of a flashing valve and then supplied to the steam network.

3.5.4 DAIRY INDUSTRY

The dairy industries also require a considerable amount of thermal energy in which a large fraction of about 70% of energy is required for process heating, which falls under the temperature range of 50–250 °C. This fraction of energy is supplied with the help of fossil fuels with a significant contribution from the biomass also in the boilers. Figure 3.8 represents the flowchart of processes involved in the typical dairy industry. The working fluid in the dairy industry is mainly in the form of steam and pressurized hot water produced at different temperatures and pressures. During dairy production, heat is required to a large extent for pasteurization and pre-heating of a boiler. Pasteurization is a process that belongs to the removal or deactivation of bacterial pathogens which is achieved by heating the raw milk below its boiling point. Such a process is essential to increase the shelf-life of dairy products.

According to a study conducted to estimate the opportunities for commercialization of solar energy in industrial sectors, India is one of the largest milk-producing countries, a fair scope of solar thermal technologies has been implemented in the dairy industry. It was suggested that the SIPH system without using any storage technology is capable of meeting 20–30% of the energy demand for milk processing in this industry. The overall energy demand of 6.40 PJ/annum was estimated, which was further reduced to 4.50 PJ per annum provided that the process heating demand is only for the pasteurization process. The corresponding energy demand

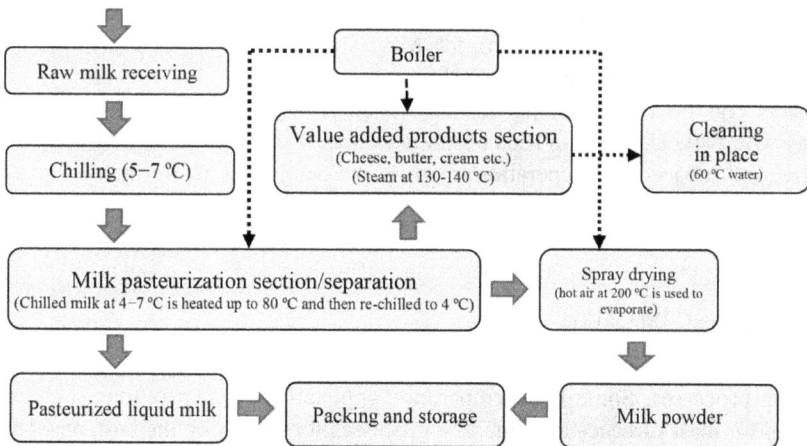

FIGURE 3.8 Flow chart of the milk processing unit [13].

is fulfilled using Arun 160 solar collector with a collector area requirement of more than 1.62 million m^2 with the average solar fraction varying between 0.18–0.32. Another advantage of the potential study estimation suggested that the system would be able to reduce about 32–144 thousand tons of CO_2 emission in a year [13].

At present many SIPH systems are in operation globally within the dairy industries. In India, Mother Dairy has installed a solar thermal system using various types of collectors with a total installed capacity of 1064 kW, accounting for 1520 m^2 of surface area and producing a temperature range of 70–90 °C. The Netherlands, having installed its largest SIPH system using FPC, produces 1680 kW and has a collector area of 2400 m^2. It also has the provision of a storage facility of 95 m^3 [11].

3.5.5 FOOD PROCESSING INDUSTRY

In the food processing industry, there are many operations such as pasteurization, sterilization, steaming, boiling, blanching, drying, cleaning, etc., which require a large amount of energy as process heat. By the year 2020, around 95 SIPH plants with a total installed capacity of 41 MW have been reported to be in use in the food industry across the world with a collector area of 69,124 m^2. Out of these plants, 38% are using FPC technology whereas 20% have installed PTC and 20% with ETC; remaining plants are also making use of other collectors such as linear Fresnels, air collectors, parabolic dishes, etc. [11, 23].

The use of solar thermal technology for food processing and preservation started many decades ago. In California, an onion drying plant employed an ETC-based solar heating system for pre-heating air and boiler feedwater. Hot water is produced by the system in the temperature range of 70–100 °C for its operation and the pre-heated air was used in a gas-fired furnace which supplied air at a temperature of 93 °C to the dryer.

In Australia, the first SIPH system using a flat-plate collector of area 94 m^2 was used in the food industry. The system produces warm water to heat the cans before packing them in a soft drink factory [24]. The USA had its largest SIPH system installed in the year 2012, consisting of flat-plate collectors with a surface area of 7804 m^2 and capable of producing 5.5 MW. The system was used to produce hot water in a turkey processing plant [25].

Many countries are utilizing air-based food drying systems using solar energy along with bioheat. In Costa Rica a similar system is installed for drying coffee, tea, maize, and tobacco. The operation involves pre-heating of air through an array of glazed and unglazed solar air collectors. This pre-heated air is then supplied to the biomass boiler for further heating and processing.

3.5.6 MINING INDUSTRY

In many mineral processing plants, thermal energy is required in metal extraction and refining processes. Some high-temperature applications for operations such as air heating for blast furnaces and smelting processes for extracting the base metal from their ore can be obtained using concentrating solar thermal technologies. Whereas small to medium temperature applications like cleaning, mining, water heating, electro-winning processes, nickel baths, degreasing chemicals, etc. are achieved by the use of flat-plate and evacuated tube collectors [9].

The world's largest SIPH system in the mining industry is installed in Oman and is integrated with PTC collectors, having a surface area of 210,000 m^2 and producing 100 MW. The second largest installed project is in Chile (2013), to refine copper in the mining industry. The system consists of flat-plate collectors with an area 39,300 m^2 with a capacity of 27.5 MW along with a storage facility of volume of 4000 m^3. The third one is used in a copper mining industry situated in Mexico, provided with FPC with a surface area of 6270 m^2 and generating 4.4 MW [26].

As suggested by Rezæi et al. [27], a solar collector array can be utilized to pre-heat the gas for the processing of iron in the CGS, which could minimize a significant amount of fuel and able to replace the conventional air heating system. It was shown that with an assembly of 450 collector modules, the fuel cost can be reduced up to 20 USD/hr which is also equivalent to the annual fuel saving of about 10678 USD; the payback period was estimated to be nearly 9 years.

3.5.7 AUTOMOBILE INDUSTRY

The automobile sector involves different process heating applications that require a tremendous amount of energy in the form of both heat and electricity. At present many countries such as South Africa, India, and Spain are implementing SIPH systems in this sector for their production systems. There are many applications in the production line of the automobile industry like paint application processes, washing of various components of the automobile vehicle, pre-treatment of body and surface preparation, etc. [9]. The heat transfer medium generally being used is hot air and, in many cases, pressurized hot water and saturated steam in the temperature range of 200 °C, which makes solar technology to be most suitable to overcome the dependency on conventional sources of energy in this sector [8].

3.6 CHALLENGES

It is clear from the previous studies that the implementation of SIPH systems in the various industrial processes has a large scope to meet the growing energy need. New solar technologies are flexible, energy efficient, and reduce pollution emissions on a large scale. Even though there exists a huge potential to convert solar energy into useful heat to fulfill the process heat load of several industries, the installed systems are limited in numbers. This is due to several challenges occurring in actual practice. Many researchers [3, 6, 12, 24, 28, 29, 30, 31] have identified the major challenges which adversely affect the integration of solar technologies, such as technical, economic, institutional, social, political issues, etc., and are summarized next.

3.6.1 TECHNICAL CHALLENGES

It can be noted that the technical difficulty, such as the dearth of design, integration, and optimization tools in the existing and optimized process heating streams, to design and construct the systems for testing and evaluating the performance results is the main barrier, which indicates weakness of the renewable energy technology. The SIPH systems can be integrated into the running manufacturing plants if the plant design has the provisions to provide sufficient land space for the installation of solar collectors, which generally have a large surface area. This land could be either open land or rooftop space. The collector area plays an important role of designing the SIPH system, which can be limited due to the infrastructure requirements in the building. Solar thermal systems are predominantly used for hot water for domestic use, drying, space heating, and heating of swimming pools, etc. However, many industrial processes may require a higher temperature range. Such a temperature could be difficult to achieve with the existing technologies, hence it requires new designs and new materials of the collector in order to reach such temperature limits. Excessive research and funding may be involved for such technological developments. An increase in fraction of solar energy share in total energy beyond a certain limit increases the size of the components, surface area of the collector, mass flow rate requirement of the heat transfer fluid, capacity of the storage device, etc., which in turn increases the cost of the SIPH. Without the use of energy storage systems, the useful surplus energy generated by the solar thermal system has to be simply thrown out. This energy can be released in absence of solar energy. Available energy storage mediums can meet this but for a few hours only, hence the development of such materials is still found to be challenging for the researchers. There is a requirement for many types of data from the actual process of heating energy use from the industries to the developers. It is also worth mentioning that the technology suffers from a lack of manufacturers, skilled technicians, and installers. Other limitations are the inadequacy of designers who are well qualified and competent enough, the requirement of software support in the integration process, etc. These technologies also suffer because the research institutes and centers which have relevant information and understanding in this field are very few. New guidelines, planning tools, and strategies must be framed for the industries. The use of solar techniques along with conventional sources must be done in such a way that it should continuously fulfill the energy demand without fail consistently. Other technical challenges include low efficiency of the collector as the solar collector suffers the loss of heat to the surroundings due to convection and conduction heat losses, limiting heat carrying capacity of the heat transfer fluids, and energy storage requirements.

3.6.1.1 Weather-Dependent Energy Source

One of the major challenges in utilizing solar-collector technology for continuous industrial process is the dependency on weather conditions. For example, cloudy, wet, and cold weather, which cannot be controlled, reduce the solar system's effectiveness, hence unpredictable weather is a challenge for solar energy. Due to Earth's rotation, solar radiation varies diurnally. Solar radiation fluctuations impact process performance [15]. The source's unpredictable behavior makes it unsuitable for continuous energy delivery [25]. Installing extra solar collectors to supply heat requirements during low sun intensity would be costly. Strong winds and tropical storms may cause damage to the solar-collector systems. Solar radiation forecasts may extract meteorological data during design. An analytical model can be developed for forecasting a system's daily performance and may be evaluated and improved through experimental investigation and validation [32]. The change in the design of the solar-collector system and thermal storage unit must be optimized due to changes in the load profile of heat demand from industrial operations. While designing a solar thermal system for industry, significant attention must be paid to the temperature requirements according to the different processes involved in the conversion of raw materials into the desired product. The type of heat supply required may vary as per the process requirements. For example, in the automobile industry hot water is required for the degreasing process and hot air for paint drying processes.

3.6.1.2 Variety of Demand Profiles in Industrial Processes

The changes in the temperature requirement are an important challenge to be accounted while designing a solar thermal system. There may be a wide variety of products and temperature requirements in an industry operation, where several products are produced. Additionally, while planning an industrial process heat application, the kind of energy to be employed must be taken into consideration. Choosing the right kind of solar collector and creating the best solar thermal storage is thus crucial.

3.6.1.3 Space Limitation for Solar Collector Installation

One of the major factors to be considered while installing the solar collector is the availability of land area. The examination of available space is critical for calculating the greatest solar collector area that may be deployed [33]. Furthermore, solar collectors must be located near demand points to reduce transmission or heat losses. As a result of the significant quantity of land necessary to install solar collectors, the industry prefers the rooftop setting. Unfortunately, not all building structures are suitable for roof installation. Few solar collection technologies such as moderate temperature solar collectors, including evacuated tube collectors, can be efficiently installed on the roof of the industries to produce steam or hot water for its requirements.

3.6.1.4 Overheating and Stagnation Problems

Sometimes, the heat transferring fluid fails to collect the absorbed solar energy from the collector surface that would lead to a high temperature at the collector. This is called stagnation. The stagnation problem can be encountered by integrating an appropriate stagnation management system to protect the solar collectors from stagnation issues [34]. The overheating issue may occur when the solar irradiance is high, with low heat demand for the process, or improper circulation of heat transferring fluid through the absorber surface. In addition to the obvious concerns associated with supporting the thermal expansion of solar array headers and pipe runs at high temperatures, further issues may occur. The abrupt pressure reduction might result in thermal shock and strains related to the rapid boiling/vaporization/condensation of the heat transfer fluid. The stagnation temperature of the collector may be controlled either by decreasing the input of solar energy or by eliminating the surplus heat that is produced by the collector [35].

3.6.1.5 Low Solar Intensity

Solar intensity variations are another barrier to solar thermal energy's widespread use. Solar energy is a preferable energy source in certain regions of ASEAN since it relies on solar radiation, which depends on geographical location [23]. Therefore, it is highly essential to conduct feasibility study to assess the possible thermal energy collection. Moreover, a solar collector can generate thermal energy for 12 hours a day, with peak output around noon, and also the conversion efficiency of the solar collectors is also very low [25]. For example, annual solar efficiency of concentrated solar power ranges by 8–10% for solar power towers, 16–18% for solar dishes, 10–15% for parabolic troughs, and 9–11% for Fresnel reflectors [8]. For utilizing, concentrated solar power systems are in need of direct sunshine, which leads to addition of tracking panels that follow the sun may extend this prime generating duration, raising installation costs.

3.6.2 ECONOMIC CHALLENGES

For small to medium scale industries, high investment cost is associated due to the design complexity and installation costs, and also the operations costs are higher as compared to the other conventional systems. One of the major barriers which makes the proposed technology less fascinating is the availability of cheaper fossil fuels. As the solar collectors suffer significant energy loss, the development of a solar heating system to produce energy at a lower price is still challenging for the researchers. If we go for high-temperature applications then the installation of tracking system becomes essential, which further increases the running and capital cost. It can be seen that industries have production targets and they have to stick to their schedules in order to fulfill those demands, and product quality also pays a significant role for them, which requires flexible, promising, and reliable technology. Many industries require less payback time specially when the initial investment is high as they may be bound under financial extremity. It is well known that the investment costs associated with most of the SIPH systems using renewable energy technology is higher; however such investments are expected to save the use of conventional energy sources. Most decision makers of the industries are very unprogressive in nature when it comes to basic needs for the adoption of such technology, such as availability of land. As far as the critical heating processes are concerned, they will first prefer the long-term proven technology in spite of thinking about the long-term damage due to these sources. Lack of competition among the manufacturers makes this technology resistible. Some of the solar thermal applications are suitable for a particular period of time such as for building space heating, crop drying, etc. Such applications become less economical as well as attractive as compared to the systems operated all around the year. Hence the demand for thermal energy must exist throughout the year. The global market is competitive enough to manufacture products at low cost. It is hard to invest large amounts of money on research and development for any manufacturers or entrepreneurs until the technology is proven to be the most economical. This is possible by taking into account the harmful effect on humans as well as the environment caused by burning the conventional sources of energy. Another reason involves the unpredictability of the investment cost as well as the expected benefits from the applications.

3.6.3 SUPPORT FROM GOVERNMENT AGENCIES

As substantial amount of energy is required in the aforementioned industries; hence, government policies and support can encourage the use of solar-collector technologies for process heating. Solar photovoltaic and solar power generation is being provided in some areas. Sometimes investors may suffer deficiency of funds; hence support may be provided in terms of financial benefits like loans, tax benefits, subsidies, etc. To encourage the investors towards SIPH, loans should be provided at a lower rate than the existing rates on commercial loans. Another good step in this direction is if the government will provide some sort of exemption in capital costs in the form of subsidies. New regulations and policies can be framed for industries to limit exhaust gas emissions parallelly with the energy being supplied by solar-assisted heating systems. Fines and other taxes can be taken from industries for high carbon emissions, which may limit the use of fossil fuels. The cost associated with the adverse effects caused by the pollutants on human bodies and the environment must be taken into account. The development of new technologies or systems is to be standardized for large-scale commercialization. Most of the new technologies suffer from the lack of manufacturing standards to ensure ease of availability and the standards are yet to be set by the regulatory authorities.

3.6.4 Environmental Challenges

Solar energy is periodic in nature and the uncertain climatic conditions lead to poor thermal performance, making it less attractive to implement. It is difficult to operate the solar-assisted industrial process heating system to achieve thermal energy at a constant temperature throughout the day as solar intensity varies with time [36]. The major difficulty is the unavailability of the sun in the nighttime, hence the industries operating in night shifts must have an auxiliary heating system. Such solar systems cannot be fruitful in industries working in many shifts. Scale and dust formation also takes place, as the system has to be kept outdoors under open sky.

3.6.5 Social Challenges

Limited capacity of the solar thermal installation for process heating in industries makes most decision makers of the relevant industries unaware about such adaptive potentials of solar thermal technologies, which makes it the key barrier to the adoption of SIPH systems for a wide range of applications in industrial sectors. As conventional technologies are already adapted and reliable, manufacturers and customers hesitate to acquire a new technology when it comes to the market. Also, there is a lack of perception among the people about using the products and services of these new technologies. Apart from these, it is also important to mention that many types of family income groups are present in the society; hence it is necessary that the products can be made affordable to each group individually. This in turns depends upon the capital cost associated with the system. Therefore, manufacturers are also not ready to initiate producing new equipment.

3.7 CONCLUSION

The industrial sector alone accounts around 35% of the total energy consumption of the world, and the majority of the energy requirement is satisfied through fossil fuels. Burning of fossil fuels releases greenhouse gases into the atmosphere, which results in undesirable effects on the environment such as global warming, climate change, etc. Solar energy has been identified as the most optimal energy resource to fulfil the thermal energy needs of the industries. As there is a lot of solar energy, installing such system will be a long-term way for industries to move toward a future with no carbon emissions. There is a huge potential for solar heating in many industries, which can reduce the dependency on fossil fuels and the environmental pollution. The selection of an appropriate solar thermal technology mainly depends on the range of temperature requirements of the process. Stationary type of collectors are the best choice for low and moderate heating loads in an industry, that is, flat-plate collectors for achieving a temperature between 30 to 80 °C, whereas evacuated tube collectors are for 50 to 200 °C. On the other hand, concentrating collectors are used for the heating applications above 250 °C or systems integrated with thermal energy storage. However, more efforts are needed from the researchers and government policy makers for successful commercialization of the system.

REFERENCES

[1] Jangde, P. K., Singh, A., & Arjunan, T. V. (2022). Efficient solar drying techniques: A review. Environmental Science and Pollution Research, 29(34), 50970–50983.

[2] Schoeneberger, C. A., McMillan, C. A., Kurup, P., Akar, S., Margolis, R., & Masanet, E. (2020). Solar for industrial process heat: A review of technologies, analysis approaches, and potential applications in the United States. Energy, 206, 118083.

[3] Kumar, L., Hasanuzzaman, M., & Rahim, N. A. (2019). Global advancement of solar thermal energy technologies for industrial process heat and its future prospects: A review. Energy Conversion and Management, 195, 885–908.

[4] Kumar, L., Hasanuzzaman, M., Rahim, N. A., & Islam, M. M. (2021). Modeling, simulation and outdoor experimental performance analysis of a solar-assisted process heating system for industrial process heat. Renewable Energy, 164, 656–673.

[5] Cresko, J., Shenoy, D., Liddell, H. P. H., & Sabouni, R. (2015). Innovating clean energy technologies in advanced manufacturing. In: Quadrennial Technology Review.

[6] Sharma, A. K., Sharma, C., Mullick, S. C., & Kandpal, T. C. (2017). Solar industrial process heating: A review. Renewable and Sustainable Energy Reviews, 78, 124–137.

[7] Allouhi, A., Agrouaz, Y., Amine, M. B., Rehman, S., Buker, M. S., Kousksou, T., . . . & Benbassou, A. (2017). Design optimization of a multi-temperature solar thermal heating system for an industrial process. Applied Energy, 206, 382–392.

[8] Uppal, A., Kesari, J. P., & Zunaid, M. (2016). Designing of solar process heating system for Indian automobile industry. International Journal of Renewable Energy Research, 6(4), 1627–1636.

[9] Farjana, S. H., Huda, N., Mahmud, M. P., & Saidur, R. (2018). Solar process heat in industrial systems–A global review. Renewable and Sustainable Energy Reviews, 82, 2270–2286.

[10] Sing, C. K. L., Lim, J. S., Walmsley, T. G., Liew, P. Y., Goto, M., & Bin Shaikh Salim, S. A. Z. (2020). Time-dependent integration of solar thermal technology in industrial processes. Sustainability, 12(6), 2322.

[11] Ismail, M. I., Yunus, N. A., & Hashim, H. (2021). Integration of solar heating systems for low-temperature heat demand in food processing industry–A review. Renewable and Sustainable Energy Reviews, 147, 111192.

[12] Kalogirou, S. (2003). The potential of solar industrial process heat applications. Applied Energy, 76(4), 337–361.

[13] Sharma, A. K., Sharma, C., Mullick, S. C., & Kandpal, T. C. (2017). Potential of solar industrial process heating in dairy industry in India and consequent carbon mitigation. Journal of Cleaner Production, 140, 714–724.

[14] Hasanuzzaman, M., Rahim, N. A., Hosenuzzaman, M., Saidur, R., Mahbubul, I. M., & Rashid, M. M. (2012). Energy savings in the combustion-based process heating in industrial sector. Renewable and Sustainable Energy Reviews, 16(7), 4527–4536.

[15] Gajendiran, M., & Nallusamy, N. (2014). Application of solar thermal energy storage for industrial process heating. Advanced Materials Research, 984, 725–729.

[16] Karki, S., Haapala, K. R., & Fronk, B. M. (2019). Technical and economic feasibility of solar flat-plate collector thermal energy systems for small and medium manufacturers. Applied Energy, 254, 113649.

[17] Sharma, A. K., Sharma, C., Mullick, S. C., & Kandpal, T. C. (2017). GHG mitigation potential of solar industrial process heating in producing cotton based textiles in India. Journal of Cleaner Production, 145, 74–84.

[18] Peters, M., Schmidt, T. S., Wiederkehr, D., & Schneider, M. (2011). Shedding light on solar technologies—A techno-economic assessment and its policy implications. Energy Policy, 39(10), 6422–6439.

[19] Ramos, C., Ramirez, R., & Beltran, J. (2014). Potential assessment in Mexico for solar process heat applications in food and textile industries. Energy Procedia, 49, 1879–1884.

[20] Sharma, A. K., Sharma, C., Mullick, S. C., & Kandpal, T. C. (2015). Potential of solar energy utilization for process heating in paper industry in India: A preliminary assessment. Energy Procedia, 79, 284–289.

[21] Sharma, A. K., Sharma, C., Mullick, S. C., & Kandpal, T. C. (2016). Carbon mitigation potential of solar industrial process heating: Paper industry in India. Journal of Cleaner Production, 112, 1683–1691.

[22] Lauterbach, C., Schmitt, B., Jordan, U., & Vajen, K. (2012). The potential of solar heat for industrial processes in Germany. Renewable and Sustainable Energy Reviews, 16(7), 5121–5130.

[23] Ismail, M. I., Yunus, N. A., Kaassim, A. Z. M., & Hashim, H. (2022). Pathways and challenges of solar thermal utilisation in the industry: ASEAN and Malaysia scenarios. Sustainable Energy Technologies and Assessments, 52, 102046.

[24] Fuller, R. J. (2011). Solar industrial process heating in Australia–Past and current status. Renewable Energy, 36(1), 216–221.

[25] Huang, J., Li, R., He, P., & Dai, Y. (2018). Status and prospect of solar heat for industrial processes in China. Renewable and Sustainable Energy Reviews, 90, 475–489.

[26] Farjana, S. H., Mahmud, M. P., & Huda, N. (2020). Solar process heat integration in lead mining process. Case Studies in Thermal Engineering, 22, 100768.

[27] Rezæi, M., Farzaneh-Gord, M., Arabkoohsar, A., & Dashtebayaz, M. D. (2011). Reducing energy consumption in natural gas pressure drop stations by employing solar heat. In: World Renewable Energy Congress-Sweden (No. 057, pp. 3797–3804). Linköping University Electronic Press.

[28] Mostafaeipour, A., Alvandimanesh, M., Najafi, F., & Issakhov, A. (2021). Identifying challenges and barriers for development of solar energy by using fuzzy best-worst method: A case study. Energy, 226, 120355.

[29] Shojaee, S. M. N., Moradian, M. A., & Mashhoodi, M. (2015). Numerical investigation of wind flow around a cylindrical trough solar collector. Journal of Power and Energy Engineering, 3(01), 1–10.

[30] Sindhu, S. P., Nehra, V., & Luthra, S. (2016). Recognition and prioritization of challenges in growth of solar energy using analytical hierarchy process: Indian outlook. Energy, 100, 332–348.

[31] Sharma, A. K., Sharma, C., Mullick, S. C., & Kandpal, T. C. (2017). Effect of incentives on the financial attractiveness of solar industrial process heating in India. Renewable Energy and Environmental Sustainability, 2, 33.

[32] Bellos, E., & Tzivanidis, C. (2018). Development of an analytical model for the daily performance of solar thermal systems with experimental validation. Sustainable Energy Technologies and Assessments, 28, 22–29.

[33] Srivastava, S. P., & Srivastava, S. P. (2013). Solar energy and its future role in Indian economy. International Journal of Environmental Science: Development and Monitoring, 4(3), 81–88.

[34] Rimar, M., Fedak, M., Vahovsky, J., Kulikov, A., Oravec, P., Kulikova, O., Smajda, M., & Kana, M. (2020). Performance evaluation of elimination of stagnation of solar thermal systems. Processes, 8(5), 621.

[35] Hussain, S., & Harrison, S. J. (2015). Experimental and numerical investigations of passive air cooling of a residential flat-plate solar collector under stagnation conditions. Solar Energy, 122, 1023–1036.

[36] Schnitzer, H., Christoph, B., & Gwehenberger, G. (2007). Minimizing greenhouse gas emissions through the application of solar thermal energy in industrial processes. Approaching zero emissions. Journal of Cleaner Production, 15, 1271–1286.

4 Solar Thermal Energy for Industrial Process Heating Applications
Medicinal/Pharmaceutical Industries

Thota S.S. Bhaskara Rao and S. Murugan

CONTENTS

4.1 INTRODUCTION

Globally, the pharmaceutical sector is an integral component of health care systems. Many organizations are actively involved in discovering, developing, producing, and distributing the medicines. There are wide variety of unit operations involved in the process of manufacturing the drugs in pharmaceutical industries; blending, granulation, grinding, coating, tablet pressing, filling, and other unit activities are all parts of the pharmaceutical processing (*Making Pharma Greener with Solar Energy—Express Pharma*, n.d.). The schematic view of the tablet manufacturing process is shown in Figure 4.1.

The pharmaceutical process is usually composed of a combination of processes based on the physical and chemical properties of the active ingredient of the drug. The variety of pharmaceutical processes are explained in Figure 4.1.

Dry granulation: The density of low-grade powder is compacted to make granules. The roller assembly process consists of screw feed, compaction, and grinding systems.

DOI: 10.1201/9781003263326-6

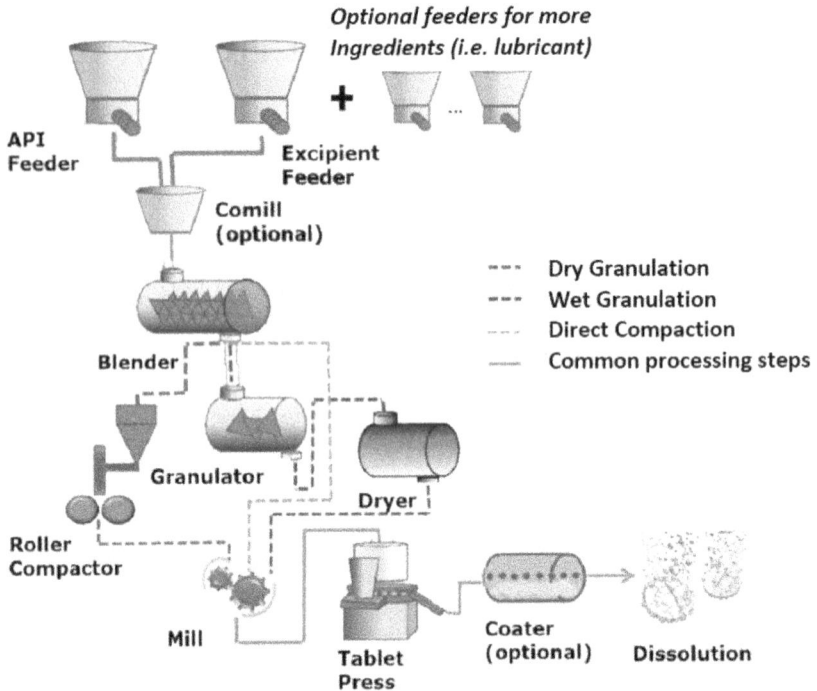

FIGURE 4.1   Tablet manufacturing process in pharmaceutical industry.

Source: *Dry Granulation Process in Pharmaceutical Industry | Production* (n.d.).

Powder blending: Powder blending is one of the oldest operations in the solid handling of the pharmaceutical industries. In this, two different chemicals are combined completely and form a new chemical that is completely different and has its own properties. Blending is an essential part in the powder/drug manufacturing industries.

High shear and wet granulation: In this, the granulation is set up on equipment with impeller blades that provides agitation or movement and apply a higher amount of shear on the powder during the process of the granulation.

Fluid bed granulation: It is also called the agglomeration process. In this, the fine powder particles are converted into a fluid by means of contact with a gas.

Hot-melt extrusion: It is the process of melting a solid product and forcing it via an orifice under extremely controlled conditions to form a new product. It is one of the promising technologies to develop a new chemical entity in a pre-developed pipeline.

Drying: Maintaining the desired moisture content in the powders of the pharmaceutical drugs is essential to keep the drugs safe in storage. The presence of water or other solvents in the drug leads to a microbiological deterioration of the product or drug. Drying is the most important pharmaceutical process to remove the moisture to a safe level by applying heated air.

Pharmaceutical milling: It is a process of reducing the size of drug particles using a tool axis rotary cutting tool.

Pressurization of powders or granules into tablets: An effective process for manufacturing and delivering a strong dosage of a solid drug.

Tablet coverings: A tablet covering or coating machine is equipment to cover the outer surface of the tablet by using a thin film of the coating material.

Pharmaceutical encapsulation: Contains a solid or liquid dosage of a drug in a soft shell or pre-formed capsule.

Drug encapsulation: Encapsulation is a technique to form the shell to a particular drug to prevent leaching before it reaches the targeted place.

Micronization: The process of reducing or downsizing the average sizes of the particles in pharmaceutical products. The techniques for micronization are milling and grinding to make the drug more stable and clinically more effective.

In most of the pharmaceutical unit processes, the thermal energy is supplied either through low-pressure steam boilers or by means of electric heaters. The pharma sector consumes around more than 930 ktoe of energy for its heating requirements. In this, 364 kilotons of coal, 2762 GWh of electricity, and 180.05 kilotons of petroleum products. The majority of the power requirements of the pharma industries is consumed by conventional energy sources and this leads to environmental degradation and increase of greenhouse gas emissions (*Identification of Industrial Sectors Promising for Commercialisation of Solar Energy*, 2011).

The applications of solar energy are mainly categorized into two, namely (i) solar photovoltaic (PV) and (ii) solar thermal. Solar PV is mainly for the generation of electricity, while solar thermal is used for heating, cooling, drying, desalination, and cooking (Hayat et al., 2019; Manchanda & Kumar, 2017).

Solar energy adoption in the pharmaceutical industry has the ability to transform the energy requirements in this sector. The pharma industries are major contributors of environmental pollution in many industry locations. Two major problems in the pharmaceutical sector are unreliable power supply and the high cost of the electrical supply consumption.

Many of the pharmaceutical industries have already started adopting the integration of solar energy technology for their heating demands. This will minimize the pollution to the surroundings and also can save on energy costs.

4.1.1 Energy Demand in Pharmaceutical Industries

The pharmaceutical industry in India ranks third in the entire world in terms of output volume, and fourteenth in domestic consumption. It also has one of the most well-organized industries. At various stages of the activities, the sector consumes both thermal and electrical energies. As a result, solar energy has a significant possibility of displacing traditional energy sources. It has been discovered that the amount of heat load accounts for 20% of the overall energy consumption in the manufacturing units. According to some empirical studies, it was concluded that solar energy could be used to replace 5% of the heating

TABLE 4.1
Energy consumption in Pharmaceutical Industries

Energy	Consumption (%)
Electricity	54
Petroleum Products	27
Coal	4
Other Fuels	15

Source: *Identification of Industrial Sectors Promising for Commercialisation of Solar Energy* (2011)

demand. As a result, solar thermal energy has significant potential in a variety of pharmaceutical industries (*ARUN Solar Boiler, Solar Thermal Technology in the Pharmaceutical Sector*, n.d.).

Generally, steam is produced in a conventional fuel-fired boiler for heating chemical substances in demand in pharmaceutical industries. Many pharmaceutical industries require a low-temperature range of steam that can be easily achieved by using non-concentrating solar thermal collectors, namely flat-plate collectors (FPC) and evacuated tube collectors (ETC) (Kumar KR, Chaitanya NK, 2021). The industry requires heat at temperatures ranging from 55 to 120 °C for the application including distillation, evaporation, and drying. The ETC system is the most preferable solar heating system to meet the requirements of heating at 120 °C. The consumption of energy by using different sources is given in Table 4.1 (*Identification of Industrial Sectors Promising for Commercialisation of Solar Energy*, 2011).

The manufacturing of pharma products in an industry may adopt either batch or continuous production. In case of batch production, after each batch the entire batch line's equipment is sterilized using low-pressure steam (4 to 5 bar). Another important process involved with the steam is multiple-effect distillation, which is primarily used for solid-liquid extraction. Importantly, electric heaters are employed for heating various fluids at different temperatures based on the process requirements. Various raw materials are mixed as per the specifications and heated using a heat exchanger. During the distillation stage of production, temperature is used to detach the numerous elements of a prepared mixture. Drying is the process of removing the fluid from the wet solid. The process of drying is adopted in many stages, especially in the dry granulation stage with or without coating. Hot air or infrared light sources are used for supplying the heat to the wet medium to cause the fluid to vaporize.

4.1.2 RANGE OF TEMPERATURE REQUIREMENTS

The pharmaceutical industry manufactures a wide range of products including ointments, tablets, powders, capsules, and liquids. The processes in these manufacturing

TABLE 4.2

Range of Temperature Requirements in the Pharmaceutical Industry Processes

Application	Temperature Range (°C)
Drying	110–120
Distillation	60–80
Evaporation	100–120

Source: Kumar et al. (2021).

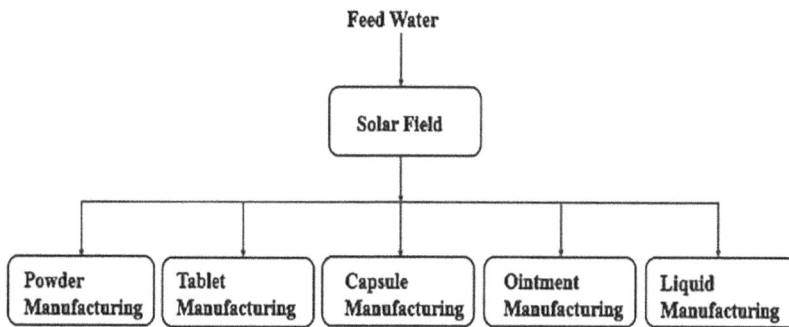

Feed Water

Solar Field

| Powder Manufacturing | Tablet Manufacturing | Capsule Manufacturing | Ointment Manufacturing | Liquid Manufacturing |

FIGURE 4.2 Process flow diagram of various pharma processes integrated with solar thermal field. Source: Kumar et al. (2021).

units are highly energy-intensive and the consumption of thermal and electrical energies varies with the type of process and product. For this, the sector requires a continuous and considerable amount of energy flow in the production units. Low-pressure steam (~5 bar, 150 °C) and hot water (60–80 °C) are commonly used to meet the thermal requirements of the products and processes. Few industries also require hot air in the temperature range of 60–80 °C in huge quantities. Two common processes, sterilization and granulation in pharmaceutical industries, require hot air in the temperature range of 60–70 °C (*ARUN Solar Boiler, Solar Thermal Technology in the Pharmaceutical Sector*, n.d.). The range of temperature requirements in the pharmaceutical industry processes are given in Table 4.2. The process flow diagram of various pharma processes integrated with the solar thermal field is shown in Figure 4.2.

4.2 SELECTION OF SOLAR COLLECTORS

The solar thermal collector plays an important role in absorbing the radiation from the sun and converting it to useful heat energy. Generally, a solar collector is a heat exchanger which converts the radiation into heat energy and then heat energy exchanges to the working fluid. The selection of the appropriate and suitable solar thermal collector is based on the temperature range, availability of the floor space, and cost. Depending on the range of working temperatures, almost every type of solar thermal collector can be used for the various heating demands of the different processes. For selecting the most appropriate and suitable solar thermal collector, there are some necessary design guidelines set up by the Solar Energy Research Institute (SERI) and The International Renewable Energy Agency (IRENA) (Tasmin et al., 2022). In general, flat plate collectors (40–60 °C), evacuated tube collectors (50–200 °C), and concentrating collectors (150–400 °C) can be used in the pharma industries. The selection of solar collectors is mainly based on the following observations:

   (i) Optical efficiency of the collector.
  (ii) Working temperature of the collector.
 (iii) Annual solar irradiation on the proposed site.
 (iv) Availability of the roof space and floor area.
  (v) Cost economics of the respective collector.

4.3 INTEGRATION OF SOLAR THERMAL COLLECTORS

The assessment methodology for the integration of solar thermal technology is mainly based on three categories: (i) pre-feasibility study, (ii) feasibility study, and (iii) final decision for further activities.

   Initially, the pre-feasibility study is a very important tool that can be used to quickly assess the possibility of the integration of solar thermal technology. If the pre-feasibility study turns in good results, then the feasibility study can be conducted for further proceedings. In this, the information is all about the integration points, production flow, floor area, and locations for the necessary storage volume tanks. After this assessment, the following parameters should be taken into consideration while establishing a plant:

  (a) Temperature range for the integration.
  (b) Heating load requirement on a daily, weekly, monthly, and annual basis.
  (c) Annual energy requirement.
  (d) Annual achievable solar radiation.

   The selection of the suitable solar thermal collector and the required volume of the storage tank can be analyzed for the best possible integration. After that, the cost analysis of the proposed plant will be calculated before it goes to the final planning for the proposed installation of the integrated solar thermal plant. This will help the plant make the final decision of whether to keep solar process heating systems. Then the company will decide, based on the results of the prior setup, if a solar heating system is desired or not. The entire assessment methodology for the integration of solar thermal technology is shown in Figure 4.3.

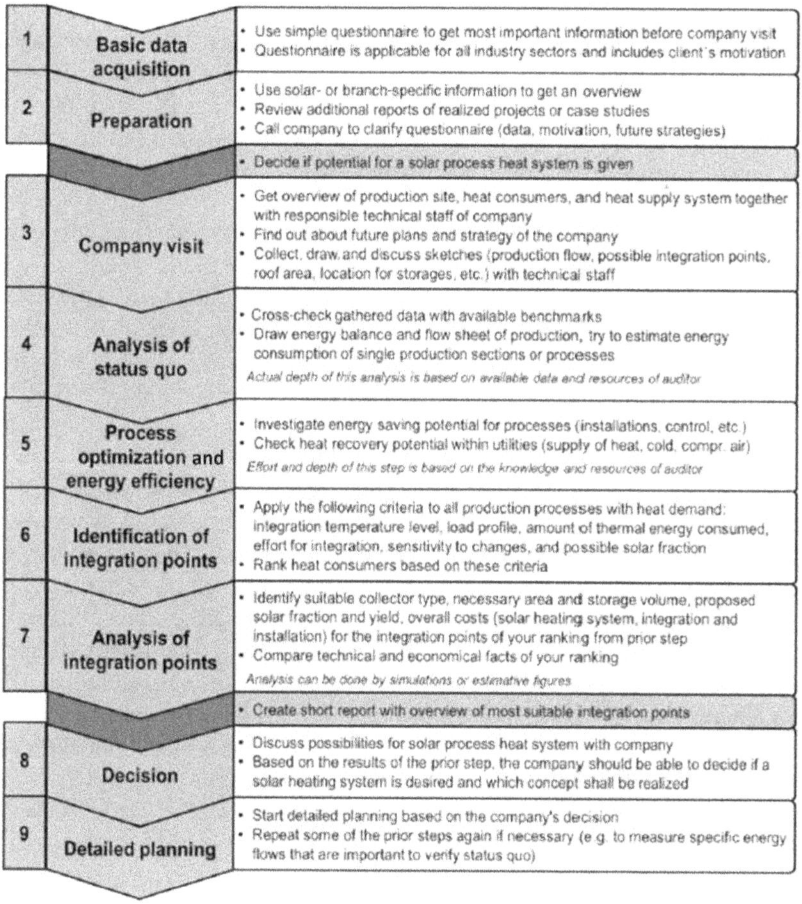

1	**Basic data acquisition**	• Use simple questionnaire to get most important information before company visit • Questionnaire is applicable for all industry sectors and includes client's motivation
2	**Preparation**	• Use solar- or branch-specific information to get an overview • Review additional reports of realized projects or case studies • Call company to clarify questionnaire (data, motivation, future strategies)
		• Decide if potential for a solar process heat system is given
3	**Company visit**	• Get overview of production site, heat consumers, and heat supply system together with responsible technical staff of company • Find out about future plans and strategy of the company • Collect, draw, and discuss sketches (production flow, possible integration points, roof area, location for storages, etc.) with technical staff
4	**Analysis of status quo**	• Cross-check gathered data with available benchmarks • Draw energy balance and flow sheet of production, try to estimate energy consumption of single production sections or processes *Actual depth of this analysis is based on available data and resources of auditor*
5	**Process optimization and energy efficiency**	• Investigate energy saving potential for processes (installations, control, etc.) • Check heat recovery potential within utilities (supply of heat, cold, compr. air) *Effort and depth of this step is based on the knowledge and resources of auditor*
6	**Identification of integration points**	• Apply the following criteria to all production processes with heat demand: integration temperature level, load profile, amount of thermal energy consumed, effort for integration, sensitivity to changes, and possible solar fraction • Rank heat consumers based on these criteria
7	**Analysis of integration points**	• Identify suitable collector type, necessary area and storage volume, proposed solar fraction and yield, overall costs (solar heating system, integration and installation) for the integration points of your ranking from prior step • Compare technical and economical facts of your ranking *Analysis can be done by simulations or estimative figures*
		• Create short report with overview of most suitable integration points
8	**Decision**	• Discuss possibilities for solar process heat system with company • Based on the results of the prior step, the company should be able to decide if a solar heating system is desired and which concept shall be realized
9	**Detailed planning**	• Start detailed planning based on the company's decision • Repeat some of the prior steps again if necessary (e.g. to measure specific energy flows that are important to verify status quo)

FIGURE 4.3   Selection and integration methodology of solar thermal technology.

Source: Brunner (2014).

4.4 CASE STUDIES

Case Study 1

Various thermal processes in the pharma plant require an enormous amount of heat energy for their heating demands. In a Swiss pharmaceutical industry, located in Switzerland, the requirement of the thermal load is around 2 TWh. Currently, around 67% of this heating demand was carried out by using fossil fuels such as oil and gas. The pollution from these fossil fuel sources causes a large amount of impact on the environment. Only 0.1% of the heating demand is supplied by using a renewable energy source, which is biogas. To meet the heating requirements of the pharma process in the plant, solar energy-based thermal collectors, namely flat plate and evacuated tube solar collectors heating systems, were integrated. For adapting and integrating this solar thermal technology, two case studies were considered for the technical feasibility and economic assessment. Both the case studies are involved in the study of drying processes (Guillaume et al., 2020).

The amount of heat supplied to the pharma processes is calculated by using Polysun simulation software. The production of heat in the first case study is about 617 MWh/year with an area of the solar collector of 1060 m² and a storage tank volume of 50 m³. Whereas, the heat production is about 382 MWh/year with 684 m² of collector area and 30 m³ of storage tank volume in the second case study.

In the first case study, among fourteen production processes in the plant, only five production processes were considered where there would be maximum suitability of the adoption of the solar thermal process with a temperature range below 65 °C. A schematic view of the integration of solar-powered heat energy for the first case study is shown in Figure 4.4. The power requirement of the five processes is shown in Figure 4.5. The red color in the graph indicates the load duration curve of the five processes, and the power production with the use of solar reaches almost 350 kW. The power produced from the solar thermal collectors can achieve 80% of the heating requirements of the five processes of the plant. Therefore, the considered capacity of a solar energy-based heating system can meet the requirements of the plant and also reduce fossil fuel consumption with a great reduction of a carbon footprint on the environment.

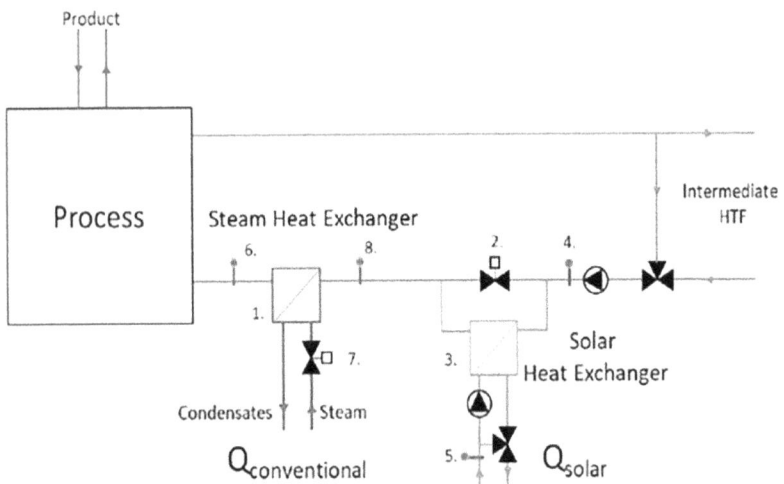

FIGURE 4.4   The case study consists of a solar thermal process in a pharma industry (Guillaume et al., 2020).

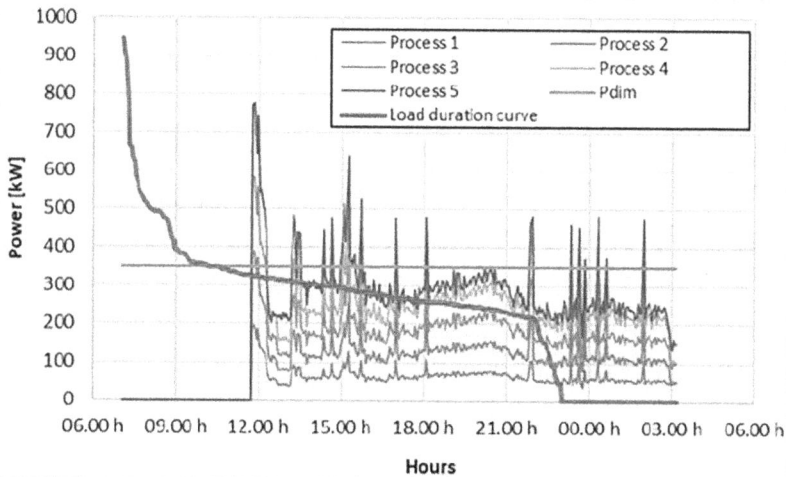

FIGURE 4.5 Variation of power requirements in the plant for the drying process (Guillaume et al., 2020).

In the second case study, the drying process occurs at four different temperatures, 50, 60, 70, and 100 °C, where the air is the heat transfer medium. Initially, the air is heated with the steam heat exchanger and then transferred to the drying chambers. The air supplied to the drying processes is pre-heated with the help of the solar heat energy system at a temperature range of 50 °C. According to the loading requirements of the drying processes, the size of the solar collectors area can be determined to a total maximum power output of 180 kW. The schematic view of the Polysun software simulation model is shown in Figure 4.6. With the help of Polysun software, the values of heat production and estimated cost of the plant were assessed. The schematic view of the integration of solar-powered heat energy for both case studies is shown in Figure 4.7.

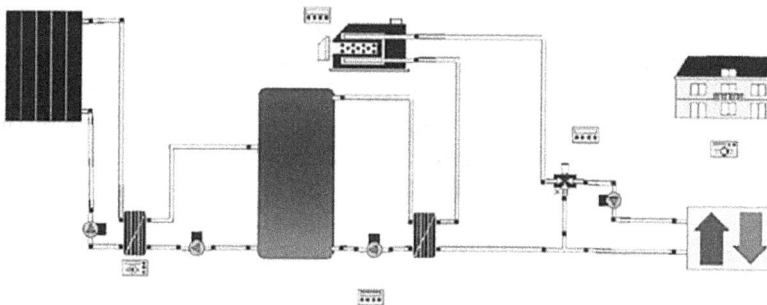

FIGURE 4.6 Polysun software used for the simulation of solar heat energy (Guillaume et al., 2020).

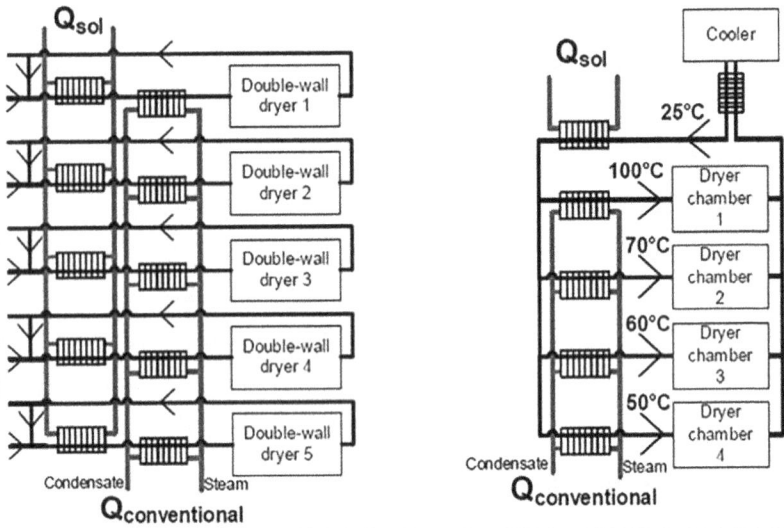

FIGURE 4.7   Schematic view of the integration of solar thermal systems in the first
case study (left) and second case study (right) (Guillaume et al., 2020).

TABLE 4.3
Variation of Different Categories of Initial Cost in Both Case Studies

Categorization	Cost in the First Case Study (USD)	Cost in the Second Case Study (USD)
Solar thermal	849256	539227
Distribution line	245839	155748
Planning	39961	25292
Others	81946	57665
Overall cost	1217002	777932

Source: Guillaume et al. (2020).

The economic study of the plant revealed that the integration of solar energy-based heating systems could cost double that of conventional heating systems. The initial investment cost of the solar thermal plant is given in Table 4.3. The plant can be operated with this fixed cost for a minimum of twenty years. Even though the initial cost is high, the adoption and integration of solar thermal powered heat energy is a viable solution to meet the Swiss Energy Strategy Reviews 2050 and also replace conventional fuels with green energy.

Case Study 2

Sunil Health Care, the world's second-largest capsule shell manufacturing company, is based in Alwar, state of Rajasthan, India, and uses hot water in the process of manufacturing capsules. Rajasthan has approximately 330 clear sunny days per year and a Global Horizontal Irradiance (GHI) of 5.5–6.0 kWh/m²/day. For the production of capsule shells, the plant needs approximately 7000 liters of distilled water at 75 °C. The plant generates 7000 liters of distilled water by using grid electricity. A diesel-powered hot water generator was used as a backup during the grid outages. To minimize the diesel consumption and reliance on the grid, a flat plate collector-based industrial solar water heating technology was established on the roof of the plant for the effective utilization of the space. The installed system has a floor area of 11,760 m² with two tanks of water storage. An additional water tank would be used as buffer storage during the cloudy days. A series of six electrical heaters were installed in the water storage tank to meet the immediate energy requirements in the plant. The developed solar energy-based water heating system in the plant has the potential to reduce the carbon footprint by 150 tons per year (*Solar Process Heat (Pharmaceutical)—IGen-Solar*, n.d.).

Case Study 3

As a part of a field case study, two pharmaceutical companies in India are studied to observe the high energy-intensive processes and to minimize fossil fuel consumption by replacing the non-conventional energy source (*Identification of Industrial Sectors Promising for Commercialisation of Solar Energy*, 2011). It was found that the boiler consumes a majority of heat energy from conventional fuels. To meet the steam requirements of the plant, the boiler was continuously supplied with the make-up water. After examining the plant heating requirements, they found a possibility to preheat the boiler make-up water by using solar energy, and by reducing the consumption of fossil fuels. This can be accomplished by employing the FPC and ETC solar thermal collectors.

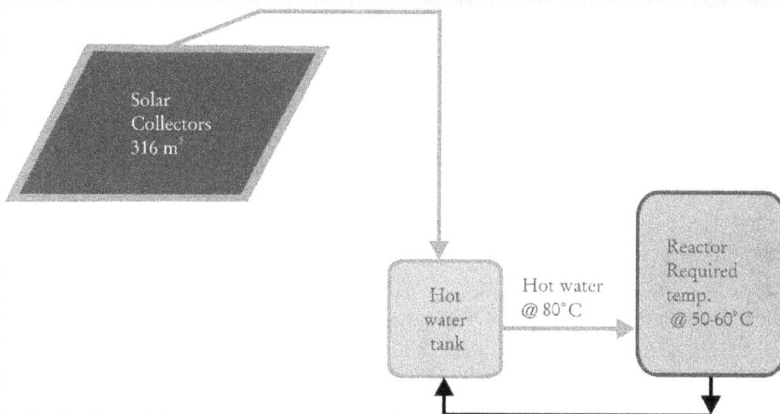

FIGURE 4.8   Schematic view of the proposed solar thermal technology for distillation application (*Identification of Industrial Sectors Promising for Commercialisation of Solar Energy*, 2011).

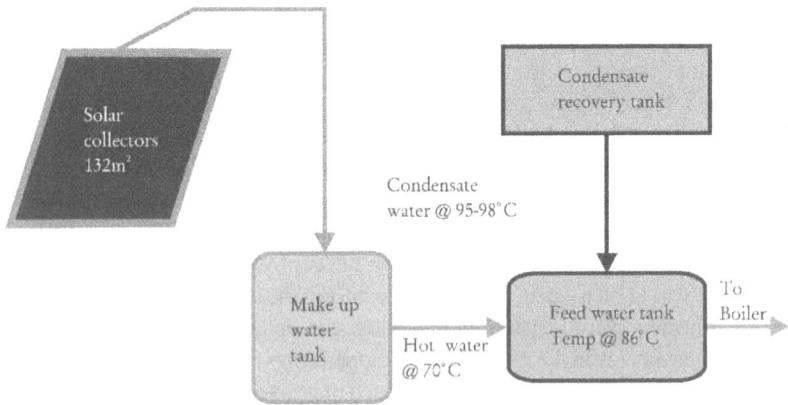

FIGURE 4.9   Schematic view of the application of pre-heating the boiler feed water by solar thermal energy (*Identification of Industrial Sectors Promising for Commercialisation of Solar Energy*, 2011).

The proposed models for the adaption of solar thermal energy to the processes include distillation and preheating the boiler feed water as shown in Figure 4.8 and Figure 4.9 respectively. By employing this solar thermal-based energy input, the estimated internal rate of return (IRR) would be in the range 5 to 30% without considering the subsidy from the government.

Case Study 4

RAM pharma is a Jordanian pharmaceutical firm that produces a wide range of medicinal goods. The plant heating requirements are met by a diesel-fired steam boiler. To reduce fossil fuel consumption and meet the heating demands of the plant, a linear Fresnel concentrating solar collector (aperture area of 22 m²/module) was installed. The schematic view of the installed linear Fresnel solar collector arrangement is shown in Figure 4.10. The collector was designed for the delivery of solar process heat and installed on the rooftop of the plant due to the high occupancy of the floor area (Haagen et al., 2015).

The schematic view of the solar-powered industrial process heat at RAM Pharma is shown in Figure 4.11. The collector modules were fitted with an entire area of 396 m² to meet a peak capacity of 222 kWth. The collector modules were linked in a U-shape and then the water passes through the collector modules and the heated water flows to the water drum. The steam-water mixture was then passed to the steam drum where separation of water and steam takes place. The developed system could reduce diesel consumption by about 30,000 liters annually.

Case Study 5

A medicinal industry in the region of Guangxi in China installed a plant for its heating demand by solar-powered heat, and recovery heat from the air

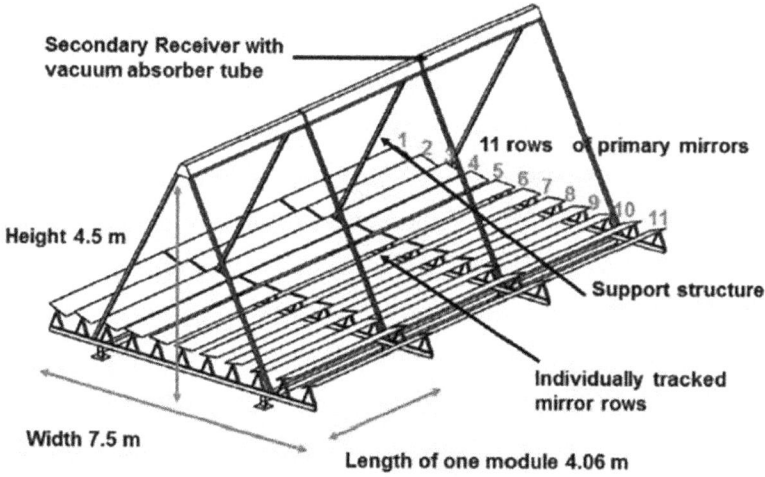

FIGURE 4.10 Schematic model of the linear Fresnel collector setup for steam generation in the pharma industry (Haagen et al., 2015).

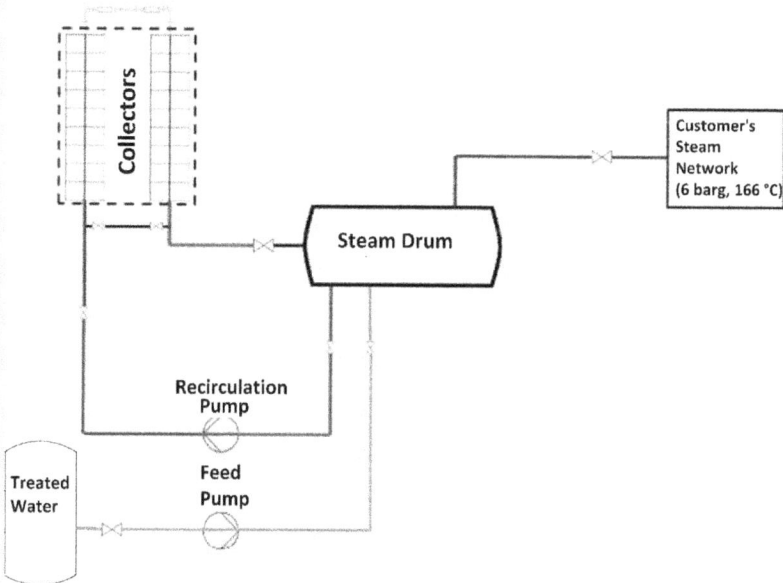

FIGURE 4.11 Schematic view of the solar-powered industrial process heat at RAM Pharma (Haagen et al., 2015).

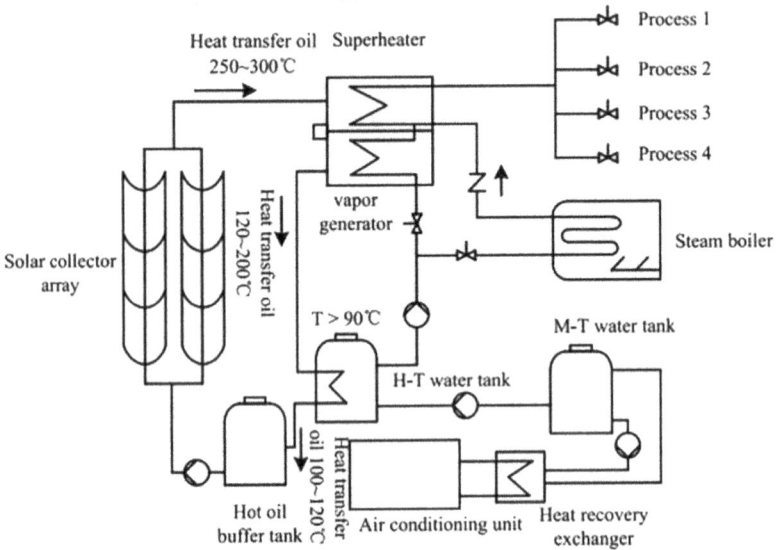

FIGURE 4.12 Schematic diagram of the solar-powered medical industry with waste heat recovery (jia et al., 2018).

conditioner to operate its auxiliary sources. The system can be used to sterilize and to clean the medical apparatus like metal nail files, razor scrappers, cuticle cutters, etc., from the high-temperature water that can be produced from the solar thermal energy. The developed project can be operated for 250 days in a year and reduce a carbon footprint of 275 tons/year for favorable economic benefits (Jia et al., 2018). The system also can save a total 480 kWh of electricity for drying every day. The schematic diagram of the developed solar-powered plant is shown in Figure 4.12.

4.5 DRUG DEVELOPMENT IN AYURVEDA

Ayurveda is one of the oldest and most traditional health care systems in India and manages the health of many people. Herbal or Ayurveda drug development is a most essential and continuous progressive process. The development of drugs in Ayurveda is different from Allopathic or chemical drug development. Ayurveda drugs should be nature friendly, have minimal side effects, be cost economical, and have high efficacy.

The procedure for developing each pharmaceutical drug involves a variety of steps, from identifying and collecting genuine raw materials or plant parts for the processing and the manufacturing of high-quality medicinal drugs. Packaging and storing the processed medicinal herbs also take a crucial role in the usage application of the corresponding drug. Generally, an herb contains more than one therapeutic property in its nature. In this scenario, different procedures may be applied for the separation of each useful constituent. Water-soluble components in the herbs are soluble in water, but some of the fats, oils, and alcohols used require solvents for extraction. Depending on the requirement of the drug, the extraction technique will be applied, and the therapeutic content will be used. Cooking methods such as heating, boiling, and frying are also used in pharmaceutical procedures because the ayurvedic plant materials can also be taken as food materials that we normally use. Both fresh and dried plant materials are used for the processing of the drugs, and various methods are used to produce them in a drug dosage form that is stable over a period of time for disease prevention. The factors to be considered before processing the plant materials in the pharmaceutical industries are provided here:

(a) Raw material type-fresh or dried.
(b) Required dosage.
(c) Plant therapeutic component solubility.
(d) Heat stability of the plant in therapeutic form.
(e) Shelf life of the dosage.

With the growing population and high cost of fuels, the energy required for the post-harvesting processes of food products has provided opportunities for the use of renewable energy sources. Among the different green energy sources, solar energy is the most abundant one, and it is available almost everywhere in the world. Among different post-harvesting processes of ayurvedic herbs, the drying process consumes a lot of energy to reduce the moisture content of the herb. Various solar thermal collectors are currently being used for the supply of heated air to dry the medicinal plants.

Solar drying is used to dry vegetables, grains, fruits, and herbs and is receiving great attention due to its lower cost, availability, and better-quality of the dried products with the inhibition of essential nutritional values. In recent years, the drying of medicinal herbs has been a focus of interest by many researchers due to increasing awareness of the public and the increased use of a variety of medicinal herbs in health care. A solar dryer is a device that reduces the moisture content of a product to a desired safe level by using solar energy (Sharma et al., 2009). The operating principle of a solar dryer is shown in Figure 4.13.

The major classifications of solar dryers are shown in Figure 4.14. The objective of this next section is to describe medicinal herb drying with solar energy in comparison with sun drying. The modeling of the entire thin-layer drying process is presented and explained with the theoretical models available in the literature.

4.5.1 AVAILABILITY OF MEDICINAL PLANTS

An herb is a plant used for medicinal and cooking purposes due to its medicinal properties. Table 4.4 gives the details of the therapeutic plants available in India and the world. India accounts for approximately 2.4% of the total land area and approximately 8% of global biodiversity. India has the largest collection of medicinal plants in the world. There are approximately 17000 high-valued medicinal plants in India, 7500 of which have medicinal values and are used as medicines.

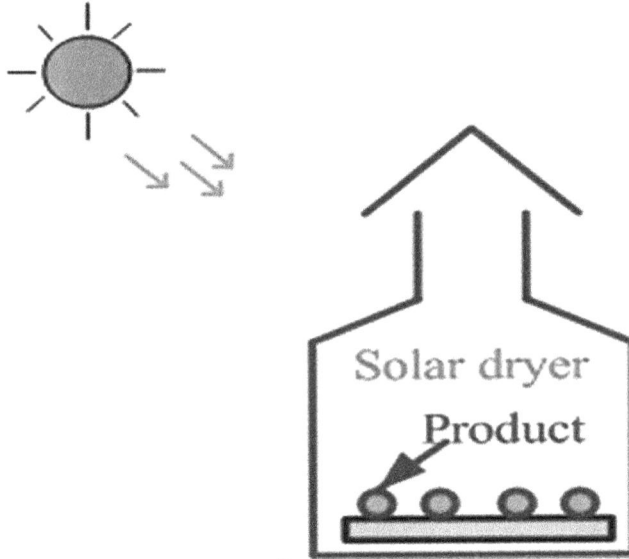

FIGURE 4.13 The operating principle of a solar dryer.

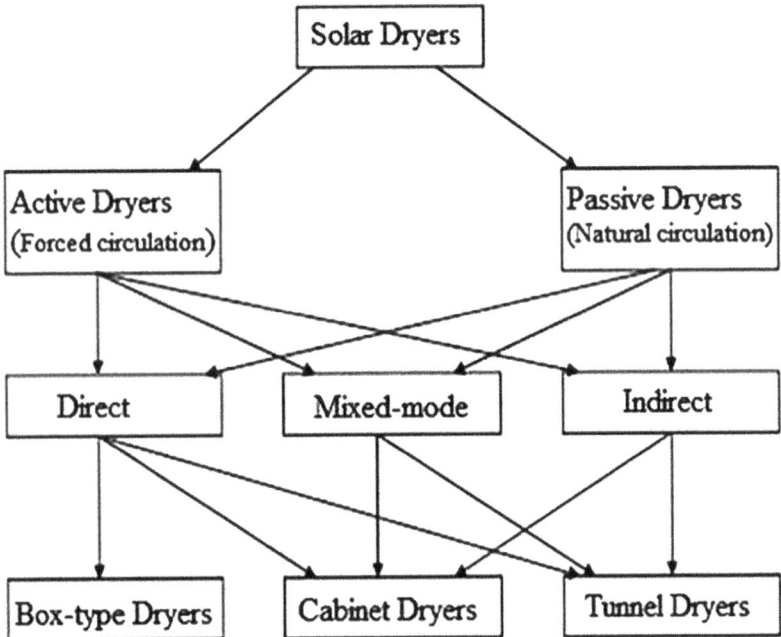

FIGURE 4.14 Classification of solar dryers.

Source: Bhaskara Rao & Murugan (2021).

TABLE 4.4
Medicinal Plants Available in the World and India

Origin	Total Number of Native Species of Flora	Number of Medicinal Plants	Percentage of Medicinal Herbs	References
World	297000	52885	10	(Schippmann et al., 2005)
India	17000	7500	44	(Shiva, 1998)
Medicinal plants in Indian Himalayas	8000	1748	22	(Samant et al., 1998)

The Indian system of medicine (ISM) prescribes the use of medicinal plants in the creation of herbal medicines in various forms. In this, Allopathic medicine is modern Western medicine that is prescribed by doctors on an evidence-based system. The medicines involved in allopathy are clinically tested drugs for better results (Kala et al., 2006). Ayurveda is an ancient form of alternative medicine, which involves the treatment of diseases with the help of herbs (Farnsworth & Soejarto, 2010; Jamshidi & Cohen, 2017). The Siddha medicinal system originated in South India is considered one of India's oldest medical systems. The Siddha system is founded based on spiritual disciplines and ancient medicinal practices (Jamshidi & Cohen, 2017). Unani medicine is Perso-Arabic traditional medicine that is practiced in South Asia and modern Central Asia. The Unani system of medicine is based on long-established knowledge and practices concerning the promotion of good health and disease prevention (Alzohairy, 2016). Tibetan medicine is an ancient, timely healing tradition from Tibet. In these medicines, the composites of various plants and occasionally minerals are used as raw materials. Some Tibetan medicines contain over 100 ingredients. Some ingredients in Tibetan medicines treat the underlying imbalance, while others treat any side effects that may occur (Nathan & Scobell, 2012). Homeopathy is a medical system based on the belief that the body can heal itself. The practitioners of homeopathy medicine use trace amounts of natural substances such as minerals and plants. They believe that these aid in the healing process (Pingale et al., 2012).

Except for Allopathic medicine, all the other medical methods use medicines made from a wide range of high-value medicinal herbs. Approximately 20% of the herbs that are available in India are used in therapeutic usage.

According to the report on export and import (EXIM), there are 880 medicinal herbs available in the Indian market (Asl Roosta et al., 2017). Some of the medicinal plants found in India are aloe vera, cinnamon, eucalyptus, hibiscus, lavender, peppermint, rosemary, dill, chives, and basil. Table 4.5 provides the botanical name and application of various medicinal leaves.

4.6 CONCLUSIONS

In this research chapter, initially, the various types of energy requirements and different processes in the medicinal and pharmaceutical industries are presented. Further,

TABLE 4.5

Some of the Potential Medicinal Herbs That Can Be Dried by Using Solar Drying and Their Availability, Color, and Applications

Herb	Color	Therapeutic Use	Availability in the World
Neem	Vibrant green	Used to treat eye disorders, intestinal worms, skin ulcers, diabetes	India, Burma, Thailand, Bangladesh, Cambodia, Indonesia, Iran, Malaysia, Nepal, Pakistan, Sri Lanka, and Vietnam
Tulsi	Green	Used to treat insect bites, respiratory problems, heart diseases	Eastern world tropics
Aloe vera	Green	Used to improve the digestive system, heal burns, relieve the anal fissures	Australia, Cuba, China, Mexico, India, Jamaica, Spain
Cinnamon	Brown	Used to treat gastrointestinal upset, diarrhea, menstrual cramps	Sri Lanka, Burma, India, South America, West Indies
Eucalyptus	Blue-green	Used as air fresheners, medicinal teas, activity against bacteria and fungi	Brazil, Argentina, South Africa, India, Galicia, Portugal
Hibiscus	Pink-red	Used to treat cancer, gallbladder attacks, skin infections, and hair growth	India, Malaysia, South Korea, China, Japan
Lavender	Medium purple	Used as a diuretic; to boost sleep; to treat depression, fatigue, insomnia; and to relieve pain	France, China, Ukraine, Spain, Morocco, Europe, South West Asia to India
Peppermint	Green	Used in cosmetics, soaps, mouthwashes, and to add flavor to foods	India, China, Brazil, Thailand, USA, Vietnam
Rosemary	Toad green	Used to stimulate hair growth, as dressing in salads, and as an anti-inflammatory	England, Mexico, USA, India, Mediterranean
Dill	Green	Used to garnish foods, and to prevent bad breath, hemorrhoids	Netherlands, North Africa, South America, China, Eastern Europe
Chives	Blue-pink	Used to improve memory and prevent cancer, osteoporosis	Asia, Europe, North America, China
Basil	Purple-green	Used to treat kidney problems, worm infections, snake and insect bites	USA, Europe, Central Africa, Southeast Asia
Ashwagandha	Dull green	Used to improve blood sugar, inflammation, memory, stress, and anxiety	Nepal, Sri Lanka, China, Oman, India
Lemon Balm	Light green	Used as a digestive tonic, consumed as a tea, and to treat skin problems	South-Central Europe, Mediterranean Basin, Iran, Central Asia, North Africa

the application of solar energy-based processes is given, and the range of temperature requirements for different processes is explained. Case studies from the various research articles about solar-powered industrial process heat are elaborated. The following important conclusions are drawn from the study:

- Integration of solar energy systems into pharmaceutical industries is technically a feasible solution. The sector requires thermal and electrical energy for the processes, including drying, sterilization, granulation, water heating, tablet preparation etc.
- The range of temperature requirements of the thermal processes in the pharmaceutical industries is 60 to 90 °C. This range of temperatures can be achieved by using flat-plate and vacuum-tube solar collectors.
- For the installation of solar panels or collectors in the pharma industries, the available floor space or rooftop space is insufficient because of exhaustive utility of the piping network over the roof.
- In the pharma sector, the closed-loop solar cycle is more feasible in the case of hot water production and the open-loop solar cycle is preferable to pre-heat the make-up water.
- Many pharma industry processes require steam; therefore installing solar thermal energy technology is a viable option to pre-heat the water before it goes to the boiler.
- The economic analysis in the few case studies indicated that integrating solar-powered energy in the industries is highly capital intensive, but it will reduce the operational cost in a daily basis.

Almost all the pharma industries emit a large amount of pollution to the atmosphere by using conventional fuels. Therefore, adopting solar thermal heat in the pharma sector can reduce the CO_2 and GHG emissions to a great extent.

REFERENCES

Alzohairy, M. A. (2016). Therapeutics role of azadirachta indica (Neem) and their active constituents in diseases prevention and treatment. *Evidence-Based Complementary and Alternative Medicine*, *2016*. https://doi.org/10.1155/2016/7382506.

ARUN Solar Boiler, Solar Thermal Technology in the Pharmaceutical Sector. (n.d.). Retrieved April 9, 2022, from www.cliquesolar.com/PharmaceuticalSolution.aspx.

Asl Roosta, R., Moghaddasi, R., & Hosseini, S. S. (2017). Export target markets of medicinal and aromatic plants. *Journal of Applied Research on Medicinal and Aromatic Plants*, *7*, 84–88. https://doi.org/10.1016/j.jarmap.2017.06.003.

Bhaskara Rao, T. S. S., & Murugan, S. (2021). Solar drying of medicinal herbs: A review. *Solar Energy*, *223*, 415–436. https://doi.org/10.1016/j.solener.2021.05.065

Brunner, C. (2014). *Solar Heat Integration in Industrial Processes*. 2. http://task49.iea-shc.org/data/sites/1/publications/IEA_SHC-Task49-Highlights-2014.pdf.

Dry Granulation Process in Pharmaceutical Industry | Production. (n.d.).

Farnsworth, N. R., & Soejarto, D. D. (1991). Global importance of medicinal plants. In: O. Akerele, V. Heywood, & H. Synge (eds.), *Conservation of Medicinal Plants* (pp. 25–51). University Press.

Guillaume, M., Wagner, G., Jobard, X., Eicher, S., & Citherlet, S. (2020). Solar thermal systems for the swiss pharmaceutical industry sector. *Proceedings of the ISES Solar World Congress 2019 and IEA SHC International Conference on Solar Heating and Cooling for Buildings and Industry 2019, 2018*, 550–559. https://doi.org/10.18086/swc.2019.12.06.

Haagen, M., Zahler, C., Zimmermann, E., & Al-Najami, M. M. R. (2015). Solar process steam for pharmaceutical industry in Jordan. *Energy Procedia, 70*, 621–625. https://doi.org/10.1016/j.egypro.2015.02.169.

Hayat, M. B., Ali, D., Monyake, K. C., Alagha, L., & Ahmed, N. (2019). Solar energy—A look into power generation, challenges, and a solar-powered future. *International Journal of Energy Research, 43*(3), 1049–1067. https://doi.org/10.1002/er.4252.

Identification of industrial sectors promising for commercialisation of solar Energy. (2011).

Jamshidi, N., & Cohen, M. M. (2017). The clinical efficacy and safety of tulsi in humans: A systematic review of the literature. *Evidence-Based Complementary and Alternative Medicine, 2017*, 1–13. https://doi.org/10.1155/2017/9217567.

Jia, T., Huang, J., Li, R., He, P., & Dai, Y. (2018). Status and prospect of solar heat for industrial processes in China. *Renewable and Sustainable Energy Reviews, 90*(June 2017), 475–489. https://doi.org/10.1016/j.rser.2018.03.077.

Kala, C. P., Dhyani, P. P., & Sajwan, B. S. (2006). Developing the medicinal plants sector in northern India: Challenges and opportunities. *Journal of Ethnobiology and Ethnomedicine, 2*(1), 1–5. https://doi.org/10.1186/1746-4269-2-32.

Kumar, K. R., Chaitanya, N. K., Sendhil, K. N. (2021). Solar thermal energy technologies and its applications for process heating and power generation—A review. *Journal of Cleaner Production, 282*, 125296. https://doi.org/10.1016/j.scitotenv.2019.135907.

Making pharma greener with solar energy—Express Pharma. (n.d.). Retrieved May 13, 2022, from www.expresspharma.in/making-pharma-greener-with-solar-energy/.

Manchanda, H., & Kumar, M. (2017). Performance analysis of single basin solar distillation cum drying unit with parabolic reflector. *Desalination, 416*(April), 1–9. https://doi.org/10.1016/j.desal.2017.04.020.

Nathan, A. J., & Scobell, A. (2012). How China sees America. *Foreign Affairs, 91*(5), 1689–1699. https://doi.org/10.1017/CBO9781107415324.004.

Pingale, S. S., Firke, N. P., & Markandeya, A. G. (2012). Therapeutic activities of Ocimum tenuiflorum accounted in last decade: A review. *Journal of Pharmacy Research, 55*(44), 2215–2220.

Samant, S. S., Dhar, U., & Palni, L. M. S. (1998). *Medicinal Plants of Indian Himalaya: Diversity Distribution Potential Value* (pp. 155–158). Gyanodaya Prakashan. http://agris.fao.org/agris-search/search.do?recordID=US201300042676.

Schippmann, U., Cunningham, A. B., Leaman, D. J., & Walter, S. (2005). Impact of cultivation and collection on the conservation of medicinal plants: Global trends and issues. *Acta Horticulturae, 676*(October), 31–44. https://doi.org/10.17660/actahortic.2005.676.3.

Sharma, A., Chen, C. R., & Vu Lan, N. (2009). Solar-energy drying systems: A review. *Renewable and Sustainable Energy Reviews, 13*(6–7), 1185–1210. https://doi.org/10.1016/j.rser.2008.08.015.

Shiva, M. P. (1998). *Inventory of Forest Resources for Sustainable Management and Biodiversity Conservation: With Lists of Multipurpose Tree Species Yielding Both Timber & Non-Timber Forest Products (NTFPs) and Shrub and herb Species of NTFP Importance* (pp. 1–405). Indus Publishing Company.

Solar Process Heat (Pharmaceutical)—IGen-Solar. (n.d.).

Tasmin, N., Farjana, S. H., Hossain, M. R., Golder, S., & Mahmud, M. A. P. (2022). Integration of solar process heat in industries: A review. *Clean Technologies, 4*(1), 97–131. https://doi.org/10.3390/cleantechnol4010008.

5 Solar Thermal Conversion Technologies for Process Heating Applications in Automobile Industries

A. Veera Kumar, D. Seenivasan, T.V. Arjunan, R. Venkatramanan, and Vijayan Selvaraj

CONTENTS

DOI: 10.1201/9781003263326-7

5.1 AUTOMOBILE INDUSTRY'S ENERGY CONSUMPTION PATTERN

The production of automobiles is a sophisticated, energy-intensive process that requires a substantial amount of raw materials and water. A huge amount of energy, either in electricity or heat, is required for several processes in the automobile industry to transform the raw materials into vehicles. The use of energy in a manufacturing plant for automobiles can be broken down into primary and secondary categories. The primary energy sources are electricity and fuels, whereas chilled and hot water, compressed air, and steam are considered secondary energy sources. In order to distribute energy throughout the plant, these secondary energy carriers are generated from main sources. The primary energy resources used in the industry and their contribution percentages are shown in Figure 5.1. Electricity is the major source used in the plant, which is consumed through the plant for various purposes such as painting (27–50%), air-conditioning, ventilation, and heating (11–20%), compressed air (9–14%), welding (9–11%), lighting (14–15%), and materials handling/tools (7–8%) [1]. Next to electricity, fuel (petroleum products—natural gas, furnace oil, etc.) is used for drying, space heating, air-conditioning, and production of steam, etc.

The total energy consumption in the automotive industry for vehicle manufacturing is separated into three areas known as the assembly shop, the paint shop, and the body shop. Figure 5.2 depicts how total energy consumption in vehicle

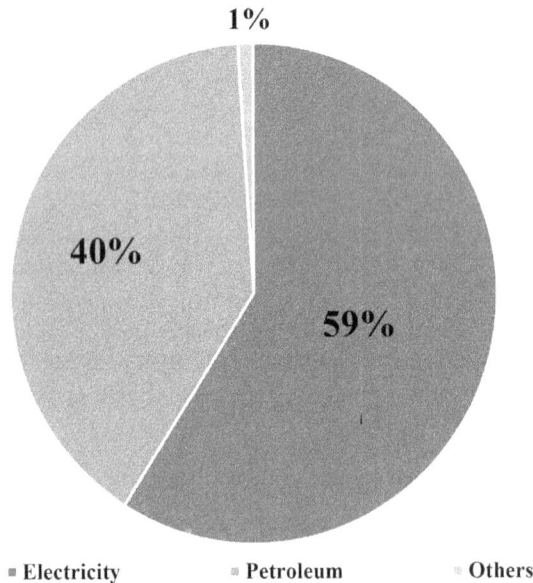

FIGURE 5.1   Energy sources contribution in the automobile industry [2].

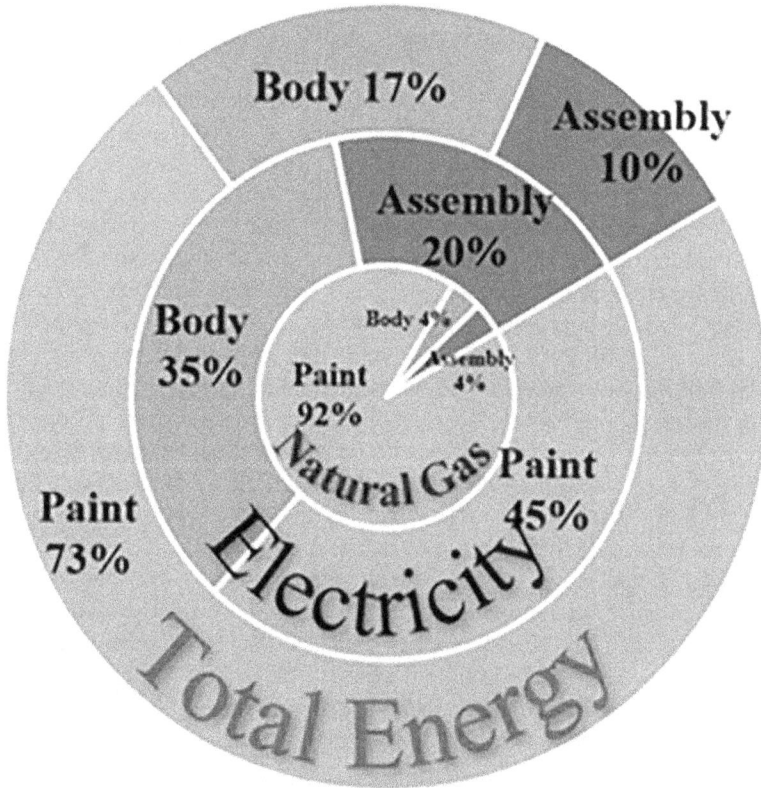

FIGURE 5.2   Energy consumption patterns in various shops in the automobile industry [2].

manufacture is divided into total energy, electricity consumed, and heat energy consumed in various shops. The paint shop consumes 73% of total energy used in the automobile industry, followed by the body shop at 17%, and assembly at 10% [3]. Overall energy usage is dominated by electricity and petroleum products, which include petrol, diesel, and natural gas in automobiles. Paint shops consume 45% of the electricity, followed by body shops at 35% and assembly shops at 20%. Natural gas is another source of energy, with a high proportion of 92% in the paint shop, a minor share of 4% in the body shop, and a small share of 4% in the assembly shop.

5.2 AUTOMOBILE MANUFACTURING PROCESSES

Automobile manufacturing is accomplished by well-designed and scheduled operations carried out in three primary production shops: the paint shop, the body shop, and the assembly shop. Metal melting furnaces, tempering furnaces, anti-corrosion coatings, and casting drying are all part of the foundry shop. Processes like press shop and zinc phosphate dip are included in the body shop. Processes such as engine component machining, component washing, and component drying are all part of the engine shop. Component machining, cleaning, and drying are all done in axle shops. The parts produced in different places are assembled in the assembly shop and before assembly, the parts are cleaned with a hot bath to remove the undesirable containments or debris like metal chips, dirt, oil, lubricant, etc. Most of the automobile manufacturing industries are equipped with electrical heaters for producing hot water in the temperature range of 80 to 120 °C. The parts manufactured in different shops are painted to improve the surface characteristics for resisting the weathering conditions (moisture, chemicals, rain, dust, etc.) and for providing an attractive aesthetic appearance. After painting, the components are exposed to a hot air environment for curing and most of the industries are using fossil fuels for producing hot air. Electrode positioning, paint baking, primer coating, paint baking, base coating, paint baking, final coating, and paint baking are all operations that are carried out in the paint shop. All of these finished items are delivered to an assembly shop, where they are put together to create a fully working automobile. Figure 5.3 depicts the manufacturing process flow chart for the automobile industry, demonstrating the direction of material flow in all key production shops [4]. Certain manufacturing processes are supplied with secondary energy carriers as a thermal energy source at moderate temperatures. The processes which require temperature less than 250 °C are indicated with asterisk (*) in the flow chart shown in Figure 5.3. The temperature range necessary for each of these six production processes, as well as the current fuel source and heat transfer medium employed, are listed in Table 5.1.

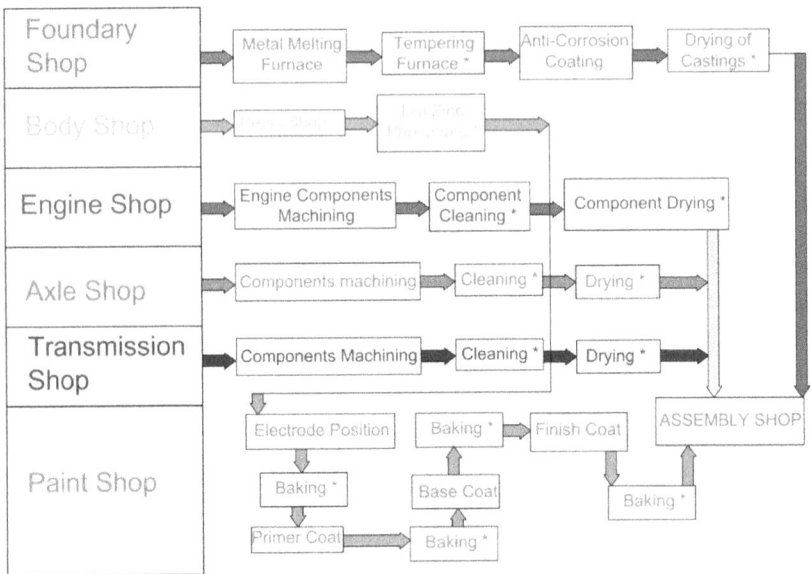

FIGURE 5.3   Flow chart of the automobile manufacturing process [2].

TABLE 5.1
Thermal Energy Is Required in the Manufacturing of Automobiles

S.No	Process Name	Fuel Source Used	Working Medium	Temperature (°C)
1	Drying of casting	Natural gas	Hot air	100
2	Dip zinc phosphating	Electricity	Pressurized hot water	80
3	Tempering furnace	Natural gas	Hot air	200
4	Paint curing process	Natural gas	Steam	200
5	Components washing	Electricity	Pressurized hot water	90
6	Components drying	Natural gas	Hot air	90

5.3 POTENTIAL FOR SOLAR THERMAL TECHNOLOGIES IN THE AUTOMOBILE INDUSTRY

Automobile industries are already starting to implement low carbon manufacturing processes to reduce carbon dioxide emissions. As part of this move, industries are focusing on improving the energy efficiency, utilization of waste energy recovery systems, renewable energy sources, etc. Solar thermal energy conversion technologies are the most promising methods for secondary energy carrier (water, steam, air) heating applications. Figure 5.2 clearly indicates that the paint shop consumes a huge amount of both electricity and fuels for its operations. The paint shop performs the painting operation on the body parts of the vehicles for providing the aesthetic and improved physical properties (mechanical and weather protection, corrosion resistance). The operations of painting and drying entail a number of different operating phases, components (scrubbers, ovens, VOC removal system, paint booths, etc.), and a large consumption of power, fuel, compressed air, and hot and chilled water. The vehicle structure from the body shop is moved to the paint shop, where a set of processes is performed to eliminate pollutants from its metal surface and to facilitate the adhesion of the next paint layer [5]. This is known as pre-treatment, and it consists of numerous steps (often 8–12), including a pre-cleaning stage, 2–3 degreasing stages, an activation stage, a phosphating stage, a passivation stage (optional), several rinse stages, and a final draining stage [5]. A simple pre-treatment process is shown in Figure 5.4. To prepare for the deposition of the next coatings, the pre-cleaning and degreasing processes remove oil, grease, lubricants, and other contaminants from the vehicle. After washing, a phosphate layer coat is applied onto the vehicle. The zinc phosphating procedure passivates the vehicle's metal surface and creates a thin non-metallic crystalline phosphate coating (2–3 μm) that chemically attaches to the vehicle's steel, for improving corrosion resistance [6]. The activation process precedes the phosphating, which aids in the development and growth of phosphate crystals on the vehicle structure. To assure the quality of the pre-treatment, it is critical to manage the pH and temperature of the phosphating bath, the concentration of chemical agents, and the time of the dipping process. The majority of the hot water is used during pre-treatment. This is due to the importance of maintaining the phosphating bath temperature at around 45 °C for the chemical

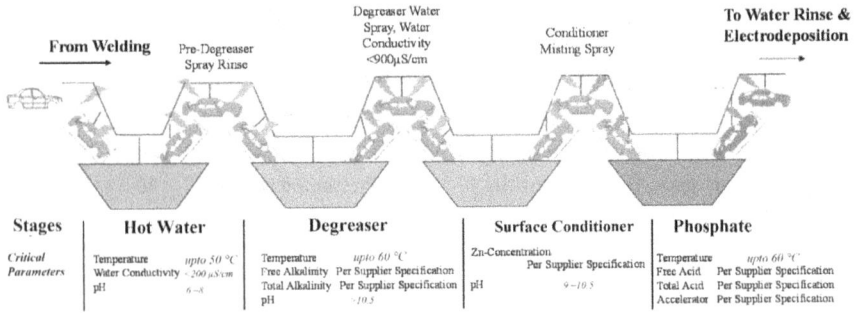

FIGURE 5.4   Pre-treatment Sequences Before Painting [8]

processes [7]. Table 5.1 summarizes the working fluid and temperature ranges required by the various pre-treatment operations.

5.3.1 PART CLEANING, WASHING, OR DEGREASING

To remove the dust content, hot water at around 90 °C is sprayed over the automotive engine components. Electrical heaters are employed to produce the hot water required for spraying.

5.3.2 DRYING OF WASHED PARTS AFTER DEGREASING

The cleaned components from the degreasing process are dried in a 90 °C hot air chamber. Electric coil heaters or natural gas-based burners are used to heat the hot air. The length of the heating cycle is determined by the size of the component to be dried.

5.3.3 DIP ZINC PHOSPHATING OR 7-TANK PROCESS

Before painting, a conventional 7-tank technique is performed to provide good paint adherence and corrosion resistance on the automotive body. Electrical immersion hot water heaters are used to heat all of the heated tanks. Table 5.2 lists the 7 steps involved in the 7-tank process, together with the temperatures involved and the required dip time.

5.3.4 MOULD DRYING

Anti-corrosion coatings are applied to casted or moulded car parts to extend their life and improve overall performance. These components are then dried with hot air blasted at a high temperature of 100 °C using a hot air blower.

5.3.5 TEMPERING OF PARTS IN A FURNACE

Hardening and tempering are commonly used to increase the wear resistance, strength, and toughness of automotive components. Low-temperature tempering

TABLE 5.2
Steps Involved in 7-Tank Process with Temperature and Dip Time Needed [2]

Step	Temperature in °C	Dip Time in Seconds
Degrease	95	15
Cold water rinse	25	5
Passivation	70	10
Cold water rinse	25	5
Derust	70	10
Cold water rinse	25	5
Zinc phosphate dip	80	5

(150–200 °C) of steel bearings, medium-temperature tempering (350–550 °C) of carbon steel, and high-temperature tempering (550–650 °C) of medium carbon alloy steel are some of the common applications of tempering in the automobile industry.

5.3.6 Paint Curing Process in Baking Ovens

The paint is baked in a baking oven using hot air generated by electrical coils or gas burners after each layer of paint is applied to the vehicle. In baking ovens, temperatures of roughly 200 °C are required.

5.4 SYSTEM INTEGRATION

Solar thermal collectors are a form of heat exchanger that absorbs solar irradiation and converts it into useful heat for use in a wide range of applications [9, 10]. However, industrial processes require a consistent source of heat energy on demand, and fluctuation is not acceptable. The difficulty of supplying continuous heat to a facility while utilizing all of the solar energy available on the collector area can be handled by integrating solar technology with an existing heating source, such as fossil fuel, electricity, or biomass. Such an integration process makes use of all the solar energy that the sun can provide while allowing for the use of conventional fuels at night and on rainy or low-sun days. For a few operations in the automotive industries, such as Liquefied Petroleum Gas (LPG) vaporization, furnace oil heating, and 7-tank dip zinc phosphating operations, the integration plan with the current heating system is proposed to satisfy the heat and temperature requirements.

Solar heat can be provided at various integration points as depicted in Figure 5.5 [11]. Solar heat may be provided at many integration levels, including the Supply (S) and Process (P) levels. In addition to these levels, the integration is further divided into direct and indirect, both of which are present in both integration levels: the process fluid is heated directly in the collectors when there is direct integration. The process fluid in the collector serves as the heat transfer fluid. The process circuit is

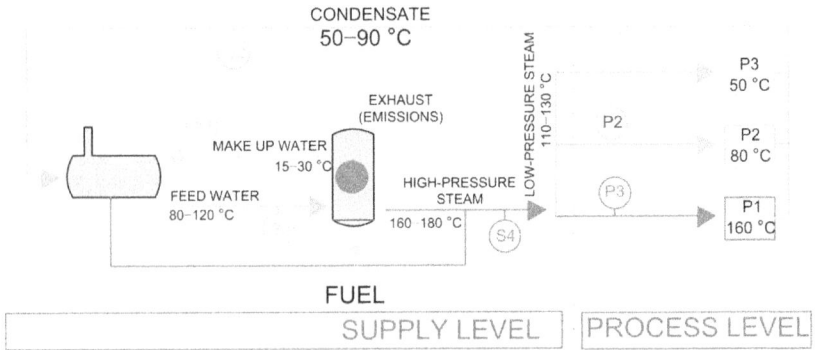

FIGURE 5.5   Potential integration points into a conventional heat-supply system [11].

separated from the solar circuit by heat exchangers in a process known as indirect integration. The heat transfer fluid and the process fluid may be different.

5.4.1 SUPPLY AND PROCESS LEVEL

The integration is accomplished at the supply level if the solar system delivers heat for the providing heat distribution network. If steam is the primary heat carrier at the process site, there are two options for incorporating solar collectors into the system. To generate steam (S4), either concentrating collectors or stationary collectors can be employed to warm make-up water (S1), condensate (S2), or feed water (S3). Water can be heated using both concentrated and stationary collectors if the primary heat transport is hot/pressurized water or thermal oil. The technique employed is mostly determined by the temperature level necessary.

The benefit of supply level integration is that the processes do not need to be fully understood. Furthermore, most organizations are familiar with their boiler's steam usage, making the load profile acquired more accurate. Moreover, supply level integration makes the system less sensible to changes in individual operations. But, at the supply level, the needed temperature is often greater than at the process level, resulting in lower system efficiencies due to larger heat losses per square metre.

Pre-heating, a concept that is frequently used in process level integration is shown in Figure 5.6. Cold water is supplied into a storage tank where it is heated by a fossil fuel boiler to the necessary temperature for the manufacturing process after being warmed in a solar field.

Direct steam generation layout is shown in Figure 5.7. In the concentrating collectors, some water is partially evaporated. Before being provided to the factory's industrial process or steam network, the solar-heated steam is first separated from the leftover water in the steam drum. The feed water, also known as treated condensate, is returned to the collector field. Indirect steam generation is another possibility. In

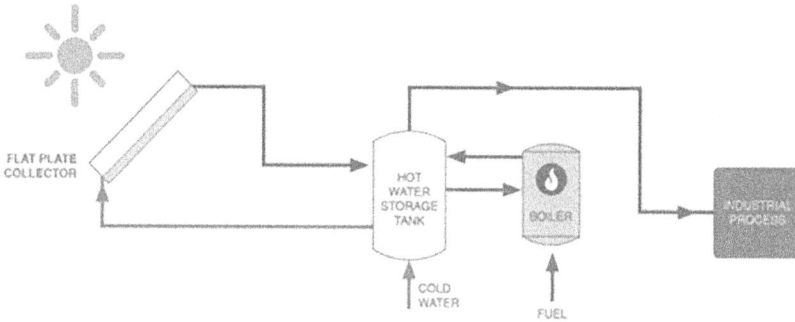

FIGURE 5.6 Scheme of solar thermal integration for pre-heating [11].

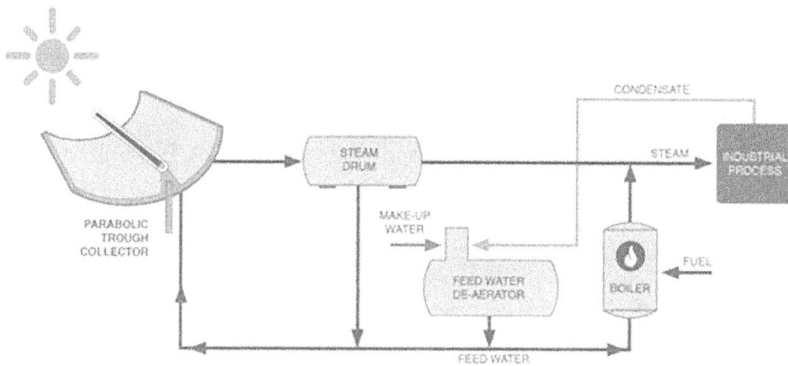

FIGURE 5.7 Scheme of solar heat integration for direct steam generation [11].

this instance, a heat exchanger uses a collector field to heat thermal oil or water in a closed circuit to produce steam.

Solar heat is used in the process, either directly or through heat exchangers. Heat is delivered at the process temperature, which is frequently lower than the standard heat-supply temperature, leading in a more energy-efficient integration. At the process level, solar integration is more variable and dependent on the process's parameters. A thorough understanding of the processes is required. Because process temperatures are often lower, integration at the process level can be more efficient. For example, concentrated solar collectors are not cost-effective in many places with high humidity and limited direct irradiation. Only stationary collectors with temperatures up to 120 °C can be utilized in this case. This temperature restriction may facilitate integration at the process level. Solar integration at the process level can be beneficial, especially if processes have stable and dependable load conditions.

Figure 5.8 shows how a bath or thermal separation process may be heated by the solar field at a certain temperature for a process heating. A boiler powered by fossil

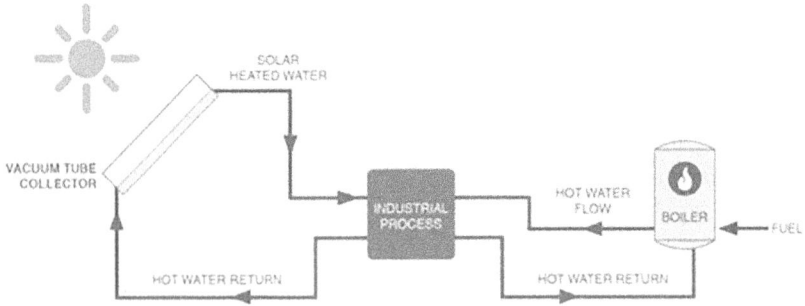

FIGURE 5.8   Scheme of solar heat integration for direct process heating [11].

fuels brings additional heat to the industrial process. The cooled-off water returns to the collector field or the boiler, respectively, because both circuits are closed.

5.4.2 Solar Energy-Based LPG Vaporization

A flammable combination of hydrocarbon gases known as LPG is used as a fuel in heating equipment and automobiles. LPG is kept liquid in pressurized tanks or containers. Liquid LPG should evaporate to turn from a liquid to a gas before being employed as a fuel for connected equipment like burners and heaters, etc. [12].

5.4.2.1 Existing System

LPG is a liquid that can be heated to well over its boiling point under a delivery pressure, and a vaporizer is designed to take that liquid and increase its temperature. By employing hot water that is between 80 and 85 °C in a closed system or hot water bath system, LPG is heated indirectly and converted into vapour. Electric heaters are used to heat the water. Figure 5.9 depicts a common hot water bath system.

5.4.2.2 LPG Vaporizer System with Solar Integration

The existing technology for vaporizing LPG incorporates a solar integration. Solar concentrators provide the necessary heat for vaporization, which lowers the equivalent electricity usage. Figure 5.10 shows an electrical heating system that is typically utilized in the automotive sector integrated with a solar system. Water is pumped in a closed loop through the solar system, gradually heating it to the required 85 °C for vaporization. LPG is heated in the liquid phase by passing water through an LPG vaporizer after the solar system has attained the required temperature. As a result, LPG will be heated and converted to vapour. The discharge water will be recirculated through the solar system at a lower temperature. When used with an LPG vaporizer, hot water or other fluid may often be kept in a separate tank at the required temperature. Due to a lack of solar energy during non-sunny hours, the temperature of the solar system lowers, and the system uses a temperature controller feedback loop to automatically activate electrical heaters.

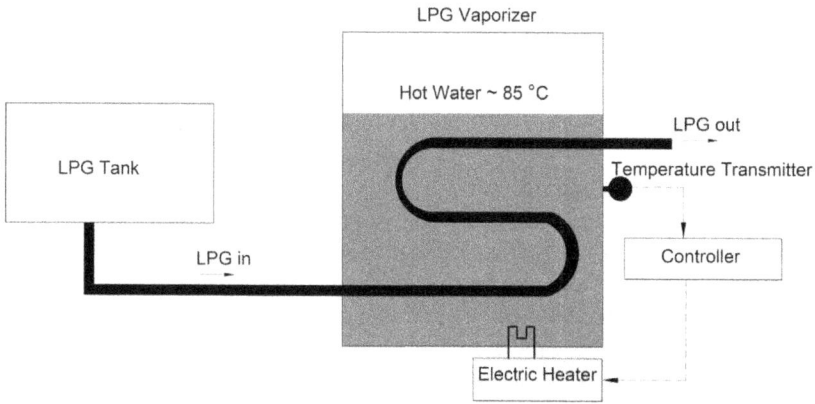

FIGURE 5.9   A simple hot water bath system.

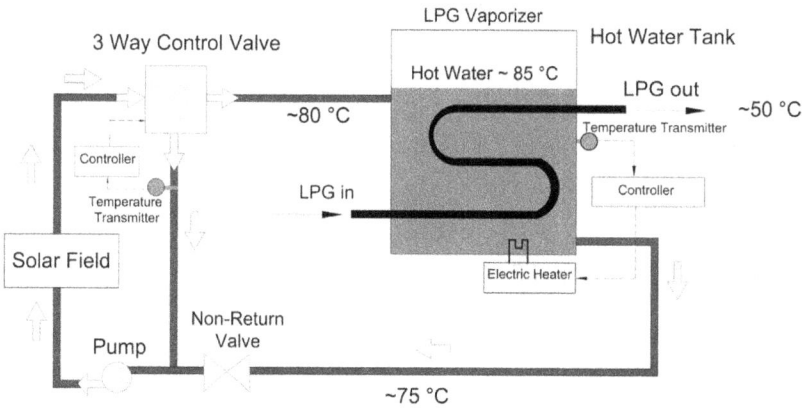

FIGURE 5.10   Solar integration with existing LPG vaporizer.

The location parameters and space availability determine the type of solar concentrating system.

5.4.3 SOLAR BASED FURNACE OIL (FO) HEATING

In the industrial sector, liquid fuel known as FO is frequently used in a variety of captive power plants, ovens, steam boilers, furnaces, etc. At ambient temperature, the fuel is quite viscous, thus pre-heating is necessary to reduce its viscosity and promote atomization at the burner tips. The proper handling and burning of fuel need atomization. So, before igniting a burner and when the FO is kept in tanks, it needs to be heated up. Before firing, oil is typically pre-heated to a temperature between 100 and 130 °C. Here, solar energy may be utilized to pre-heat instead of steam or electricity, which is the optimum use of solar energy [13].

5.4.3.1 Existing FO Heating System

There are primarily two types of systems utilized for FO heating, electrical heating and steam heating. The bank that provides electrical heating to the coil heaters is situated near the FO tank and close to the burner. Electrical heating provides all of the heat necessary for FO heating. The use of temperature feedback allows for the control of electrical heating.

The FO is heated using steam coils and supplied at desired hot condition to the burner or any other application. During startup, only electrical heating is used. After that, steam is supplied through a heat exchanger for pre-heating FO. Using a steam valve control, the desired temperature is maintained. Accordingly, depending on the quality of the fuel, the FO temperature at the burner must be regulated between 100 and 130 °C by utilizing either steam or electricity. A FO pump is often used to transfer fuel from the FO tank to the FO heater and subsequently to the burner in a typical setup. By controlling the FO heater, it is made sure that the burner receives FO at the required temperature.

5.4.3.2 Solar Integration with FO Heating System

The heat load for FO heating may be satisfied by installing a solar concentrator system to produce low pressure steam. In India, solar energy may be used 270 days out of the year for 6–7 hours every day. When solar energy is insufficient to produce enough heat, as is the case during overcast days and hours, a typical steam or electrical heater system is offered as a backup. When solar energy exceeds the actual requirements, the pressure in the steam drum rises, operating as a buffer storage device for hours when solar energy is unavailable. Steam will always be supplied to the application owing to the operation of control valves. Figure 5.11 illustrates how the solar system is integrated with the existing FO heating system.

Depending on the location and space conditions, any suitable concentrated solar thermal technology may be employed in the solar system. Steam can be produced using a parabolic trough collector and parabolic dish to heat FO.

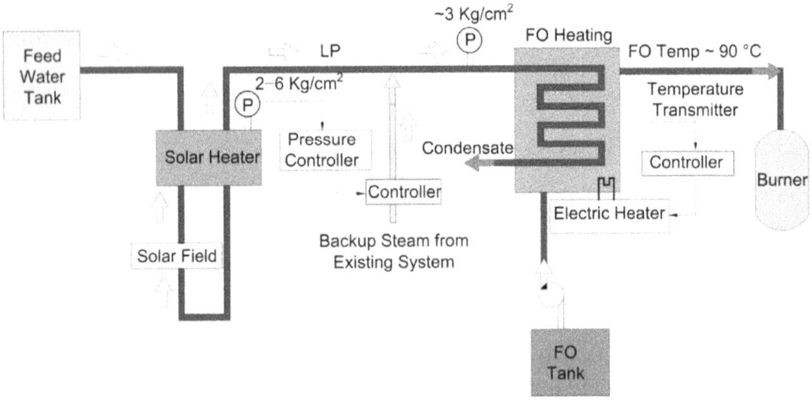

FIGURE 5.11   Solar system integration with existing FO heating system.

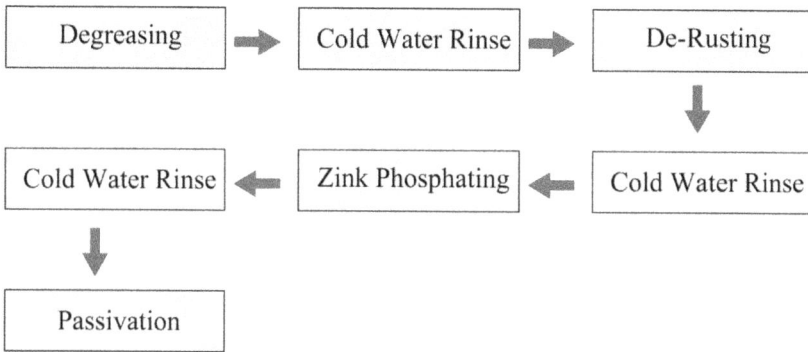

FIGURE 5.12   Process flow of 7-tank process.

5.4.4 UTILIZING SOLAR ENERGY TO 7-TANK DIP ZINC PHOSPHATING HEATING APPLICATION

A standard 7-tank dip zinc phosphating technique is used in an automobile industry paint shop to provide superior paint bonding and resistance to corrosion. Figure 5.12 depicts the process flow. Depending on the technology used in this pre-treatment process, the needed temperature for each stage varies and typically ranges between 60 and 80 °C. The tanks are heated using electrical immersion heaters, direct fire, thermal oil, or closed steam coils.

5.4.4.1 Standard 7-Tank System

All tanks are heated directly using electric immerse heaters or steam coils. Depending on the temperature needed for the different bath sizes, hot water generators or electrical heaters are utilized to heat the bath. The bath is filled with the electrical immersion heaters and heated to the desired temperature. The use of closed hot water heating coils, which may be placed at the tank's sides and allow the sludge to settle with a minimum clearance of 1 foot, is another heating technique. Hot water is produced using a variety of fuels, including natural gas, FO, LPG, and high-speed diesel. Table 5.3 depicts the typical hot water application temperatures without the use of a pre-conditioner for various processes [14].

TABLE 5.3
Usual Range of Temperatures for Using Hot Water in Different Processes

Tank	Step	Temperature in °C	Dip Time in Minutes	Tank Material
1	Degrease	60–70	1–2	Mild Steel (MS)
2	Cold water rinse	–	1	MS
3	De-rusting	60–70	1–2	Acid-resisting material
4	Cold water rinse	–	1	Acid-resisting material
5	Zinc phosphate	70–80	1–2	MS
6	Cold water rinse	–	1	MS
7	Passivation	70	1	MS

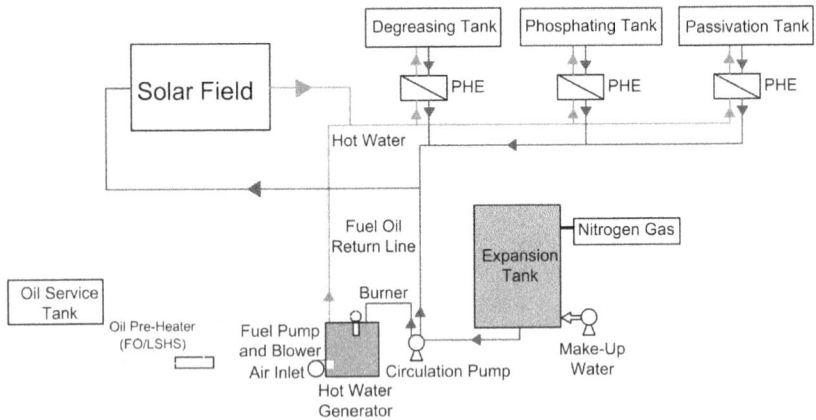

FIGURE 5.13   Solar system integration with current system for various heating processes.

5.4.4.2 7-Tank System Integrated with the Solar System

Hot water for low-temperature applications may be produced using solar energy. To produce hot water, the current system may be effectively incorporated with the solar system. When solar energy is insufficient, the existing system serves as a backup for steam generation. Figure 5.13 depicts the solar integration with an existing heating source for different processes.

5.5 SELECTION OF SOLAR THERMAL COLLECTORS

Selection of an appropriate solar thermal collector for the integration in industrial process heating applications primarily depends on the temperature range and type of working medium that is employed. There are a wide range of solar thermal collectors in practice, and generally they may be categorized into two, concentrating and non-concentrating collectors. Solar thermal collectors can provide energy for manufacturing processes operating in the low (100 °C) and medium (100–400 °C) temperature range. Non-concentrating collectors such as flat plate collectors (FPC) [15, 16] and evacuated tube collectors (ETC) [17, 18] are commonly used in low-temperature applications. Concentrating collectors such as parabolic trough collectors, parabolic dishes, and central receivers are used for medium-temperature applications. Table 5.4 depicts the process and its integration with appropriate solar thermal technology, based on the temperature range achieved by various solar collectors and the temperature required for the process given by heat application media.

5.6 CASE STUDIES

Solar thermal conversion technologies in automobile industries are successfully implemented in many automobile manufacturing industries situated in different parts of the world [19]. Depending upon the temperature attainable, solar collectors may be divided into low-, medium-, and high-temperature collectors. The industrial case

TABLE 5.4

Solar Thermal Technology Is Used at Several Stages of the Vehicle Manufacturing Process

Process Steps	Energy or Fuel Being Used	Application Media	Temperature Need °C	Recommended Solar Technology
Degreasing of automobile parts or part cleaning	Natural gas	Pressured hot water	90	ETCs
Dip zinc phosphating process	Electricity	Hot water	80	FTCs
Paint shop air-conditioning for wet paints	Natural gas	Hot or cold air supply	50	ETC based chillers
Mould drying	Electricity	Hot air	100	Solar air heaters
Paint shop with paint baking ovens for curing vehicle paint	Natural gas	Hot air	200	linear Fresnel concentrating collectors
Drying of cleaned parts	Natural gas	Hot air	90	PTCs with air as working medium
Paint shop evaporation drying	Natural gas	Hot air supply	100	Solar air heating
In a paint shop, the body of a vehicle is rinsed with hot water	Electricity	Hot water	40	FPCs
Tempering in furnace	Natural gas	Hot air	200	Solar air heaters with concentrating collectors

studies are explained in the following section under the classifications of low and medium-temperature applications.

5.6.1 Low-Temperature Applications

Wheels India Limited, situated in Chennai, is a vehicle wheel manufacturer owned by the TVS Group. It employs evacuated tube collectors to heat and maintain bath temperatures for the preparation of vehicle components prior to painting. The furnace oil burner provides the heat essential for hot water. In April 2013, a 1365 m^2 solar water heating system with evacuated tube collectors (in Figure 5.14a) was constructed to heat tanks for washing and polishing metal vehicle wheels [20].

The ETC tubes are filled with a heat transfer fluid (water-glycol combination), and the solar heat is transmitted to the demineralized (DM) water loop through a heat exchanger. The DM water acts as a heat transfer fluid, circulating in closed loops to transmit solar heat to the degreasing tanks via submerged coils that operate as heat exchangers. Because the tanks are used as storage, no additional storage is required. The solar system is not pressurized, and the DM water expansion is balanced using

FIGURE 5.14A ETC water heating system at Wheels India Ltd [20].

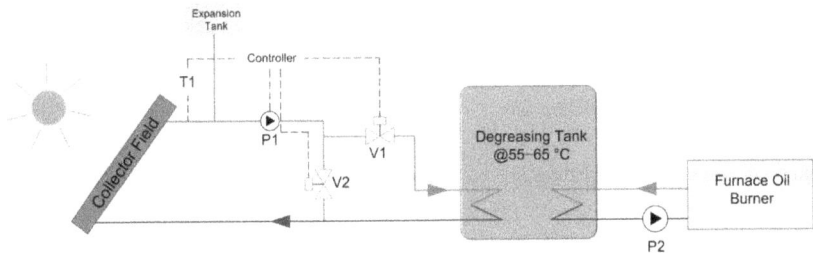

FIGURE 5.14B Hydraulic scheme of solar water heating system at Wheels India Ltd.

an open expansion tank. Figure 5.14b depicts the hydraulics of a solar water heating system. Hot water at 55 °C in tank 1, knock-off degreaser at 70 °C in tank 2, Degreaser 1 at 60 °C in tank 3, and Degreaser 2 at 60 °C in tank 4 are used to clean and wash metal wheels manufactured via casting.

The company utilizes around 3,000 litres of furnace oil per day, which costs INR 165,000 (EUR 2000) per day. They have saved roughly 380 litres per day since installing solar water heaters, saving around INR 20,000 (EUR 250) every day. With subsidies and tax incentives, the total capital cost for the solar systems was Rs 21,000,000/-. The payback time was around two years and reduction in carbon dioxide emissions by 280 tonnes per year by replacing furnace oil.

Sona Koya Steering Systems is a leading producer of steering elements for cars and light motor vehicles in India [21]. This company installed a solar water heating system which replaced the conventional water heating system operated by diesel. The solar system was capable of generating 1260 KWh of solar thermal energy in 24 hours. This resulted in savings of 36,000 litres of diesel per annum. Sona Koyo also supplied the hot water to their employees' cafeteria by using this solar water heater. A pictorial view of the solar assisted water heating system is shown in Figure 5.15.

Harita Seating Systems Ltd is the largest manufacturer of seating systems for automobile sectors in India. The company employed a Liquefied Petroleum Gas

FIGURE 5.15   Rooftop solar water heating system in Sona Koya, Chennai, India [21].

(LPG) based furnace heating system for supplying the thermal energy needed for the pre-treatment process. Nearly 20 to 25 tonnes of LPG per year were required for the pre-treatment heating process. A 360 kW rooftop solar thermal system (evacuated tube type collector) with an absorber area of 501 m^2 was installed to deliver the required thermal energy to the process tanks, which has reduced their expenditure about 20 to 25 tonnes of LPG per annum. Figure 5.16 indicates the solar process heating system in Harita Seating Systems Ltd [21].

In Nagpur, a CPC-based process heating system for automobile component washing was installed in 2013 [22]. The solar CPC has a surface area of around 442 m^2. Solar CPCs are used to produce high-temperature water at 85 °C. The solar power plant is comprised of solar collectors, valves, a water storage tank, a circulation pump, controls, etc. The schematic diagram of the system is depicted in Figure 5.17 (a). The temperature of the water is raised in the system by several passes through the storage tank. After the sun rises and the temperature of the water in the collector rises, the collector begins to absorb solar energy. The temperature sensor detects this and activates the circulation pump, forcing the primary circuit water through the solar collectors. Solar collectors gather heat, which is then transported to a storage tank. The process is supplied with water when the temperature in the storage tank reaches the target temperature of 85 °C. On the component washing tanks, electric heaters are installed, which turn on when the appropriate temperature cannot be reached using the solar system owing to low radiation levels or during non-sunny hours/days.

Water from the storage tank is cycled through heat exchangers, which indirectly maintain the proper temperature in the process tanks, and the water is returned to the storage tank after releasing its heat, and the cycle continues. This solar process heating system produces process heating water at temperatures as high as 90 °C. It also connects smoothly with the current system, ensuring its stability while

FIGURE 5.16   Solar assisted process heating system used in Haritha seating systems, Hosur India [20].

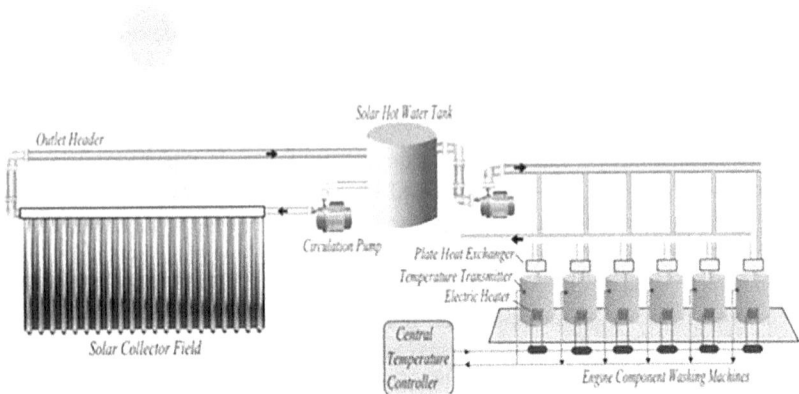

FIGURE 5.17A   Layout of the CPC heating system at Nagpur [22].

lowering power use. The cost of a system based on a CPC was Rs. 8,444,000/-. In 2014, total units of electricity saved were 174,910 kWh, equal to 164 metric tonnes of CO_2 emissions. The payback period is around 5 years when considering the benefits provided by the Indian government in the form of subsidies and financial

FIGURE 5.17B   CPC based solar installation at an automobile industry [22].

FIGURE 5.18   Demonstration plant "Hammerer Transport Company", Austria [23, 24].

rewards for renewable energy. Figure 5.17 (b) is a photograph of a solar water heating system.

For the Hammerer transport firm in Austria, a transportation sector system was developed [23, 24]. A 126 kW system (180 m^2 collector array) was constructed to provide hot water for cleaning trucks' transport containers as well as space heating in the offices. It is shown in Figure 5.18.

Solar thermal energy is employed in the bodyworks pre-treatment line at Nissan Motors Company, where the pre-treatment for grease removal, phosphatizing, and electro-coating processes are installed [25]. The solar heating system covers around 530 m^2 and is made up of 42 batteries with six units each. The system is installed on the roof of the building where the activities are taking place, at a 23° angle to the horizontal and 30° to the south. An NG system was installed before the construction to give heat to the exchangers and meet the process's energy requirements. According to estimates, the new solar system will deliver 479,990 kWh per year, reducing energy

FIGURE 5.19   Rooftop FPC solar water heating system in Nissan, Spain [25].

provided by GAS by 1,314,216 kWh per year and saving 42,823 € per year. CO_2 emissions are reduced by 217 tonnes per year in the atmosphere. The overall investment calculated is around 280,000 €; however, Nissan's contribution was up to 140,000 €, owing to regional government incentives and a tax credit for the first year's entire investment. A pictorial of a FPC solar water heating system is shown in Figure 5.19.

Lackiererei Vogel is a paint shop located in Zwickau, Saxony. This company's main areas of operation include car painting, industrial part varnishing, car and part maintenance, labelling, and transportation services. The solar process heating system was installed in August 2010 at Lackiererei Vogel [25]. Lackiererei Vogel uses solar process heat from 43 m³ vacuum tube collectors to heat the painting cabins to 22 to 24 °C during the painting. It also generates hot air for drying processes at 60 to 70 °C. As a result, a solar thermal plant was developed to lower the company's natural gas demand. Figure 5.20 shows a pictorial view of an ETC solar heating system.

Vaporizados Palencia, located in Villamuriel de Cerrato, is a company that specializes in washing truck containers for carriers of various items like chemicals, food, and industrial oils [25]. Depending on the materials delivered, each washing procedure is different. A solar thermal system of 140 m² was installed with two buffer tanks of 5,000 litres each. With 20 trucks washed every day, the average water need is around 300 litres per truck. The total project cost around 85,000 €. The payback period was 7 years. A picture of solar water heating system is shown in Figure 5.21.

In the FASA Valladolid, the solar thermal energy is used in the bodyworks sheet surface preparation for the welding process and before the painting [26]. The solar heating system has a surface area of around 244 m² and is made up of five batteries. Due to a shortage of space above the roofs, the system is mounted on the ground (in Figure 5.22). The NG heating system produced the needed energy, 254 MWh/year, but solar heat currently provides 128 MWh/year, providing over 50% of the requirements. The solar heat is used in the degreasing tanks, which have a temperature of 50 °C

FIGURE 5.20 ETC solar heating system in Lackiererei Vogel at Zwickau, Saxony [25].

FIGURE 5.21 Solar water heating system in Vaporizados Palencia at Villamuriel de Cerrato [25].

FIGURE 5.22   FPC solar water heating system in FASA Valladolid, Spain [26].

and require water to maintain the level. As a result of this activity, the solar heat prevents the operation from requiring additional thermal input. Due to the significant discounts available to large fuel customers, the solar system saves 9,200 € per year, placing the payback time at 10 years. The entire investment calculated is around 150,000 €; however, Renault only invested just over 100,000 € since the project got incentives from local governments and a tax reduction of the whole expenditure for the first year.

Lackiercenter Schulte is a medium-sized paint company that specializes in car body part restyling and it installed a solar process heat system in April 2009 [27]. The solar heating system has a surface area of 136 m² installed with vacuum tube collectors. Two 5,000 litre buffer storage tanks are used to store solar heat gains. The heat source for the painting chamber is provided by one of the storage tanks, which requires a constant temperature of 23 °C, while the heat supply for the drying chamber is provided by the other storage tank, which requires a constant temperature of 70 °C. Solar heat is transferred from the buffer to the cabin supply air via a water/air heat exchanger. The cost of a vacuum tube collector system, which includes the heat recovery system, was 116,000 €. The bank supported the funding at 30% of the investment cost. The payback period was expected to be 7 to 8 years. A pictorial of an ETC solar water heating system is shown in Figure 5.23.

5.6.2 MEDIUM-TEMPERATURE APPLICATIONS

Mahindra's 700-acre Greenfield manufacturing plant in Chakan, Pune district, was established in 2007 to produce multipurpose vehicles (MPV), sport utility vehicles

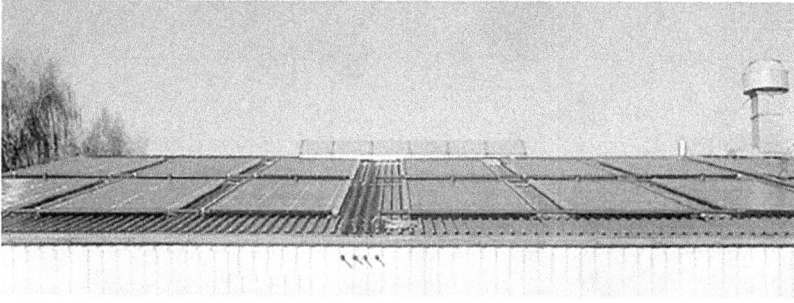

FIGURE 5.23   ETC solar water heating system in Lackiercenter Schulte, Meppen, Germany [27].

(SUV), and vehicles [28]. Mahindra Vehicle Manufacturers Limited (MVML) has deployed an ARUN dish, a dual-axis tracked Fresnel Paraboloid Solar Concentrator (FPSC), to generate hot water for washing engine components in washing machines. The system provides pressurized hot water at 120 °C for 7 hours per day, which is used in the degreasing of engine components. Layout of the plant is shown in Figure 5.24 (a).

It is mounted on a 3 m × 3 m column with an 8 m height, and the region beneath the dish is used as a vehicle test facility, reducing space needs. The dish is a 100% indigenously developed FPSC with a point focus. In comparison to any other solar concentrator, the revolutionary dish design and automated 2-axis tracking technology assure the maximum thermal energy output per square metre of collecting area. The ease of use, along with strict safety requirements, guarantees that minimal maintenance is required over a long length of time.

The ARUN dish system at MVML was implemented at a cost of Rs. 39,00,000/-, which included the rest of the system's expenditures such as plumbing, civil construction, etc. The Ministry of New and Renewable Energy (MNRE) sanctioned and granted a total subsidy of Rs. 10,14,000/- for this system. As a result, the total project cost, excluding the MNRE subsidy, is Rs. 28,86,000/-. The unit owner can also take advantage of the IT Act's extra benefits of accelerated depreciation of up to 80% of the cost of the project. This aids in the reduction of tax outflows in the first year of expenditures. Due to the reductions in power purchase prices, the project payback period was roughly 2 years, and the project life cycle was 25 years. As a result, the FPSC system serves as a model case study for other vehicle companies and startups interested in implementing CST-based systems.

The temperature of heated air essential for curing paint in all vehicle production facilities ranges between 80 °C to 180 °C, and most petroleum fuels such as furnace oil, diesel, or kerosene are used to generate hot air that is blown within ovens by combustion. The unit established a solar air heater with a surface area of 292 m² to heat the air to a temperature of 112 °C [29].

A V-shaped corrugated unique aluminium absorber, silica sealant, aluminium extrusions, EPDM rubber, mineral wool insulation, and thick toughened glass were

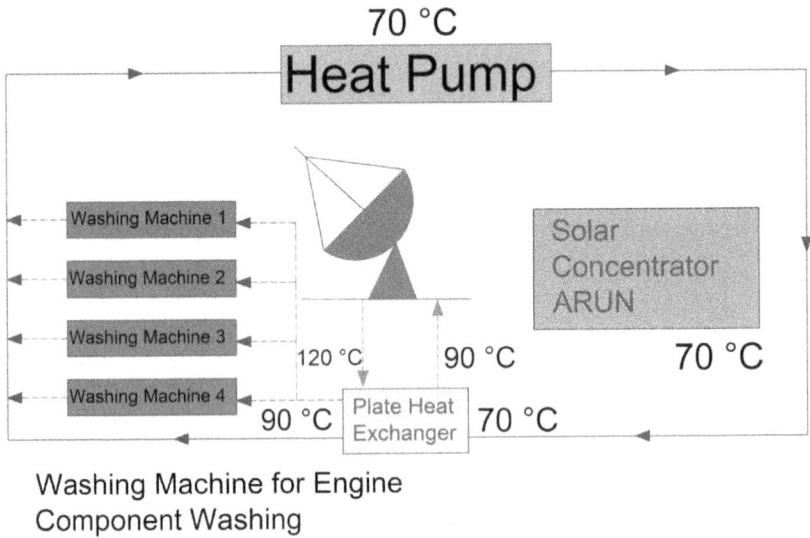

FIGURE 5.24   Layout of the FPSC system at MVML, Chakan, Pune [28].

FIGURE 5.25   Solar hot air for paint shop manufacturing unit, Chennai, South India [29].

used to construct the solar air heater (Figure 5.25). Suction blowers deliver the hot air collected from solar air heaters through an insulated metal duct to the oven's fresh air inputs. With the MNRE subsidies, tax savings under the accelerated depreciation method, and fuel savings achieved against the investments, the payback period was estimated to be roughly 1.3 years.

In Mysore, SKF Technologies Private Limited maintains a production site for automobile sealing systems. A solar concentrating system based on a parabolic trough collector (PTC) has been constructed for process heat application at SKF Mysore [30]. The system consists of 40 SolPac P60 PTCs, each with a 6.41 m^2 aperture area, and are

FIGURE 5.26A   Pictorial view of the phosphating plant [30].

coupled in a series-parallel configuration to produce pressurized hot water at 130 °C. The system has been placed on the establishment's terrace, with a total trough collecting area of 256.4 m². With the help of a backup generator tracking system and a support structure, this collection of collectors provides hot water. The highest working temperature that can be achieved with this technique is 210 °C, which is heated by the sun's rays focused on the receiver tube. They used a parabolic trough-based solar concentrating system to create pressurized hot water at 130 °C for process heat. The collectors have a total area of 256 m² and are positioned on the plant's rooftop (Figure 5.26a).

There are 11 tanks in the phosphating facility, and there are 5 key stages. The oil and grease from the stampings are removed in the first stage. Another round of cleaning is performed in the second step to entirely remove the chemical and oil from the stampings. The stamping is then coated with a phosphate layer in the third stage. The rinse procedure is the fourth stage, during which all dirt and pollutants are eliminated. The last procedure entails drying the stampings in ordinary dryers before sending them for further processing. The first, second, and third steps of the procedure all necessitate the use of hot water. A diesel-powered backup generator is included in the system. The water utilized for circulation through treatment tanks for the phosphating process is heated by parabolic trough collectors, which maintain a temperature of 95 °C. Phosphating metal components reduce the likelihood of rust development. Layout of plant is shown in Figure 5.26b.

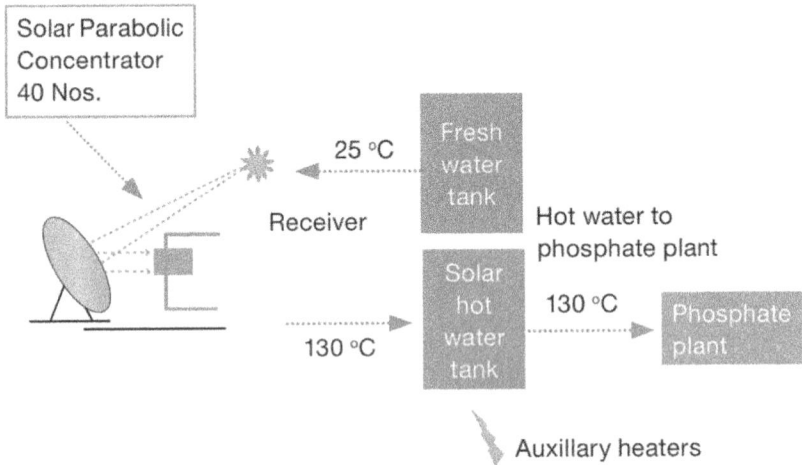

FIGURE 5.26B   Layout of the phosphating plant at SKF Technologies Pvt Ltd., Mysore.

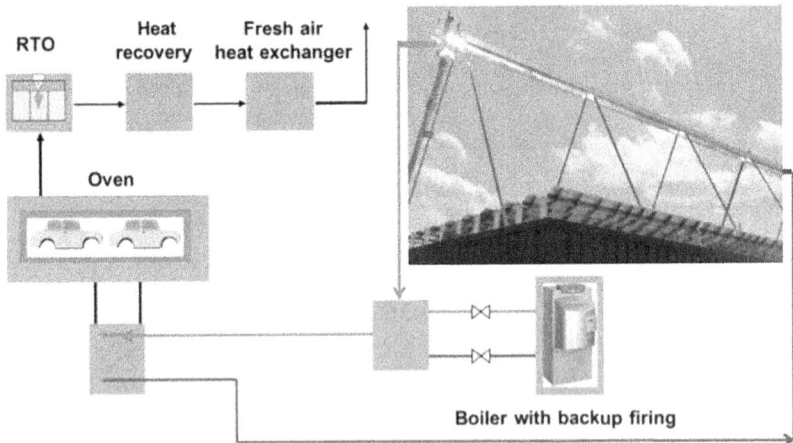

FIGURE 5.27   Fresnel collector-based heating system at Dürr Campus [31].

The cost of a system based on a parabolic trough collector was roughly Rs 7,050,000/-, which included the rest of the system's expenditures such as pipe and civil construction. MNRE authorized and gave a total subsidy of around Rs 1,384,500/-. As a result, the unit owner incurred a total project cost of around Rs 5,670,400/-. In addition, the additional benefits of the 80% accelerated depreciation plan under the IT Act will result in lower tax outflows in the first year. As a result, the complete project had a 4-year payback period due to the owner's significant savings over diesel fuel previously used for heating.

On the Dürr Campus in Bietigheim-Bissingen near Stuttgart, Germany, a pilot system of a solar-powered convection oven was constructed in February 2012

TABLE 5.5
SIPH Used in Automobile Industries by Country [21, 33]

Country	Name of the Industry	Solar Thermal Applications	Type of Collector	Temperature Range (°C)
India	Mahindra Vehicle Manufactures	Hot water required to clean the components of engines	FPSC	120
	Wheels India Limited	Clean and wash metal wheels	ETC	60–70
	SKF Technologies Private Limited	Hot water is used for phosphating process	PTC	130
	Harita Seating Systems Ltd	Pre-treatment heating process	ETC	–
	Sona Koya Steering Systems	Hot water is used for washing the steering elements	ETC	–
	Indian Automobile Industry	Process heat for automobile component washing	CPC	90
Spain	Nissan Avila	Hot water required for degreasing and electrocoating process	FPC	45
	FASA Valladolid	Thermal energy is required in bodywork section for surface treatment process before painting and welding	FPC	50
	Vaporizados Palencia	Hot water is used for washing truck containers	FPC	–
Germany	Lackiercenter Schulte	–	ETC	70
	Lackiererei Vogel	Car painting	ETC	60–70
South Africa	Bayerische Motor Werke Manufacturing	Need hot water in the paint shop section, to remove dust particles from engine components	ETC	90

[31]. A 6-module Industrial Solar Fresnel collector with a total aperture area of 132 m² heats pressurized water to a temperature of 180 °C at the output. To supply the convection oven's thermal power demand, a fossil-fuelled backup boiler is used in addition to the solar thermal system. Finally, a secondary pressurized water-air heat exchanger in the convection oven dissipates the heat. In the paint shop, hot air is required at the range of 80 °C to 100 °C for pre-treatment in the paint shop [32]. Figure 5.27 shows the schematic view of the Fresnel collector-based heating system.

Table 5.5 shows the countries which are using solar industrial process heat (SIPH) systems in their automobile industry, as well as the production specifications.

5.7 CONCLUSION

Automobile industries consume a huge amount of thermal energy for many of its low-temperature processes including degreasing, washing, drying, etc. Solar hybridization with current systems can partially fulfil the automotive industry's need for thermal energy. Most of the automobile industries are facilitated with adequate rooftop area on their factories to install the stationary type of collectors and they can supply moderate temperature requirements. The challenging task in the integration of solar thermal energy systems along with the conventional heating system requires more effort and research to ensure the smooth integration to ensure the maximum utilization of solar energy without any interruption in the process.

REFERENCES

[1] Giampieri, A., Ling-Chin, J., Ma, Z., Smallbone, A., & Roskilly, A. P. (2020). A review of the current automotive manufacturing practice from an energy perspective. Applied Energy, 261, 114074.

[2] Uppal, A., & Kesari, J. P. (2015). Solar industrial process heat in Indian automobile industry. International Journal of Latest Technology in Engineering, Management & Applied Science-IJLTEMAS, 4(10), 117–123.

[3] Automotive manufacturing solutions for paint shops, www.automotivemanufacturingsolutions.com/processmaterials/solar-process-heat-for-paintshops.

[4] Clique solar solution for automobile, www.cliquesolar.com/AutomobileSolution.aspx

[5] Streitberger, H. J., & Dossel, K. F. (Eds.). (2008). Automotive paints and coatings. John Wiley & Sons.

[6] Talbert, R. (2007). Paint technology handbook. CRC Press.

[7] Painting makes the difference; 2012

[8] Akafuah, N. K., Poozesh, S., Salaimeh, A., Patrick, G., Lawler, K., & Saito, K. (2016). Evolution of the automotive body coating process—A review. Coatings, 6(2), 24.

[9] Veera Kumar, A., Arjunan, T. V., Seenivasan, D., Venkatramanan, R., & Vijayan, S. (2021). Techno-Economic evaluation of an evacuated tube solar air collector with inserted baffles. Proceedings of the Institution of Mechanical Engineers, Part E: Journal of Process Mechanical Engineering, 235(4), 1027–1038.

[10] Kumar, A. V., Arjunan, T. V., Seenivasan, D., Venkatramanan, R., & Vijayan, S. (2021) Thermal performance of an evacuated tube solar collector with inserted baffles for air heating applications. Solar Energy, 215, 131–143.

[11] SW-Solar. (2020). Solar Payback. Abgerufen am August 2020 von Solar Payback: www.solarpayback.com

[12] www.altenergy.com/Default.htm Alternate Energy Systems, Inc

[13] https://beeindia.gov.in/sites/default/files/2Ch1.pdf Fuels and Combustion

[14] **www.thinchemie.com**Dip zinc phosphating-Thin Chemie Formulations

[15] Vijayan, S., Arjunan, T. V., Kumar, A., & Matheswaran, M. M. (2020). Experimental and thermal performance investigations on sensible storage based solar air heater. Journal of Energy Storage, 31, 101620.

[16] Matheswaran, M. M., Arjunan, T. V., & Somasundaram, D. (2019). Analytical investigation of exergetic performance on jet impingement solar air heater with multiple arc protrusion obstacles. Journal of Thermal Analysis and Calorimetry, 137(1), 253–266.

[17] Kumar, A. V., Arjunan, T. V., Seenivasan, D., Venkatramanan, R., Vijayan, S., & Matheswaran, M. M. (2021). Influence of twisted tape inserts on energy and exergy performance of an evacuated Tube-based solar air collector. Solar Energy, 225, 892–904.

[18] Venkatramanan, R., Arjunan, T. V., Seenivasan, D., & Kumar, A. V. (2022). Parametric study of evacuated tube collector solar air heater with inserted baffles on thermal network for low-temperature applications. Journal of Cleaner Production, 367, 132941.

[19] Mayyas, A., Qattawi, A., Omar, M., & Shan, D. (2012). Design for sustainability in automotive industry: A comprehensive review. Renewable and sustainable energy reviews, 16(4), 1845–1862.

[20] Sopro India, www.soproindia.in/wheelsindia-proj01.html

[21] http://ship-plants.info.

[22] Pathak, A., Deshpande, K., & Jadkar, S. (2017). Application of Solar Thermal Energy for Medium Temperature Heating in Automobile Industry. IRA-International Journal of Technology & Engineering (ISSN 2455–4480), 7(2 (S)), 19–33.

[23] www.estif.org/fileadmin/estif/content/policies/downloads/D23-solar-industrial-process-heat.pdf

[24] Yuan, G., Hong, L., Li, X., Xu, L., Tang, W., Wang, Z. (2015). Experimental investigation of a solar dryer system for drying carpet. Energy Procedia, 70, 626–33. doi:10.1016/j.egypro.2015.02.170

[25] www.solar-process-heat.eu/

[26] https://www.solar-process-heat.eu/fileadmin/redakteure/So-Pro/Installations/ESCAN_Fasa.pdf

[27] https://www.solar-process-heat.eu/fileadmin/redakteure/So-Pro/Installations/Gertec_Schulte.pdf

[28] Concentrated solar heat India, www.cshindia.in/images/pdf/M&M_1.pdf

[29] Solar heat conference 2012, http://cms.shc2012.org/program/displayattachment/48e7fc94032?filename=26617.pdf&mode=abstract.

[30] Concentrated solar heat India, www.cshindia.in/images/pdf/SKF.pdf

[31] Zahler, C., & Iglauer, O. (2012). Solar process heat for sustainable automobile manufacturing. Energy Procedia, 30, 775–782.

[32] Kumar, K. R., Chaitanya, N. K., & Kumar, N. S. (2021). Solar thermal energy technologies and its applications for process heating and power generation—A review. Journal of Cleaner Production, 282, 125296.

[33] Farjana, S. H., Huda, N., Mahmud, M. P., & Saidur, R. (2018). Solar process heat in industrial systems—A global review. Renewable and Sustainable Energy Reviews, 82, 2270–2286.

6 Solar Energy in Food Processing Industries

An Overview

Jasinta Poonam Ekka

CONTENTS

6.1 INTRODUCTION

The use of processed food has increased globally over time due to urbanization, higher income, and change in consumers' mindsets regarding the quality of food (Regmi and Gehlhar, 2005). The growth in the global population will have increased food demand by 60% by the year 2050 (Ladha-Sabur et al., 2019) to fulfill the needs of 9 billion people (Gowreesunker et al., 2017). The food processing industry

DOI: 10.1201/9781003263326-8

149

transforms raw edible materials into high-valued consumable products (flavor, aroma, and texture). The food processing industry throughout the world is one of the largest consumers of energy consumption. The food production industries are taking opportunities to reduce costs without affecting the output, product quality, and profitability. The sales value of processed food is three times higher than that of fresh products. All the food processing industries are either partially or fully dependent on burning fossil fuels to fulfill the energy needs for power and heat energy. The use of energy has been the primary concern during the last few decades with globalization and escalation in prices. Several efforts have been made to reduce fossil fuel consumption, which also creates environmental issues like global warming and climate change by the emission of GHGs. The food industry consumes about 30% of the world's total energy (Ismail et al., 2021). In India, the food processing industry has mainly increased during the last two–three decades. The food processing and beverage industry involves several processes that require energy. About 13% of the energy is consumed for producing thermal heat. Food processing industries use nearly 68% of fuel for process and space heating; while 16% is used as electrical energy for running motors, pumps, etc.; 6% used in refrigeration; and 2% for compressors (Wu et al., 2013). Different food processing industries include fruits and vegetables, milk and milk products, beer and alcoholic beverages, meat and poultry, marine products, grain processing, packaged or convenience food, and packaged drinks. Most energy is consumed during the various processes involved like cooking, washing and cleaning, sterilization, boiling, cooling, pureeing, baking, pasteurization, brewing, drying, and dehydration. These industries are the major users of resources like water, energy, and packaging materials and create waste and harmful emissions. Due to this, these industries face many pressures from the government and international organizations. Wu et al. (2013) reported that in UK the food industries emitted about 13 MtCO2e with a total energy consumption of 42 TWh. Figure 6.1 indicates the share of different food process industries. The major sectors of food processing are:

- Fruits and vegetables.
- Meat and poultry.
- Fisheries.
- Milk and dairy.
- Food grains and cereals.
- Plantation crops.
- Confection items.

6.2 PROCESSING METHODS IN FOOD PROCESSING INDUSTRIES

"Food processing is described as a systematic series of actions directed to add value to a food product," as defined by the Food and Drug Administration (FDA) in 1999. These actions include drying, reducing the size (cutting, peeling, grinding), heat transfer (cooking, cooling, freezing, pasteurization), and separation (washing, sorting, grading) (Bowser, 2019). Food processing aims to increase the shelf life and quality (color, texture, flavor). The processing technique is different for different types of products.

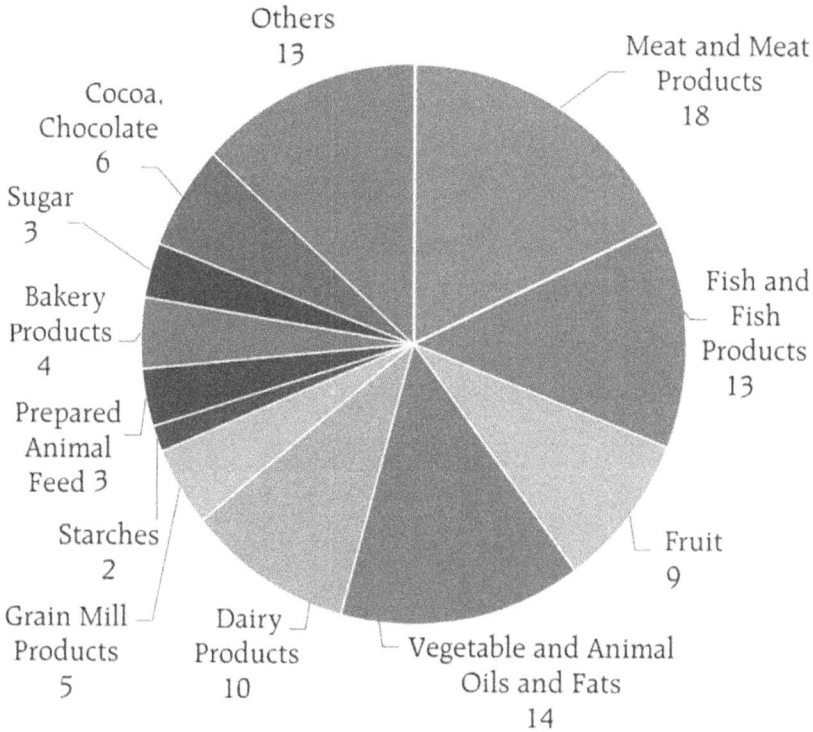

FIGURE 6.1   Share of different types of processed food.

Source: Dhanya et al. (2020).

6.2.1 FRUITS AND VEGETABLES

Fruits and vegetables are essential to diets and contain minerals, vitamins, fibers, etc. Fruits and vegetables are highly perishable and seasonal. Therefore, for long-term usage, processing is required. India is one of two of the world's largest producers of fruits and vegetables after China. It contributes 15% of the total world's agriculture production. As per the data of (CIPHET) Ludhiana, post-harvest losses in India are 18% of the production. Various processing techniques have been introduced to reduce the post-harvest losses of fresh fruits and vegetables and to add value to the products, such as dehydration, canning, pickling, sauces, and preservations. In India processing of fruits and vegetables accounts for only 3%, which is presently less when compared to China (23%), the US (65%), and the Philippines (78%). Some of the fruits and vegetables processing units are:

1. Tomato puree, sauces, ketchup.
2. Mango pulp and other fruit juice.
3. Frozen vegetables.
4. Pickles, chutney.
5. Potato fries.

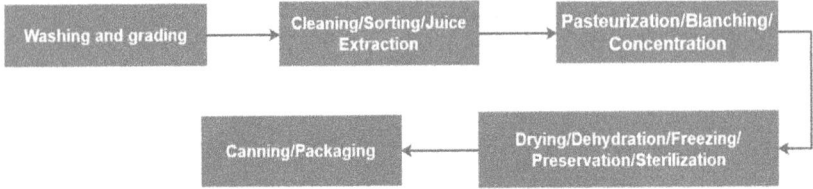

FIGURE 6.2 Fruits and vegetable processing techniques.

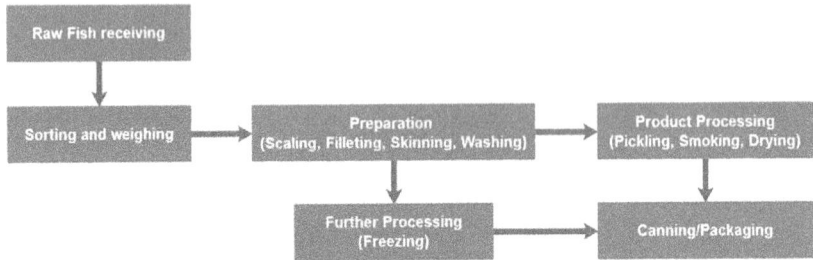

FIGURE 6.3 Methods of processing raw fish.

The processes used in fruit and vegetable processing are reported in Figure 6.2.

6.2.2 MEAT, POULTRY, AND FISH PROCESSING

Meat and fish products are highly nutritional and contain fats, proteins, vitamins, and minerals (zinc and iron). These products have become an essential part of the diet and are consumed worldwide. The main purpose of processing is to avoid microbial growth. The various steps involved in processing raw fish are given in Figure 6.3.

6.2.3 MILK AND DAIRY PROCESSING

As milk and its by-products (yogurt, butter, cheese) are highly perishable, they are processed for value addition and increased shelf storage life. India is the fastest-growing nation in milk and dairy products; it produces about 123.7 million tons of milk per year, and the same amount is processed in processing plants. Figure 6.4 provides the details of the milk processing plant (Mane, 2013). Pasteurization is the main process in the milk industry; it to helps destroy the disease-causing bacteria, protozoa, and yeast and enhances the shelf-life of milk.

6.2.4 BAKERY

Bakery is the most vital industry in the food processing industry throughout the world. The value of the world bakery industry was $216 billion in 2020. Bakery

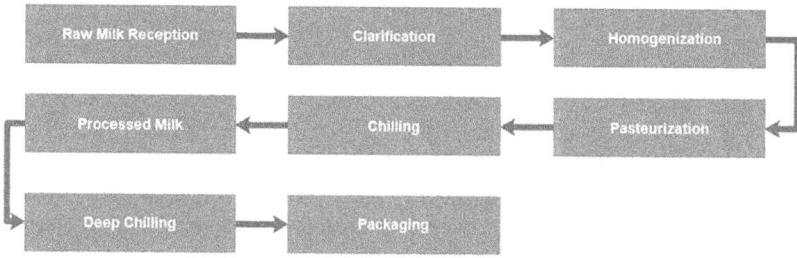

FIGURE 6.4 Details of the milk processing plant.

Process	Products			Equipment
	Bread	Cake	Minor Pastry	
Weighing	x	x	x	- scale
Kneading	x	x	x	- kneader - mixer
Forming	x	x	x	-
Leavening	x	-	-	- fermentation chamber
Baking	x	x	x	- electric ovens

FIGURE 6.5 Methodology of processing different bakery products.

Source: Briceño-León et al. (2021).

processing includes the making of cakes, bread, biscuits, doughnuts, truffles, brownies, etc., performed in different stages such as mixing and kneading of dough which in later stages are subjected to baking. Baking is the final process accomplished at a very high temperature in an electric oven with a temperature range between 190–250°C (Briceño-León et al., 2021). The process involved in the bakery is discussed in detail in Figure 6.5.

6.2.5 CONFECTIONERY PRODUCTS

Globally, confectionery items are greatly enjoyed as a major food delicacy because of their strong flavor and texture. Confectionery means "food products which are rich in sugar and carbohydrates." In India, the confectionery market produced about 138,000 metric tons of chocolates, mints, chewing gums, hard candy toffees, and other sugar-filled candies for around Rs. 20,000 million in 2005. The basic processes involved in confectionery products are mixing sugar or glucose, cooking, kneading, cooling, molding, sorting, wrapping, weighing, and packaging.

6.3 ENERGY CONSUMPTION PROFILE

Food processing industries are the largest consumers of energy in the manufacturing sector. The most energy is consumed in process heating, cooling/refrigeration, and boiler losses; it accounts for 29%, 15.5%, and 22%, respectively. The energy consumption in food processing industries is governed by various parameters like the thermophysical properties of the raw material, the technology used, equipment, range of processing temperatures, production capacity, and production process organization (Wojdalski et al., 2015). Most food processing units require thermal heat for various processes such as drying, boiling, sterilization, pasteurization, cleaning, cooking, extracting, brewing, etc. The food industry in the UK consumes approximately 11.5% of the industrial energy, which is 36.8 TWh both in thermal and electrical energy. Due to the escalation in fuel prices and associated environmental problems, major food processing industries are shifting toward an alternative energy source (Hall and Howe, 2012).

6.3.1 MILK AND DAIRY INDUSTRY

Milk and dairy processing are highly energy-intensive among the different food processing industries. The demand for milk and dairy products has increased by 6% in the last few decades. Dairy products are manufactured with condensed milk. Electrical energy is used for refrigeration, pumping, and separation, whereas thermal energy is required to clean vessels, sterilize, evaporate, and pasteurize. Figure 6.6 briefly shows the energy consumed in manufacturing different by-products of milk. Cheesemaking requires high energy (13.85 MJ/kg), whereas milk powder has the highest energy usage. High-temperature heat is required for sterilization, and the pasteurization of milk consumes 17–26% of the total energy (Ladha-Sabur et al., 2019). The highest amount of energy is consumed in making powdered milk, which is performed by the spray drying technique (Drescher et al., 1997).

6.3.2 ENERGY NEEDS IN BAKERY

Modern bakeries are commercial and highly automated; they are able to produce up to 136,000 kg of 100 varieties of bread every day. The amount of energy consumed in the baking of bread is 5.2 MJ/kg. In US, the total energy needs in bakeries are fulfilled by 63% electrical supply and 37% by natural gas. The major electrical energy is

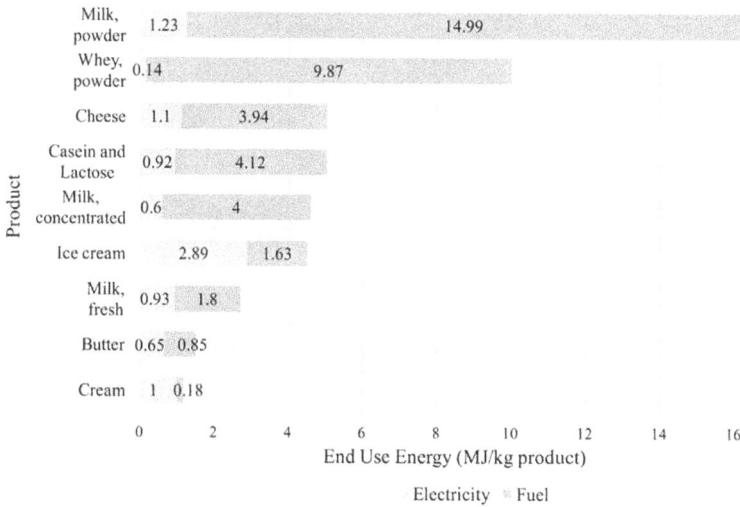

FIGURE 6.6 Energy demand in milk and dairy processing.

used in boilers to produce steam for fermentation, and in ovens, freezers, coolers, and for cleaning and washing. Energy use also depends on the type of product manufactured. Baking and freezing consume a significant amount of energy. Since baking is performed at high temperatures in the oven, the energy consumed is high due to air's low convective heat transfer coefficient. Figure 6.7 elaborates the amount of energy consumed by each product in the bakery industry.

6.3.3 Fruits and Vegetables

The process of cooling vegetables in cold storage is the most energy-intensive segment. A large amount of refrigeration and heating are required to reduce spoilage and maintain the freshness of fruits and vegetables. Cooling fresh products before processing helps maintain moisture content, vitamins, and sugar content. Blanching helps keep the texture and color of various vegetables like cauliflower, broccoli, etc. Drying is another method of processing; chips and French fries consume a massive amount of energy as the products are dehydrated up to 2% of moisture content. Figure 6.8 shows some data on energy consumed in processing fruits and vegetables.

6.3.4 Energy Requirements for a Meat Processing Unit

The meat processing industry is one of the largest food and beverage industries. A 9% rise in meat and poultry consumption was seen in Europe from 2007 to 2014. Poultry processing is more energy-intensive than meat processing due to feather and hair removal and singeing operations (Ladha-Sabur et al., 2019). Commonly, fossil fuels are used to process meat, while electrical energy is used for refrigeration. The

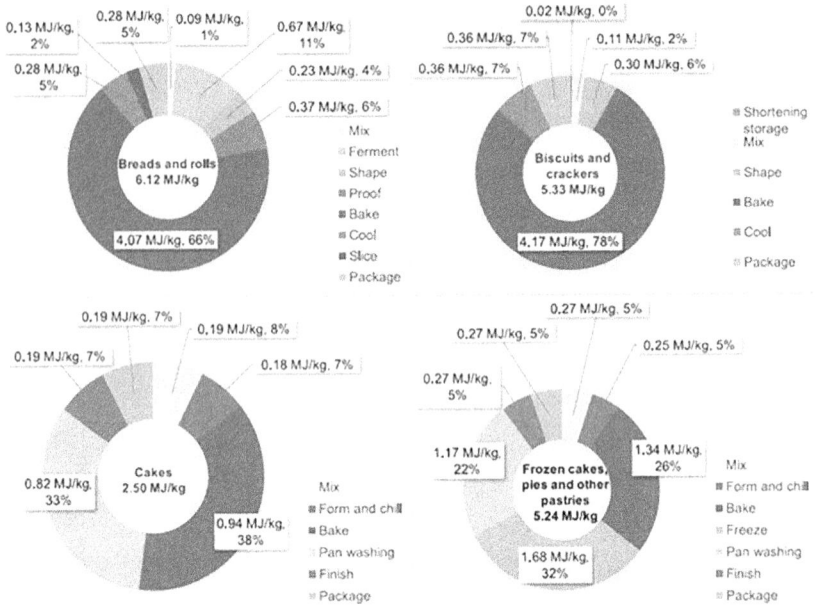

FIGURE 6.7 Energy consumed in production of different bakery items.

Source: Ladha-Sabur et al. (2019).

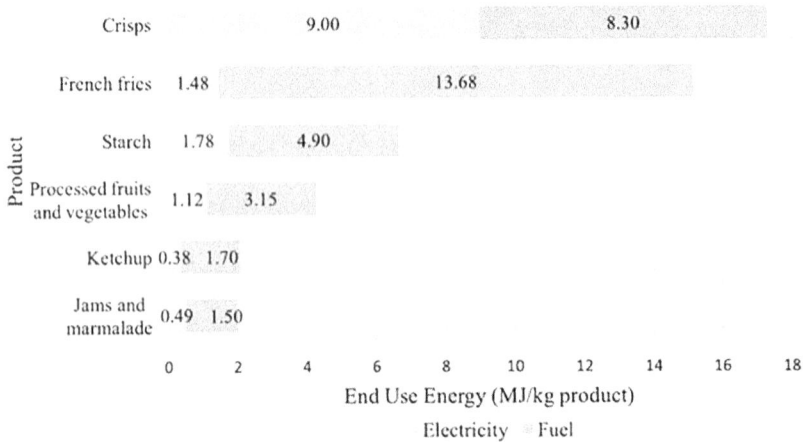

FIGURE 6.8 Energy consumed in the processing of fruits and vegetables.

amount of energy required can be decoupled as thermal energy and electrical energy. Figure 6.9 shows the energy demand in the meat processing unit. The processes which involve significant energy consumption in the production of meat are, broadly given as:

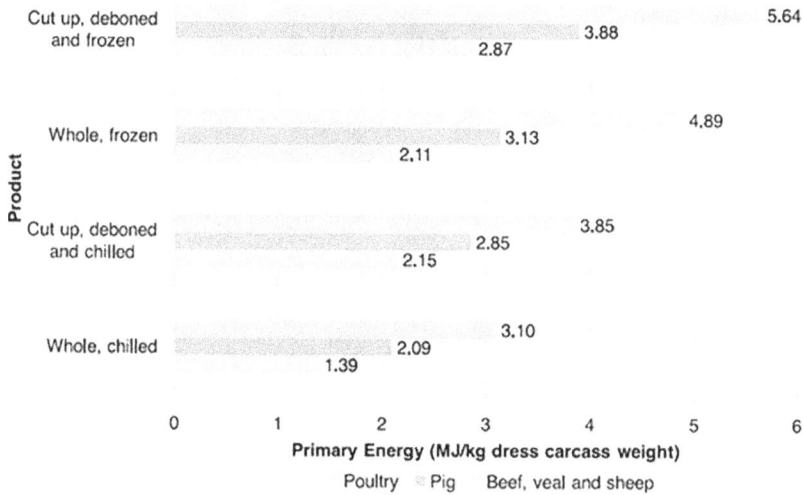

FIGURE 6.9 Energy demand in a meat processing unit.

TABLE 6.1
Characterization of Energy Required in Cold Processing

Process	Description	Temperature Range (°C)
Slow cooling	Cold air passed through the forced draught cooling unit, passing carcasses in the chilled room.	0–4
Fast cooling	Meat is chilled at a temperature of −1°C/5 hours	−20 to −15
Spray chilling	Intermittent stage of cooling by spraying water	2
Super chilling	Carcass cooled to its initial freezing point in which 5 to 30% water was frozen.	−4 to 3

Source: Iten et al. (2021).

1. Refrigeration of meat for storage and cooling of the building.
2. Thermal energy is required for scalding, cooking, sterilization, and pasteurization. The details of the thermal energy requirement are given in Table 6.1.

In most food industries, there is a simultaneous demand for cooling, hot water, and electricity; the cooling is required at a temperature of 12–15°C, and the refrigerator is kept at 0–4°C; electrical chillers fulfill this energy need. The total power required annually to run this chiller is 38.180 kWh (Herrando et al., 2021). Table 6.2 reports the annual energy requirement in terms of cooling, hot water, and electrical energy.

TABLE 6.2
The Annual Energy Requirement of Hot Water, Cooling, and Electrical Energy (Herrando et al., 2021)

	Energy Consumption (kWh/year)	Energy Requirement (kWh/year)
Hot water	21949	10316
Electricity for equipment	26,057	26,057
Cooling		
Rooms (12–15°C)	38,180	90,374
Refrigeration chamber (0–4°C)	95,820	226,811

6.3.5 CONFECTIONERY

The energy consumed in the confectionery industry is based on the type of product being manufactured and the processes employed. Sugar is the main ingredient in the confectionery industry. In Japan, about 65% heat energy is used in evaporative crystallizers, while melting and centrifugal drying consumes 25% and 22% of total electrical energy, respectively (Ladha-Sabur et al., 2019).

6.4 INTEGRATION OF SOLAR ENERGY IN FOOD INDUSTRIES

Solar heating and cooling technologies are considered an alternative option to reduce the dependency on fossil fuels and their associated emissions. The demand for energy in bulk occurs during the day hours. Therefore, it's a good match for solar energy generation in the food processing sector. The energy demand in these industries is characterized by cooling, hot water, and electricity. Several processes require solar process heat in the food and beverage industries. A low-temperature level up to 95°C is used in washing, cleaning, blanching, scalding, smoking, and tempering. The medium temperature level is from 95 to 250°C, supplied in the form of steam and is applicable for sterilizing, pasteurization, and drying. Figure 6.10 illustrates the thermal energy requirement in food industries for different processes. Table 6.3 briefs the use of different types of solar collectors with their operating temperature range. The cooling demand can be met by solar absorption cooling and refrigeration system. The application of solar thermal process heat is widely used in food processing industries for quality food at a low price. Solar energy conversion systems are used to generate thermal energy and electrical power. Solar thermal conversion systems convert solar radiations into heat energy using a flat plate and concentrating collectors. The heat collected is transferred to the working fluid. Flat plate collectors are used for low-temperature ranges and concentrating collectors (parabolic trough, dish collectors) with tracking mechanisms for higher temperatures above 250°C. Most of the food processing sectors have installed FPC (38%), PTC (20%), and ETC (20%) because these solar collectors can operate at a wide range of temperatures. Figure 6.11 reports the different types of solar collectors installed in various food industries. In contrast, photovoltaic cells convert solar radiation directly into electricity. Solar energy can be substituted to

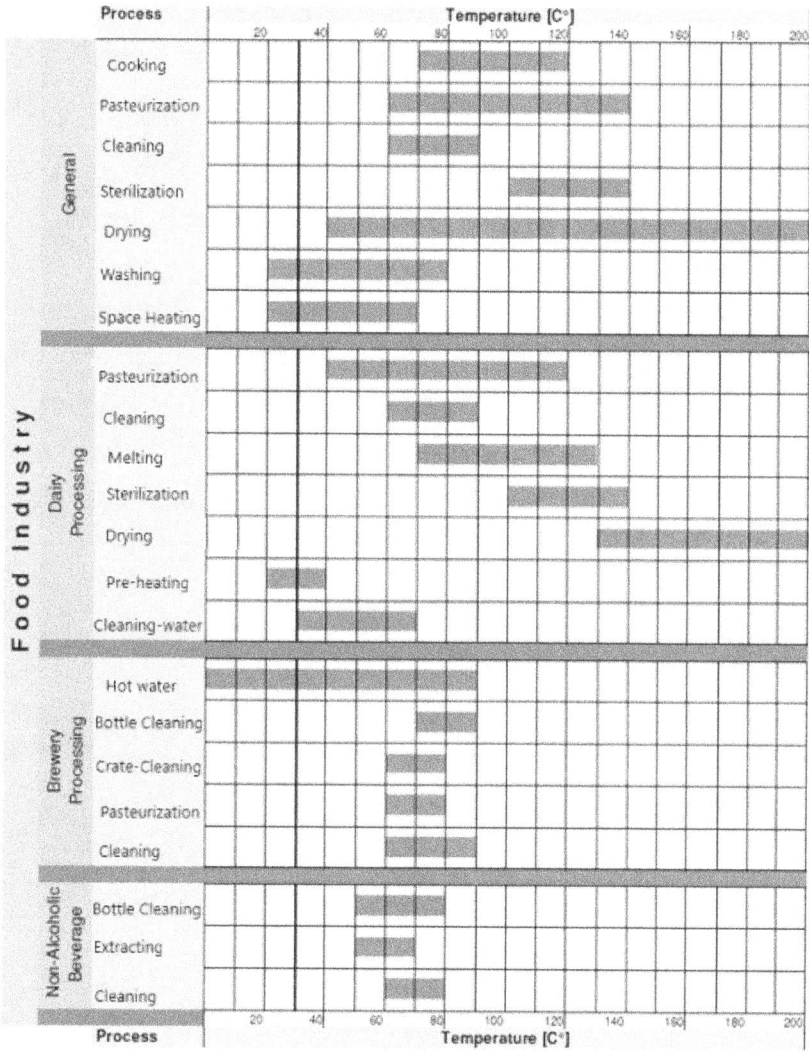

FIGURE 6.10 Chart represents the temperature range of various processes in the food industry.

supply the energy required in the food industries in terms of electrical and thermal energy. Photovoltaic panels can be an alternative to diesel generators and grid electricity. About 28% of the total 337 units use solar process heat in the industrial food sector, and about 95 solar projects are for solar thermal energy (Ismail et al., 2021). Several studies have been carried out on the solar process heat to fulfill the needs of thermal energy requirements in food processing industries. Drying and dehydration is the most commonly used process to enhance the storage life of fruits and vegetables. The dried products can be consumed during the off-seasons. Fudholi and Sopian (2019) performed the review of FPC for the drying application; it was found that the collector's efficiency ranges from 28% to 62%. Several research works were performed on the dehydration process under

TABLE 6.3
Type of Solar Thermal Conversion Systems for Food Processing

Type	Concentration Ratio	Temperature Range °C
Flat plate collector (FPC)	1	30–80
Evacuated tube collector (ETC)	1	50–200
Compound parabolic collector (CPC)	1–5	60–240
Fresnel lens collector	5–15	60–300
Parabolic trough collector (PTC)	10–40	60–250
Cylindrical trough collector	10–50	60–300

Source: Mekhilef et al. (2011).

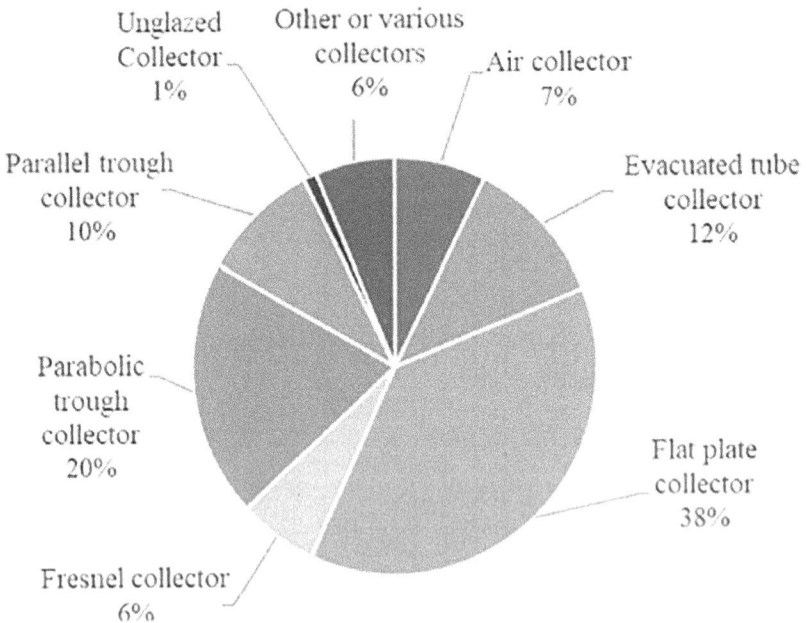

FIGURE 6.11   Types of solar collectors with their percentage contribution to food industry.

different operating parameters to control product drying temperature and quality. Solar concentrators were used for boiling Indian gooseberry for making Amla candy, pickles, etc. The various processes like boiling, blanching, frying, and roasting are well-performed under 200°C (Eswara and Ramakrishnarao, 2013). Different types of solar collectors are installed worldwide to meet the energy needs for processing fruit and vegetables. Table 6.4 depicts various types of solar collectors installed in fruit and vegetable processing industries with the power generated and area required. Fathey Mohamed Atia et al. (2011) designed a low-cost milk pasteurization unit using a flat plate solar air collector (FPC). It was developed for no grid supply areas. Figure 6.12 shows the working model of milk pasteurization using a flat plate solar collector. The milk stored in the tank flows due to

TABLE 6.4

Solar Collectors Installed for Fruit and Vegetable Processing

Name	Collector	Temperature, °C	Collector Area, m²	Power, kWh
Krispl fruit juice (Austria)	FPC	80	112	78
CPC Juice Manufacturer	ETC	75–150	225	158
KEAN soft drinks	ETC	75–150	225	158
Ostervang Greenhouse (Denmark)	FPC	-	15680	9880
Agrolibano (Honduras)	FPC	60–70	44	31
ZACATECAS (Mexico)	Air collectors	55–120	120	84
Papes safor S.L (Spain)	PTC	200–250	175	134
Stapleon Spence fruit packing Co. USA	Unglazed collector	-	2637	1846

Source: Ismail et al. (2021).

1- Fresh milk tank
2- Solar flat plate collector
3- Holding tube
4- Cooling unit
5- Pasteurized milk outlet

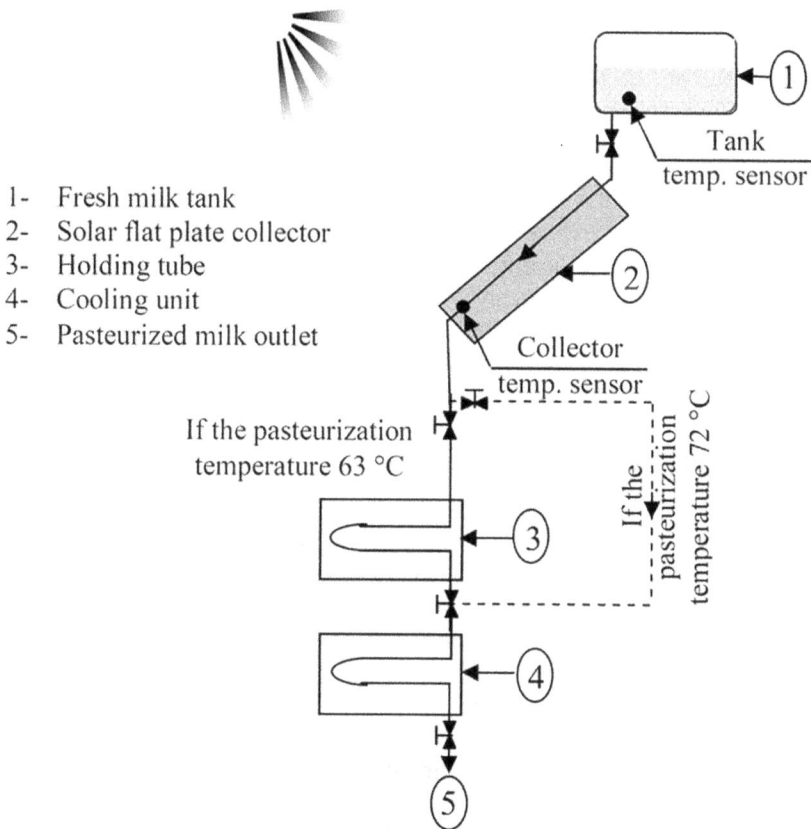

FIGURE 6.12   Flat plate solar collector used for the pasteurization of milk.

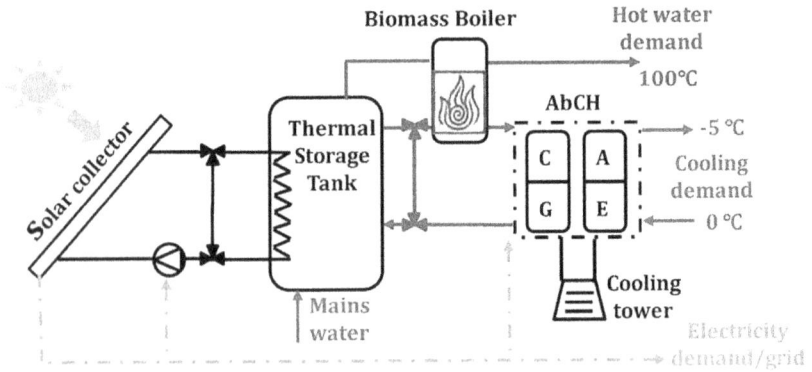

FIGURE 6.13 Schematic of PVT integrated with hot water storage for the absorption chiller.

gravity and is controlled by the valve arrangement. The milk flows in the serpentine path of the solar collector to reach the pasteurization temperature and then flows to the cooling unit. Herrando et al. (2021) proposed a system in which the thermal output of the PVT collectors is used to heat water in the storage tank to fulfill the thermal energy demand used in the NH_3-H_2O absorption system to provide the cooling. An auxiliary heater is required to attain a hot water temperature of 100°C for food processing and to feed water at 80°C to the absorption chiller. A biomass boiler replaced the conventional boiler to act as an auxiliary heater to increase performance and reduce carbon emissions (see Figure 6.13).

Kizilkan et al. (2016) replaced the conventional energy system in an ice cream plant in Isparta, Turkey, with a solar energy system. The various processes in ice cream making were analyzed, and the process heat requirement in the factory changed from grid electricity to process heat from parabolic trough solar collectors. The shifting of the grid to solar energy saves 98.56% of energy. The developed system has a high initial investment cost with a payback of 8.5 years.

6.5 CASE STUDIES

6.5.1 Solar Pond for Dairy Industries

A solar pond of 6000 m² was designed to supply about 15000 m³ of hot water at a temperature of 99.88°C to Gujarat Energy Development Agency, Gujarat Dairy Development Corporation Limited (Figure 6.14). The hot water supplied to Kutch dairy, located near the pond, saves around 935 MT of lignite annually. The Bhuj solar pond was developed by the collaborative effort and Tata Energy Research Institute under the National Solar Pond Programme of the Ministry of Non-Conventional Energy Sources (Kumar and Kishore, 1999).

6.5.2 Milk Pasteurization Using Solar Concentrators

The milk pasteurization process is performed at a temperature range of 65–75°C and is highly energy-intensive. The use of solar concentrators reduces fossil fuel and GHGs emissions. The system comprises a Fresnel paraboloid concentrating solar

FIGURE 6.14   The photographic view of the solar pond located in Bhuj, Gujarat.

Source: Kumar and Kishore (1999)

collector for a medium temperature range. The dish is automatically oriented toward the sun, and the working fluid attains 200°C temperature at the outlet (Figure 6.15). Daily, about 20,000 to 30,000 liters of milk is processed, saving approximately 80 liters to 100 liters of furnace oil, and saving the environment from burning fuel. This project was funded by MNRE for the milk pasteurization process for Mahanand Dairy, Latur, Maharashtra, India (Eswara and Ramakrishnarao, 2013).

Tyras S.A. Dairy industry in Greece uses solar collector arrays with an area of 1002 m², and thermal power of 730 kW capacity is utilized for the generation of hot water of 50,000 L stored in the buffer tank. The system was designed for economical operation, and this system produces an average of 700 MWh of thermal energy. The solar process heat contributed about 7% of the total energy demand (see Figure 6.16).

6.5.3 Solar Chips and Frito Lays

A chips manufacturing company (Frito Lays) has installed a solar concentrator in Modesto, California. The plant was commissioned in 2008 with the collaborative effort of the California Energy Commission. Solar concentrators were installed in a five-acre field with 54,000 sq. ft of a concave mirror to capture the solar radiation intensity (shown in Figure 6.17). A total of 384 solar collectors are used to generate steam that is used for heating and cooking oil for frying. The utilization of solar energy has reduced the use of natural gas, and the plant produces around 145,000 bags of chips every day. The use of solar potential has reduced the emission of greenhouse gases, reducing 1.7 million pounds of CO_2 into the atmosphere. The company has reduced 22% of electrical energy per bag.

FIGURE 6.15 Fresnel parabolic concentrator designed for milk pasteurization.

FIGURE 6.16 Solar arrays installed for process heat in Tyras S.A. Dairy industry in Greece.

6.5.4 Solar Plant "Berger" for Processing Meat

Fleischwaren Berger is located in lower Austria and produces about 80–90 tons of meat and sausage daily. The raw material for meat processing was supplied from abattoirs, a flat plate solar collector with an effective area of 1067 m^2 mounted on

FIGURE 6.17 Frito lays manufacturing plant at Modesto, California (concave mirror collectors).

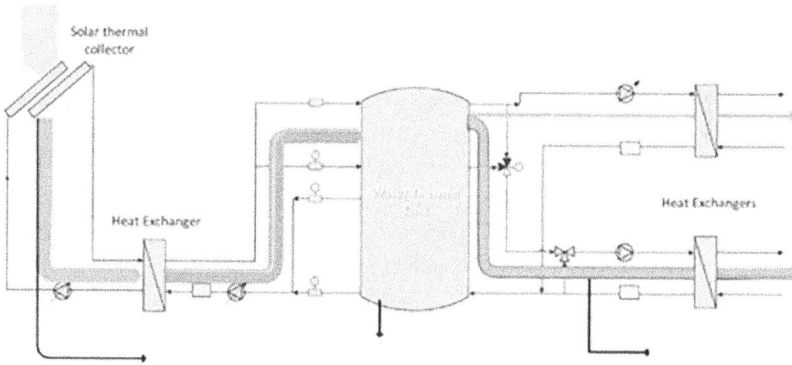

FIGURE 6.18 Process heat for a meat processing unit.

the rooftop of the factory. The thermal energy generated is used to preheat the water for the boiler. The water is preheated up to 60°C, and the steam is utilized for ham cooking. The hot water used in the maturation room helps in the long storage life of the sausages. Figure 6.18 gives the detail of the system used in a meat processing unit. The total heat generated by the solar collector is about 475 MWh, of which 21% is utilized for drying air conditioning and 78% of thermal energy is given for feed water heating (Cotrado et al., 2014).

FIGURE 6.19   SEED developed a solar cabinet dryer with PV for solar drying of fruits and vegetables.

6.5.5 SOLAR CABINET DRYER DEVELOPED BY SEED

To reduce the post-harvest losses in the rural areas, SEED introduced solar drying technology to help the villagers. This food processing method increases the shelf life of high moisture content fruits and vegetables. SEED designed a solar cabinet dryer for drying Gum karaya, a forest product, and benefited 10,000 tribal families of India. The developed dryer dries the product efficiently in 2–3 days, while it takes about 10–15 days to dry in the open sun.

A commercial dryer was fabricated using the solar thermal and solar photovoltaic combination for making a mango bar. The project was initiated with the help of United Nations Development Programme (UNDP), the Department of Science and Technology (DST), the Government of India, and the Renewable Energy Project Support Office—the United States Agency for International Development (REPSO-USAID) (Figure 6.19). About 150 SEED solar dryers have been installed in 13 states of India. It has successfully developed 70 products using solar dryers for fruits, vegetables, spices, and herbs on a commercial scale. Food processed in solar dryers is superior to open sun drying in hygiene, color, texture, and higher in nutritional content (Eswara and Ramakrishnarao, 2013).

6.6 CONCLUSIONS

The chapter presented an overview of energy usage in the various food industry sectors and the integration of solar technology in the food processing industries to overcome the energy crisis and environmental issues. The implementation of solar energy in the food processing industries will create new opportunities in the near future. Currently, the food processing industries require electrical energy and fossil fuel to meet the energy demand to accomplish various processes. The use of conventional energy sources like fossil fuel and electricity in food processing can at least in part be replaced by solar technology. Solar energy has the potential to produce solar process heat to fulfill the needs of thermal energy in the food industry. Solar collectors such as flat plate collectors, parabolic trough collectors, and evacuated tubes solar collectors are used to generate low- to high-temperature process heat. It has been reported that both solar thermal and PVT systems can be used equally in food industry process applications. The PVT collector is used for the direct conversion of solar energy to electricity and thermal energy. Apart from this, a detailed analysis has been performed to study the energy consumption pattern followed by various products in the processing unit. Major energy is consumed in manufacturing milk powder, potato chips, and French fries, which require freeze-drying and drying, respectively. Additional energy is required in maintaining hygiene and cleaning, where a large amount of thermal energy is required. The replacement of conventional fossil fuels with solar energy is an efficient method to save energy and the environment.

REFERENCES

Bowser, T.J., 2019. Food processing facility design. In *Handbook of Farm, Dairy and Food Machinery Engineering* (pp. 623–649). Elsevier Inc. https://doi.org/10.1016/B978-0-12-814803-7.00024-5.

Briceño-León, M., Pazmiño-Quishpe, D., Clairand, J.M., Escrivá-Escrivá, G., 2021. Energy efficiency measures in bakeries toward competitiveness and sustainability—case studies in Quito, Ecuador. Sustainability (Switzerland) 13. https://doi.org/10.3390/su13095209.

Cotrado, M., Dalibard, A., Söll, R., Pietruschka, D., 2014. Design, control and first monitoring data of a large scale solar plant at the meat factory Berger, Austria. Energy Procedia 48, 1144–1151. https://doi.org/10.1016/j.egypro.2014.02.129.

Dhanya, V., Shukla, A.K., Kumar, R., 2020. Food processing industry in India: Challenges and potential. RBI Bulletin March 2020.

Drescher, S., Rao, N., Kozak, J., Okos, M., 1997. Review of energy use in the food industry. Proceedings ACEEE Summer Study on Energy Efficiency in Industry 29–40.

Eswara, A.R., Ramakrishnarao, M., 2013. Solar energy in food processing—A critical appraisal. Journal of Food Science and Technology 50, 209–227. https://doi.org/10.1007/s13197-012-0739-3.

Fathey Mohamed Atia, M., Mostafa, M. M., El-Nono, M.A., Abdel-Salam, M.F., 2011. Solar energy utilization for milk pasteurization. Misr Journal of Agricultural Engineering 28, 729–744.

Fudholi, A., Sopian, K., 2019. A review of solar air flat plate collector for drying application. Renewable and Sustainable Energy Reviews 102, 333–345. https://doi.org/10.1016/j.rser.2018.12.032.

Gowreesunker, B.L., Mundie, S., Tassou, S.A., 2017. The impact of the UK's emissions reduction initiative on the national food industry. Energy Procedia 123, 30–35. https://doi.org/10.1016/j.egypro.2017.05.093.

Hall, G.M., Howe, J., 2012. Energy from waste and the food processing industry. Process Safety and Environmental Protection 90, 203–212. https://doi.org/10.1016/j.psep.2011.09.005

Herrando, M., Simón, R., Guedea, I., Fueyo, N., 2021. The challenges of solar hybrid PVT systems in the food processing industry. Applied Thermal Engineering 184. https://doi.org/10.1016/j.applthermaleng.2020.116235.

Ismail, M.I., Yunus, N.A., Hashim, H., 2021. Integration of solar heating systems for low-temperature heat demand in food processing industry—A review. Renewable and Sustainable Energy Reviews 147, 1–15. https://doi.org/10.1016/j.rser.2021.111192.

Iten, M., Fernandes, U., Oliveira, M.C., 2021. Framework to assess eco-efficiency improvement: Case study of a meat production industry. Energy Reports 7, 7134–7148. https://doi.org/10.1016/j.egyr.2021.09.120

Kizilkan, O., Kabul, A., Dincer, I., 2016. Development and performance assessment of a parabolic trough solar collector-based integrated system for an ice-cream factory. Energy 100, 167–176. https://doi.org/10.1016/j.energy.2016.01.098

Kumar, A., Kishore, V.V.N., 1999. Construction and operational experience of a 6000 M2 solar pond at kutch, India. Solar Energy 65, 237–249. https://doi.org/10.1016/S0038-092X(98)00134-0

Ladha-Sabur, A., Bakalis, S., Fryer, P.J., Lopez-Quiroga, E., 2019. Mapping energy consumption in food manufacturing. Trends in Food Science and Technology 86, 270–280. https://doi.org/10.1016/j.tifs.2019.02.034.

Mane, S.R., 2013. Energy management in a dairy industry. International Journal of Mechanical and Production Engineering 1(4), 27–32.

Mekhilef, S., Saidur, R., Safari, A., 2011. A review on solar energy use in industries. Renewable and Sustainable Energy Reviews 15, 1777–1790. https://doi.org/10.1016/j.rser.2010.12.018.

Regmi, A., Gehlhar, M., 2005. New directions in global food markets. Agricultural Information Bulletin No 794. February.

Wojdalski, J., Grochowicz, J., Drózdz, B., Bartoszewska, K., Zdanowska, P., Kupczyk, A., Ekielski, A., Florczak, I., Hasny, A., Wójcik, G., 2015. Energy efficiency of a confectionery plant—Case study. Journal of Food Engineering 146, 182–191. https://doi.org/10.1016/j.jfoodeng.2014.08.019

Wu, H., Tassou, S.A., Karayiannis, T.G., Jouhara, H., 2013. Analysis and simulation of continuous food frying processes. Applied Thermal Engineering 53, 332–339. https://doi.org/10.1016/j.applthermaleng.2012.04.023.

7 Large-Scale Solar Desalination System

Guna Muthuvairavan and
Sendhil Kumar Natarajan

CONTENTS

7.1 INTRODUCTION

Water is a critical component of the planet's ecosystem. It is the most fundamental and necessary condition for all living creatures to exist, including humans, animals, trees, and birds. When it comes to economic development and public health, water is a natural resource that has a significant impact. Water scarcity is a significant problem in both developing and underdeveloped countries worldwide. Over 96% of the world's water remains in the oceans. The remaining 2% is freshwater, 71% of which is trapped in mountains in the form of ice, glaciers, and so on. Despite its widespread abundance, it is less readily available for direct human consumption. Water used for drinking and potable purposes makes up less than 1% of the total volume of water on the planet. As a result, those who live in small island communities in the ocean face an even greater challenge because they have no access to freshwater. Nowadays, the majority of health issues can be traced back to a lack of access to clean drinking water. Water salinity has increased dramatically in recent decades as a result of low rainfall. Furthermore, as a consequence of the increasing population, rapid urbanization, industrialization,

DOI: 10.1201/9781003263326-9

and other causes, the water has become increasingly polluted. These activities harm water quality in agricultural and rural areas. It is estimated that women collect these openly polluted water sources every day for nearly 200 million hours. Water-borne diseases claim the lives of approximately 3.575 million people each year. Many villages in developing and underdeveloped countries lack even the most basic medical facilities (K. Sampathkumar, Arjunan, Pitchandi, & Senthilkumar, 2010). According to the ranking of water-stressed countries for 2040, Gulf countries such as Bahrain, Kuwait, and Qatar will top the list when industrial, residential, and agricultural sectors are considered, whereas India will rank 40th out of 161 countries in terms of water shortage (Luo, Young, & Reig, 2015). However, India has a lengthy coastline of 7,517 kilometres, which is a significant advantage for desalination to address the growing water scarcity problem in the country in the near future (Manju & Sagar, 2017a). Recently, it was discovered that nearly 40% of the people worldwide are suffering from an extreme shortage of water, with this figure expected to rise to around 60% by 2025. Desalination of water has been highlighted as one of humankind's top ten concerns, and it is projected to receive even greater attention in the next decades, as the oceans and seas provide a steady water resource in times when conventional energy sources seem to be less reliable. Such facts tend to make desalination an appealing alternative to dwindling freshwater supplies (Ibrahim, Rashad, & Dincer, 2017).

Numerous desalination technologies, including desalination of all forms of water resources, are now accessible. The most widely used desalination process is reverse osmosis (RO), which accounts for approximately 51% of total capacity globally. Additionally, thermal desalination facilities such as MSF and MED plants provide 40% of the total, with a relative contribution of 32% and 8%, respectively. Desalination processes on a smaller scale, such as VCD, humidification and dehumidification (HDH), and electrodialysis (ED), account for the remaining 9% (Latteman, 2010).

The World Health Organization (WHO) reports a salinity limit of 500 ppm is permissible, with a maximum limit of 1000 ppm for some specific cases. The majority of readily accessible water has a salinity ranging up to 10,000 ppm, whereas seawater seems to have a salinity of 35,000–45,000 ppm due to the presence of salt content, as seen in Table 7.1.

TABLE 7.1
Types of Water Sources and Their Total Dissolved Solids (TDS)

Classification of Water Source	TDS (in ppm)
Freshwater (drinkable)	Less than 500
Freshwater	Less than 1,000
Brackish water (mild)	1,000 to 5,000
Brackish water (moderate)	5,000 to 15,000
Brackish water (heavy)	15,000 to 35,000
Seawater (standard average)	35,000
Seawater	35,000 to 45,000

Source: Kucera (2014).

Several desalination techniques have been successfully used to reduce the salinity level of water resources to an allowable limit of less than 500 ppm, after various improvements, making it economically viable today (Tiwari, Singh, & Tripathi, 2003). In this chapter, various desalination techniques, particularly thermal-based high-capacity desalination methods such as MSF, MED, VCD, low-temperature thermal desalination (LTTD), solar integrated desalination, hybrid desalination, and recent improvements in these systems, as well as development directions for large-scale desalination, are discussed.

7.2 DESALINATION

The sources of feed or saltwater include brackish water, the sea, wells, wastewater, rivers and streams (surface water), process waters, and industrial feed waters. These water sources may contain dissolved salts above the WHO's allowable limit, and the dissolved salts can be removed using various methods of desalination to meet rising demand. Desalination is the removal of excess dissolved salts from feed/saltwater or, in some situations, the chemical process of obtaining potable water from saline water. Desalinated water can indeed be utilized for a variety of industrial, commercial, and municipal applications. In general, any method of desalination sends salt/feed water to the desalination device, which produces two distinct water streams as shown in Figure 7.1. The first is potable water or freshwater, which eventually has less salinity (lower salinity than feed/saltwater). Another stream is concentrated brine, which is highly salted and mineral concentrated water (higher salinity than feed/saltwater).

The potable or freshwater obtained after desalination should be re-mineralized in accordance with the requirements for human consumption. The concentrated brine stream produced by the desalination processes must be disposed of properly without affecting the ecological system.

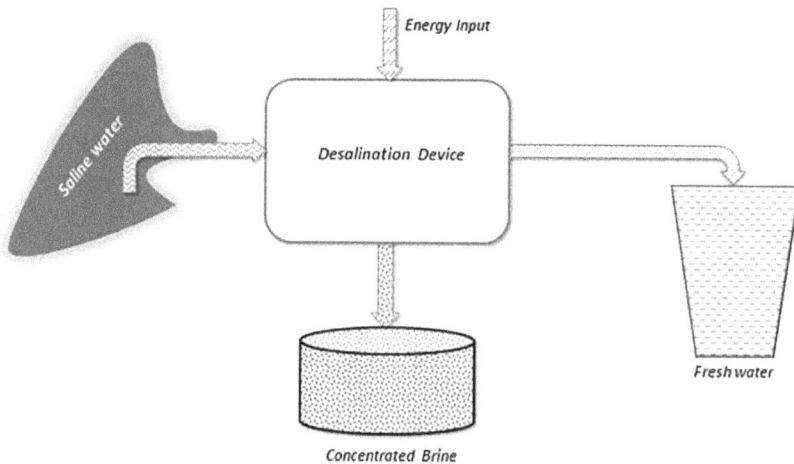

FIGURE 7.1   Desalination streams.

7.3 TYPES OF DESALINATION

Several well-established desalination techniques as shown in Figure 7.2 have become available in recent years, with some of them still in the research and development stages. Two major categories can be distinguished when it comes to desalination procedures:

- Desalination methods that remove salt by the use of heat (phase change process).
- Desalination methods based on membranes.

In order to remove salt from a supply of saline water, the desalination process requires a significant quantity of heat or electrical energy, and in some circumstances

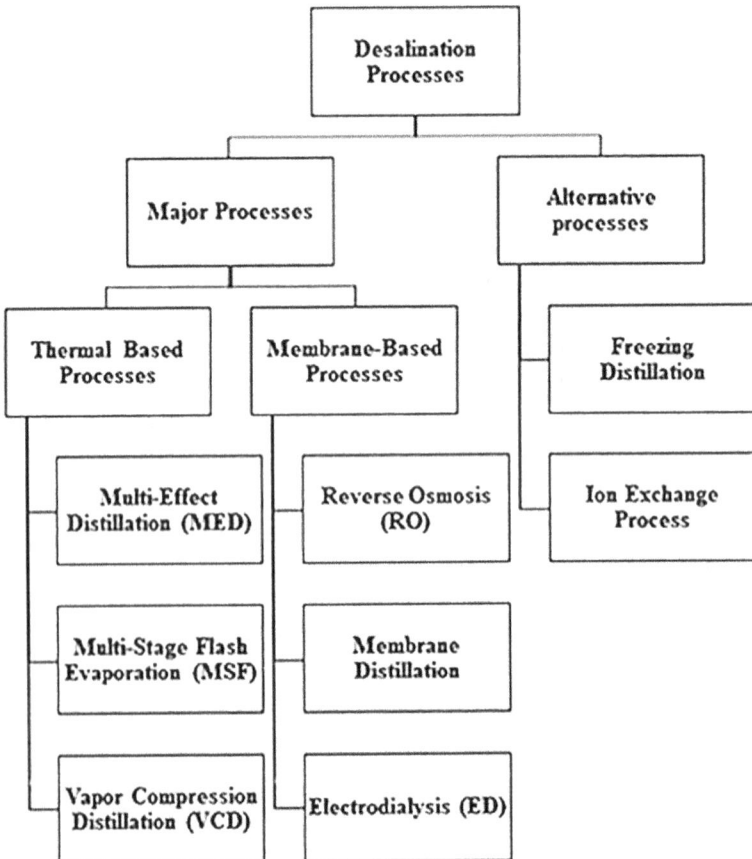

FIGURE 7.2 Types of desalination processes.

Source: Shatat & Riffat (2014).

TABLE 7.2

Specific Energy (Thermal and Electrical) Consumption and Water Output Quality for Conventional Desalination Processes

		MSF Evaporation	MED	VCD		RO
				MVC	TVC	
Freshwater Capacity (m3/day)		50,000 to 70,000	5,000 to 15,000	100 to 2,500	10,000 to 35,000	24,000
Thermal Energy Consumption	Specific Thermal/heat Energy Consumption (kJ/kg)	190 to 390	230 to 390	Nil	187 to 227	Nil
	Equivalent Consumption of Electrical Energy (kWh/m3)	9.50 to 19.50	5 to 8.50	Nil	14.50	Nil
Specific Electrical Power Consumption (kWh/m3)		4 to 6	1.5 to 2.50	7 to 12	1.80 to 1.60	2 to 5.55
Quality Level of Water (ppm)		Around 10				200 to 500

Reproduced from Manju and Sagar (2017b).

both. Table 7.2 summarizes the energy requirement and water output quality depending on the desalination plant's capacity and the selected technique. Except for MVC and RO, desalination methods such as MSF, MED, and TVC require thermal energy from fossil fuels or renewable energy sources to heat the feed water to the requisite temperature prior to introducing it into the plant's operation. Despite the fact that all desalination procedures require electrical energy to run certain components, such as pumps and other auxiliary equipment, desalination may be accomplished without electricity. In addition, the water quality derived through procedures other than RO is around 10 ppm, and it must be further treated for human use (Manju & Sagar, 2017b).

Aside from these major processes, there are a few other desalination procedures that are rarely used, such as ion exchange and freeze desalination. The primary focus of this chapter is on the large-scale thermal-based desalination technologies that are now in use.

7.4 THERMAL-BASED DESALINATION

In the thermal-based desalination process also known as distillation, this method of removing contaminants from water involves evaporation followed by condensation. Solar still desalination is one of the low-cost and sustainable thermal desalination approaches. In addition to desalination of seawater, the solar energy still can be utilized for water treatment of industrial effluent, ethanol, brackish water, and reject water from reverse osmosis systems. Nevertheless, the freshwater productivity from solar stills is only about 1.5 to 2 litres per square metre per day. Hence, many researchers across the world are working on different design constructions (Suraparaju & Natarajan, 2021c, 2021d), materials (Suraparaju, Dhanusuraman, & Natarajan, 2021; Suraparaju & Natarajan, 2020, 2021a), energy storage systems (Suraparaju & Natarajan, 2021b), micro and nano natural materials as surface enhancers in absorber basins (Natarajan, Suraparaju, Elavarasan, Pugazhendhi, & Hossain, 2022; A. Sampathkumar & Natarajan, 2021b, 2021a) to enhance the daily freshwater productivity of solar stills. Although the modifications enhanced the productivity of solar stills up to 3.5 to 5 litres per square metre per day, the application of solar stills is restricted only to serving small-scale freshwater requirements. Hence, the other thermal desalination systems can be effectively used for large-scale freshwater requirements with huge capital and maintenance costs relative to solar stills. However, these costs can be compensated by the freshwater supply, and hence these thermal desalination systems can be recommended for medium- or large-scale freshwater requirements. However, due to the high cost and widespread use in large-scale water supply projects, this method of desalination is usually not recommended when desalinating brackish water.

A general classification for thermal-based desalination can be found in the following primary categories:

- Multi-stage flash (MSF) evaporation.
- Multi-effect distillation (MED).
- Vapour compression distillation (VCD).
- Low-temperature thermal desalination (LTTD).

The capability to be driven by any form of thermal source, as well as the fact that they utilize relatively less electricity, are the two most significant benefits of thermal-based desalination processes (Wang, Christ, Regenauer-Lieb, Hooman, & Chua, 2011). MSF, MED, and MVC are the three most common commercial thermal desalination methods, accounting for 87.3%, 12.5%, and 0.2% of commercial desalination facilities, respectively (Feria-Díaz, López-Méndez, Rodríguez-Miranda, Sandoval-Herazo, & Correa-Mahecha, 2021).

7.4.1 MULTI-STAGE FLASH EVAPORATION

MSF evaporation plants were first built in the 1950s. These plants can be either once-through or recycling plants, depending on their configuration. After passing through the heater as well as flash chamber, the incoming feed water is discharged in a once-through plant. Because it's already been warmed up, the cooling water from a recycled plant can also be used to pre-heat the feed water. These days, a co-generation system is employed to meet the energy requirements of the system.

Multi-stage evaporators with 19–28 stages are employed in modern MSF plants. However, some authors report that some MSF systems include 40 stages, allowing them to provide freshwater outputs varying between 10,000 to 40,000 m³/day (Feria-Díaz et al., 2021). However, expanding the number of stages in the MSF system results in a more complicated and expensive system (Saidur, Elcevvadi, Mekhilef, Safari, & Mohammed, 2011).

The majority of the world's desalination capacity (65%) appears to be concentrated in the Gulf countries of the Middle East, which relies on its oil reserves to fuel primarily MSF desalination (Alsehli, Choi, & Aljuhan, 2017). Despite its reliability, this type of desalination is a highly energy-intensive method that demands both mechanical and thermal energy to produce potable water. There are no moving parts in these plants, except for pumps that circulate the water. The water quality is between 2 and 10 ppm. Consequently, the water that is deemed safe for human consumption is re-mineralized before consumption.

Multiple stages make up the MSF plant. As used in this context, the term "stage" refers to a flash chamber unit that can operate under a range of pressures. Low-pressure stages are used for each subsequent stage of the system. The saline water is heated externally and then fed at high pressure into the first stage of the flash chamber unit. The pressure is released in the flash chamber's first stage, causing the incoming feed water to boil and accelerate the flashing or evaporation process. Further stages of the flash chamber unit continue the same process with progressively lower pressure than the preceding stage. The heat exchanger part of the system, which carries the cooling water through its inlet tubes, condenses the saltwater that has evaporated after flashing. The feed water is now pre-heated before it enters the brine heater using the condenser outlet water, reducing the system's energy consumption.

Figure 7.3 depicts the MSF evaporation process, which includes a brine heater component to heat saline water to a temperature just below saturation temperature. As it moves through the flash chamber units, the water boils quickly due to the low environmental pressure, which causes vaporization. Incoming feed water nearly vaporizes as a result of the sudden introduction of heated saltwater into a lower atmospheric pressure, causing the "flashing effect." The amount of saltwater evaporated depends primarily on the chamber pressure at each stage of the evaporation process.

The MSF plant is capable of operating at temperatures over 115°C. Typically, the MSF facilities operate at top brine temperatures (TBT) between 90 and 120 degrees Celsius, depending on how the scale is controlled (Saidur et al., 2011). Even though

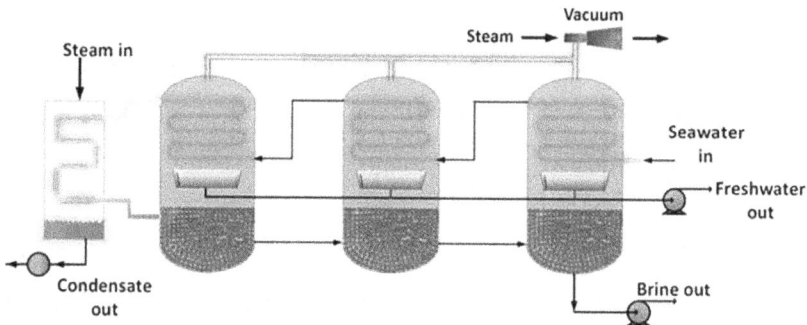

FIGURE 7.3 Schematic of MSF distillation system.

Source: Feria-Díaz et al. (2021).

the higher temperature improves the system's efficiency, the precipitates of calcium sulphate that form on the tube's inner surface cause scaling, which leads to mechanical and thermal issues like tube clogging. It is important to note that the temperature of the top brine, the salinity of the input water, and the number of stages in the distillation process as well as the fouling barrier of the brine heater all have an impact on distillate production at the MSF facility (Baig, Antar, & Zubair, 2011; Ettouney, 2002; Fath, Abbas, & Khaled, 2011).

In the Arabian Gulf, the possibility of an MSF system using solar energy has been investigated. A pilot plant using a flat plate collector-driven single-stage flash desalination technology. It was tested by changing various conditions and variables. Experiments with varying flash temperatures, valve metering, and flow rates were used to optimize the conditions. The system is feasible and more efficient at high flash (equilibrium) temperatures and relatively high input water flow rates. Furthermore, increasing the flash temperature increases the amount of heat that can be reclaimed from the vapour that is condensing, which reduces the heat input demand. As a result, the system requires less space for solar collectors. Both the reduced heat input and the greater freshwater production make this system more economically viable. In this investigation, the flow rate was the primary input variable. Higher flow rates produced more freshwater than lower flow rates. The operation was optimized to generate 11 litres of freshwater at 0.7 litres per minute of feed water flow rate, 80°C temperature, and −14.3 psi beginning vacuum pressure. Because the vacuum pressure dropped dramatically in the first 30 minutes, the majority of the freshwater was created during this time. To enhance the process's reliability and efficiency, the study might be conducted using solar (80%) and heating elements (20%) as a hybrid unit. Instead of corrosive copper collectors, special polymers can be employed (Abutayeh, Humood, Alsheghri, Al Hammadi, & Farraj, 2013).

Process waste water (PWW) was modelled to validate the feasibility of saline PWW in a flowsheet environment and optimize MSF distillation operating parameters such as water yield pressure, flow rate, temperature, NaCl, performance ratio, consumed steam, and thermal energy to distinguish patterns in flash stages for entire desalination plants. The resulting MSF model computation was thoroughly validated using data supplied by industries. Both simulators lower the 4.5V/V% from the fine chemical industry process seawater or 45000 ppm NaCl content to 0.05V/V% or 500 ppm saline content. The production of RO ranged from 3.0% to 12.6%, and the yield of MSF was 11.7%. A steam flow rate of 15 m³/h is recommended for the efficient operation of the heat input portion (Toth, 2020). The distillate yield at the MSF plant could be increased by raising the temperature difference between hot brine and incoming saltwater. (Moustafa & Jarrar, 1985).

7.4.2 Multi-Effect Distillation

The MED method is the first large-scale desalination procedure that was used. The first MED plants were built in Saudi Arabia in 1930, and they had only two or three effects at the time. Vapour flows through the tubes in this system. Outside of the tubes, seawater is used to cool the evaporated saline water. The action takes place entirely in two side chambers. It's not in line with MSF.

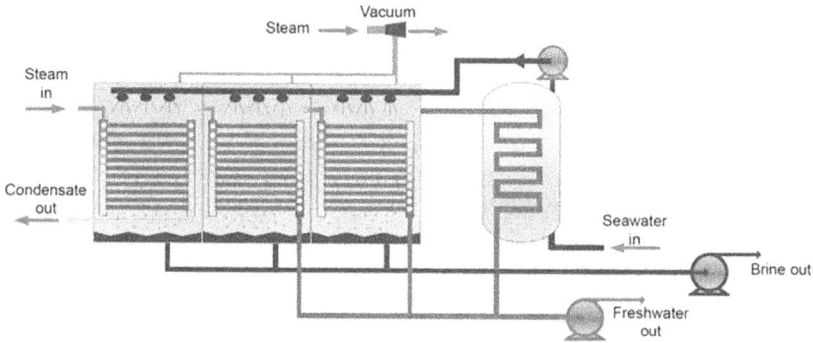

FIGURE 7.4   Schematic of MED system.

Source: Feria-Díaz et al. (2021).

To produce the desired results in the same way that MSF does, MED is carried out in a sequence of chambers (effects) and makes use of the evaporation and condensation processes at lowered ambient pressure in the various effects. As indicated in Figure 7.4, the evaporator and condenser are the two major components of the unit's overall construction. It is typically made of stainless steel. Water is heated by sending it through many pre-heaters (evaporators) in each effect before being sent to the condenser, which is pre-heated first by the condensation heat of the vapour. After passing through the last of these, a steam boiler or other source of heat is utilized to elevate the temperature of the supply water to its saturation point for the "top" effect.

Before the heated feed water enters the evaporator, it is sprayed thinly over the evaporator tubes to accelerate the boiling and evaporation processes. Only a portion of the seawater that was pumped into the tubes in the first effect evaporates out of them. First, seawater is used to warm the entering feed, and then some of it is utilized to power the next effect, that runs at a reduced pressure and draws its feed out from brine produced by the first effect. This second effect produces product water, whereas the next effect uses heat to vaporize some of the leftover feed water. This is a cycle that is repeated all over the facility. Eventually, the effects will be implemented even at lower temperatures. Maintaining the effects at lower and lower pressures is how this is done (or via higher vacuums utilizing an air ejector). The evaporator condenses the heating steam, which is then recirculated back to the boiler for regeneration. The bottom of the unit is filled with concentrated brine, which is then blown out as a waste product.

MED plants have fewer effects than MSF stages. Due to the link between the number of effects and the performance ratio, the number of effects utilized in big plants is typically between 8 and 16. The number of effects less 1 is a rough approximation of the MED plant's performance ratio (N - 1). In this scenario, the required number of effects in a MED facility for an 8:1 performance ratio facility would be 9. This is much less than the cost of a comparable MSF plant. When compared to MSF plants, MED plants have fewer effects, which results in capital cost savings. However, when more effects are added, the performance ratio increases.

The advantages of MED over MSF are as follows (García-Rodríguez, 2003; Thimmaraju et al., 2018):

- Improved utilization of the temperature differential.
- A better heat exchange coefficient.
- Less pumping power requirement.
- A lower working temperature (less susceptible to scaling and corrosion).
- More adaptable to partial load operation.
- Less expensive.
- Better suited to low capacity.

Compared to MSF, MED has higher thermal efficiency to generate freshwater because the latent heat form of thermal energy is used several times, allowing the system to operate near thermal equilibrium (Wang et al., 2011).

Due to the significant consumption of fossil fuels, MED technology is an extremely energy-intensive desalination technique, resulting in high desalination prices. Two novel methods based on MED technology, namely boost hot water MED (Boost-HW-MED) and hot water MED (HW-MED), were presented and compared to conventional MED technology using 90°C wastewater from the petrochemical, iron, and steel industries as a heat source. The Boost-HW-MED system enables pre-heating some of the effects of seawater before spraying and evaporation to the highest temperature possible, which reduces the temperature difference between each evaporator by using the hot water residue as a heat source, increasing system efficiency and reducing the cost and area required for pre-heating the evaporators (Qi, Lv, Feng, Lv, & Xing, 2017).

The organic Rankine cycle (ORC) was proposed to be integrated with the MED for simultaneous electrical energy and freshwater generation. For the 20, 30, 40, and 50 kW MED/ORC power capacities, the suggested integration boosts freshwater output by 8.8, 13.6, 17.65, and 22%, respectively. Furthermore, in order to achieve improved MED efficiency and potable water production, this MED/ORC integration does not claim a significant increase in overall heat transfer area. When a 10 kW rise in electrical power generation is combined with a 3.95% improvement in performance ratio (PR), a 1.57% increase in overall heat transfer area is needed. Only a 6.9% increase in overall heat transfer area is necessary for 50 kW electrical energy production, which is 22% more efficient than the MED system without ORC integration. Because of the first effect's large water vapour production, the inclusion of MED/ORC allows for a less specific region and higher MED's PR (Aguilar-Jiménez et al., 2020).

MED's operating temperature is around 70°C, which ensures that scaling and corrosion on the tube's surface are minimized. Pre-treatment of incoming saline water is requisite for the reverse osmosis method of desalination to work, but the MED method doesn't require that the feed water is of high quality. MED generates freshwater more efficiently than MSF because it uses latent heat energy multiple times, allowing the system to run near thermal equilibrium. These plants use less energy and have better thermal efficiency than MSF plants. For this reason, MED water production may be more efficient and cost-effective for high-quality distillate and medium-capacity water production.

7.4.3 VAPOUR COMPRESSION DISTILLATION

Vapour compression distillation is a work-driven process. Rather than heating directly from the boiler unit, the VCD method heats by compressing the vapour. Electrical energy is used to generate the majority of the heat in this process. Vapour's temperature rises as a result of its rapid compression. The supply water is evaporated through a network of tubes in a low-pressure atmosphere. Typically, this technology is utilized to meet small-scale freshwater needs. To generate enough heat to evaporate the entering seawater, two separate devices are developed in different configurations. If the compressor is mechanical, the process is known as mechanical vapour compression (Figure 7.5); if it is a steam jet, the process is known as thermal vapour compression (Figure 7.6).

The mechanical compressor is generally driven using diesel or electricity to yield water evaporation. Outside of the heated tube bundle, feed water is sprayed. Here, the water is boiled and partially evaporates, producing more vapour. In the case of vacuum compression evaporation using a steam jet, a venture orifice located at the steam jet produces water vapour, resulting in a reduced atmospheric temperature. The steam jet pressurizes the withdrawn water vapour, which then condenses on the tube walls, providing the heat needed to evaporate the saltwater on the other side of the evaporator. In general, the freshwater production per day from MVC (100–3000 m³) is lesser than that from TVC (10,000–30,000 m³) (Rabiee, Khalilpour, Betts, & Tapper, 2018).

With an operating temperature of less than 70°C, the VCD system is a power-efficient and simplistic operation. Additionally, corrosion and scale formation on the tube is significantly reduced as compared to MSF and MED.

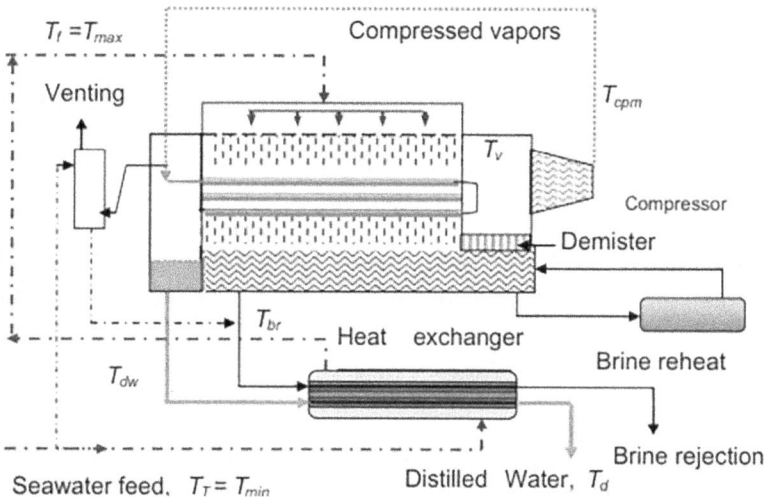

FIGURE 7.5 Schematic of MVC distillation system.

Source: Feria-Díaz et al. (2021).

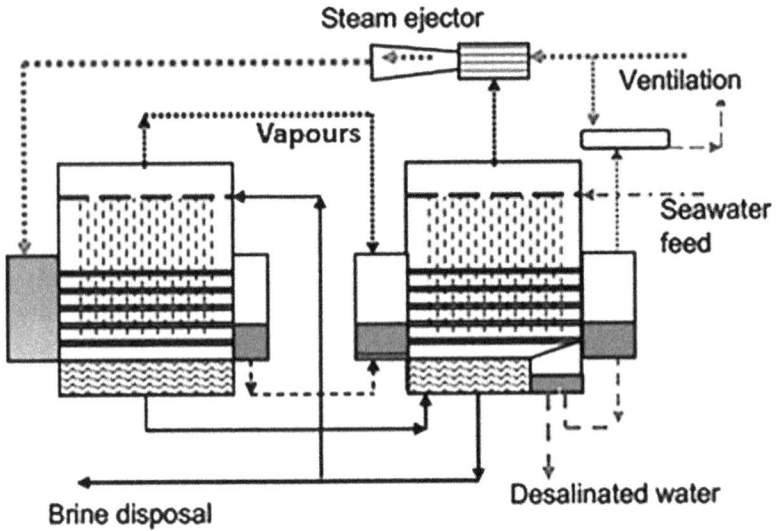

FIGURE 7.6 Schematic of TVC distillation system.

Source: Feria-Díaz et al. (2021).

Membrane distillation (MD) is less energy efficient than thermal desalination. This increases the cost of the MD's freshwater output. Pure water separation happens in MD as a result of the vapour pressure differential. Thus, the MD process is possible at atmospheric pressure and low temperature. As a result, combining renewable energy sources or low-temperature waste heat energy makes the MD system economically viable.

The MD is employed to substitute the pre-heater heat exchanger that is used in MVC to recover thermal energy from the brine stream. By substituting an MD module for the costly heat exchanger equipment, free heat energy is efficiently utilized and capital costs are lowered.

The MVC-MD connected system generates water at a cost that is about 6% less than a standalone MVC. Increases in MVC's operating temperature, a reduction in MVC's recovery ratio, and a reduction in MD's capital cost all contribute to larger savings. At a specific cost of approximately US $40/m² for an MD system operating at 70°C MVC with 50% recovery, the hybrid conductive gap, and air gap MD systems save approximately 4–8% (Swaminathan, Nayar, & Lienhard V, 2016).

It is possible to use VCD systems on islands, maritime bases, and ships, because of their great efficiency in turning heat into work, the much smaller volume required compared to conventional distillation systems, and the fact that the compressor can be powered by electricity or diesel engines. However, the intricate boiler design and the limited water production due to the compressor's capacity are drawbacks (Zheng, 2017).

7.4.4 Low-Temperature Thermal Desalination

When it comes to industrial areas where steam is commonly accessible, MSF and MED are generally chosen, which utilize both low- as well as high-pressure steam of power plants, depending on the application. To endure the high temperatures requires expensive materials such as Cu-Ni alloy steel, high-grade stainless steel, titanium, or duplex steel. Pre-treating the input seawater using expensive and potentially harmful chemicals is required in order to prevent scaling. High-temperature effluent from power stations and/or desalination systems has the potential to produce negative effects on the environment and ecosystem. This can result in a reduction in the incubation rate of eggs as well as a significant reduction in the maturation of larvae. The MSF and MED facilities currently demand 20 to 40 kWh of thermal energy for each cubic metre of freshwater delivered (Wakil, Choon, Thu, & Baran, 2014). In addition to being expensive and energy-intensive, present conventional desalination plants also dump highly salinized water at elevated temperatures into the environment, creating a significant environmental hazard (J. Hou, Cheng, Wang, Gao, & Gao, 2010).

Because of the greenhouse gasses (GHGs) and environmental degradation associated with traditional energy sources, as well as their finite availability, fossil fuels cannot be relied on as a long-term energy source. As a result, it is necessary to seek out unpolluted, renewable sources of energy for the supply of drinking water through desalination (Gnaneswar, Nirmalakhandan, & Deng, 2010). Oceans collect the most solar energy due to their larger surface area. The ocean's upper surface collects solar energy and keeps it as sensible heat. With exception of sunlight, ocean thermal gradients are available 24/7 (Muthunayagam, Ramamurthi, & Paden, 2005). Large amounts of heat are expelled as "waste" from nuclear and thermal power plants, steel factories, petroleum refineries, and petrochemical industries in the form of condenser reject effluent. Thermal power plants had progressed much in terms of design and performance, but waste heat that travels through discharge water would continue to be an issue (Bendig, Maréchal, & Favrat, 2013).

The LTTD process is based on the principle of lowering the boiling point of water in a vacuum. In other words, incoming seawater's boiling point is lowered by flashing it in a flash chamber that is in a vacuum by an external vacuum pump as shown in Figure 7.7. Thus, in contrast to MSF, MED, and VCD techniques, the LTTD process does not require a large quantity of thermal energy to heat the brine since it operates with a temperature gradient of even 10°C to 20°C. However, about 1% of the yield rate can be achieved with this thermal gradient (Abraham & Robert Singh, 2006).

There are numerous advantages to the LTTD process over other large-scale desalination technologies, including the following:

- Less prone to rust and scaling (due to low thermal gradient).
- Reduced use of fossil fuels for brine heater.
- Improved thermal efficiency.
- Feasibility of using renewable energy sources to operate the plant.
- Blowdown of concentrated brine is kept under control.

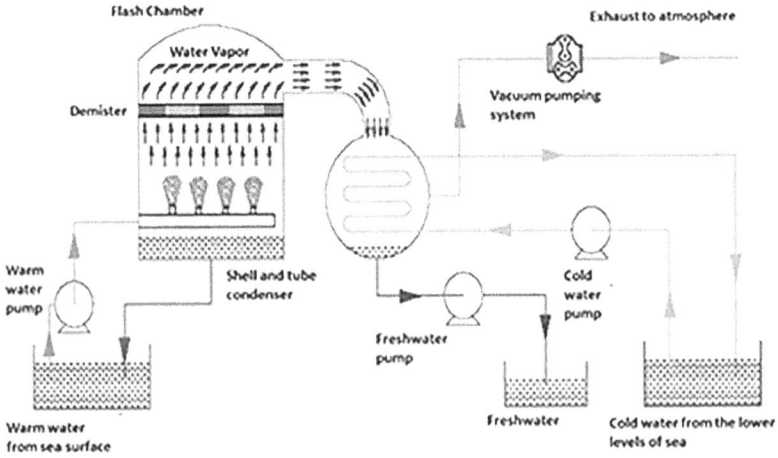

FIGURE 7.7   Schematic of ocean thermal energy-based LTTD system.

Source: Balaji (2016).

Recent years have seen the development of some successful LTTD systems based on ocean thermal energy (OTE) gradient (Abraham, 2007; Abraham & Robert Singh, 2006; Chik et al., 2015; Jin & Wang, 2015; Mutair & Ikegami, 2014; Senthil Kumar, Mani, & Kumaraswamy, 2007; Sistla, Venkatesan, Jalihal, & Kathiroli, 2009), nuclear power plant waste heat (Saari, 1978), and thermal power plant (Chandrakanth, Venkatesan, Prakash Kumar, Jalihal, & Iniyan, 2018; Low, 1991; Tay, Low, & Jeyaseelan, 1996; Venkatesan, Iniyan, & Jalihal, 2014, 2015) waste heat to fulfil the growing need for pure water on a large scale. Aside from that, the geothermal energy-driven LTTD process is still in the research and development stage (Chandrasekharam, Lashin, Al Arifi, Al-Bassam, & Chandrasekhar, 2020).

7.5 SOLAR ENERGY INTEGRATED LARGE-SCALE DESALINATION SYSTEM

Conventional desalination systems that use fossil fuels have also contributed to the emission of GHGs. Due to this, researchers have been pushed to investigate other methods of supplying renewable energy to desalination units (Subramani, Badruzzaman, Oppenheimer, & Jacangelo, 2011). The production of 1000 m^3/d of seawater desalination necessitates the use of around 10,000 tonnes of oil each year (Kalogirou, 2005).

Desalination plants driven by renewable energy sources are especially well suited for use in arid, semi-arid, and isolated places where every conventional source of power is not available (Chaibi, 2000). Solar, geothermal, and wind energy are examples of renewable energy sources that are often explored for desalination, with solar energy accounting for almost 57% of the renewable energy-based desalination industry (Eltawil, Zhengming, & Yuan, 2009). Most desalination units need either thermal or electrical energy input, both of which can be delivered by solar energy. As a result, solar-powered desalination systems have received a great deal of attention. Solar thermal collectors have advanced to a very high state of development (García-Rodríguez, Palmero-Marrero, & Gómez-Camacho, 2002). Countries with abundant fossil fuel reserves, such as the Middle East and Gulf states, have shifted their attention to solar energy in order to supply pure water sustainably (Sharon & Reddy, 2015). Several solar thermal collector types, such as the parabolic dish collector, parabolic trough collector (PTC), linear Fresnel reflector (LFR) and power tower, have been investigated for many decades as means of generating the thermal energy for several applications (Sahu, K, & Natarajan, 2021). The modular advantage of the solar parabolic dish collector allows it to create heat at temperatures ranging from 150 to 2000°C, making it one of the approaches used to generate thermal energy from solar radiation (Sahu, Arjun Singh, & Natarajan, 2020).

It is apparent that increasing plant capacity reduces the freshwater cost for all the high-capacity desalination procedures. However, RO systems and MSF evaporation generate low-cost freshwater for low capacity and large capacity demand accordingly. Table 7.3 highlights the cost of freshwater generated from the various desalination procedures for varied capacity range (Ullah & Rasul, 2019).

Many solar collector technologies, including CPCs, vacuum tubes, and flat plate collectors, have been incorporated to heat the input water and thereby minimize greenhouse gas emissions. Table 7.4 illustrates the energy demand for solar collectors combined with MSF, MED, and TVC as well as PV integrated RO systems (Vassilis, Soteris, & Emmy, 2016).

The typical solar-powered MSF and MED plant are shown in Figure 7.8 (a) and (b) respectively. A solar-powered MSF system with 10 m^3/d water production

TABLE 7.3
Freshwater Cost of Desalination Processes

Process	Freshwater Capacity (m^3/day)		Freshwater Cost ($/$m^3$)
MSF Evaporation	Low	< 100	2.20–8.80
	Medium	12,000–55,000	0.84–1.31
	High	> 91,000	0.46
MED	Medium	23,000	1.54
	High	5,28,000	0.46
VCD	Low to Medium	1,000–12,000	1.77–2.34
RO	Low	< 100	1.5–18.75
	Medium	15,000	1.62
	High	1,00,000–3,20,000	0.45–0.66
Solar Pond + MED	Medium	20,000	0.89
	High	2,00,000	0.71
Solar CSP + MED	Medium	> 5,000	2.40–2.80

Source: Adapted from Ullah and Rasul (2019), Manju and Sagar (2017b).

TABLE 7.4
Energy Demand of Solar Collector Technology Integrated Desalination Methods

Solar Collector Integrated Desalination Method	Temperature Range (°C)	Freshwater Capacity (m³/day)	Energy Requirement (kWh/m³)
MSF + CPC	90–120	> 10,000	60–70
MED + CPC	110–160	> 5,000	60–70
MED (Low Temperature) + Flat Plate/Waste Heat	70–80	> 5,000	60–70
TVC + Vacuum Tube	Up to 110	> 5,000	8–10
RO + PV	Up to 35	< 100	4–5 (for seawater) 0.5–1.5 (for brackish water)

Source: Reproduced from Vassilis et al. (2016).

capacity including brine recycling was created in 1974 by Mexico and the Federal Republic of Germany. Plant components include a double tube flat plate and PTC that use water as a heat transfer fluid, storage tanks that can supply thermal energy for 24 hours, and a desalination system (Manjarrez & Galván, 1979). When compared to standard solar stills, the performance ratio of MSF units powered by solar energy was observed to be 3–10 times higher, and also the cost of water production and capital investment for the same generation capacity was found to be lower for solar MSF plants (Signh & Sharma, 1989). It is worth noting that the cost of producing water from a multi-stage flash desalination unit can be reduced by expanding the plant's capacity and by combining the desalination plant with such a solar pond that allows for both solar energy collection and storage (Szacsvay, Hofer-Noser, & Posnansky, 1999). A real-time study of the solar MSF unit confirmed that water production could be enhanced by employing water as the working fluid in the solar collectors and then by expanding the number and volume of storage tanks (Abdallahan, 1991). Laboratory prototype trials have demonstrated that roughly 15 m³/d of water can be produced by combining ten flash desalination modules with a 1 m² surface area working at 0.9 bar with a 70°C solar pond (Safi, 1998). The utilization of 3160 m² PTCs can deliver roughly 76% of the energy consumption required by an MSF system (Al-Othman, Tawalbeh, El Haj Assad, Alkayyali, & Eisa, 2018). For MSF plants, using PTC to produce enough energy to yield high volumes of potable water at $2.72/m³ represents an enormous potential for alternative sources of energy in optimizing and minimizing the operational expenses of such thermal desalination systems, which are currently in use (Alsehli et al., 2017; Luqman et al., 2020). Operation of the solar MSF unit throughout a wide temperature range, as well as discharge of the distillate at the final stage, could improve its gained output ratio (S. Hou, Zhang, Huang, & Xie, 2008). MSF facilities connected with solar ponds and driven in part by electricity are more cost-effective than other solar-powered desalination

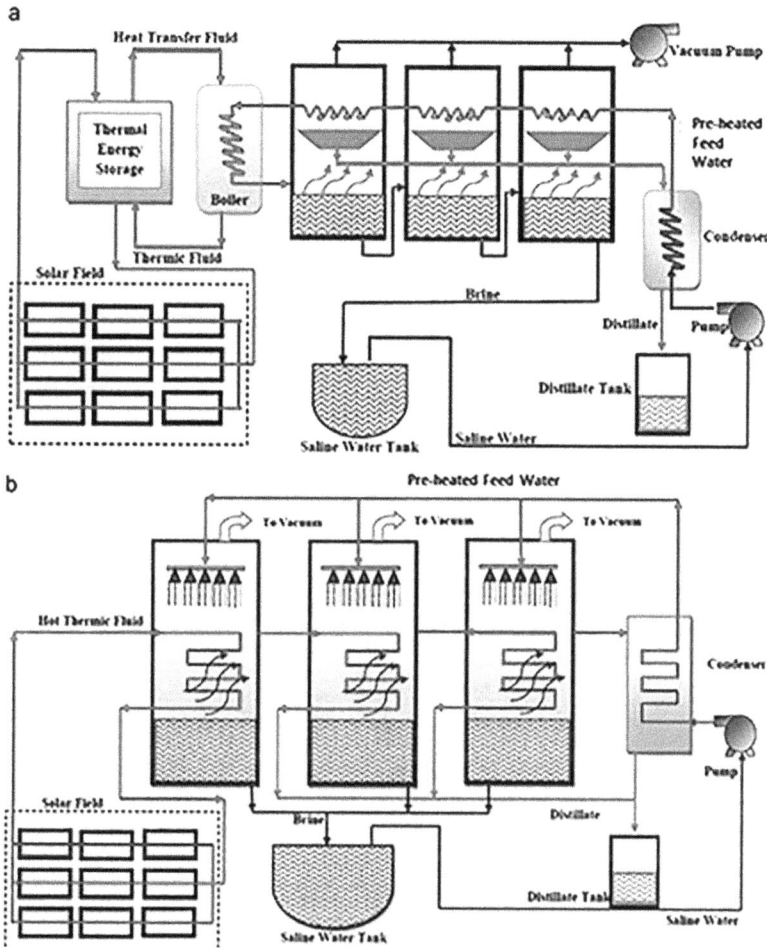

FIGURE 7.8   (a) Solar-powered MSF system and (b) solar-powered MED system.

Source: Sharon and Reddy (2015).

methods available (Suri, Al-Marafie, Al-Homoud, & Maheshwari, 1989). Because of their superior feasibility with solar thermal desalination, MED systems are a rising market share (Fath et al., 2011).

An innovative supercritical ORC design integrating evacuated tube collectors (ETCs) as a low-grade solar energy source powers MEDs coupled with mechanical vapour compression (MVC). Typical solar-powered MVC and TVC distillation system are shown in Figure 7.9 (a) and (b) respectively. The effect of increasing the number of effects on the performance ratio (PR), specific power consumption, solar collector area, and system efficiency is also investigated. The efficiency of supercritical ORC was around 10% at 5 MPa and 150°C with R152a working fluid. The performance of the low temperature (LT)-MED-MVC system was found to

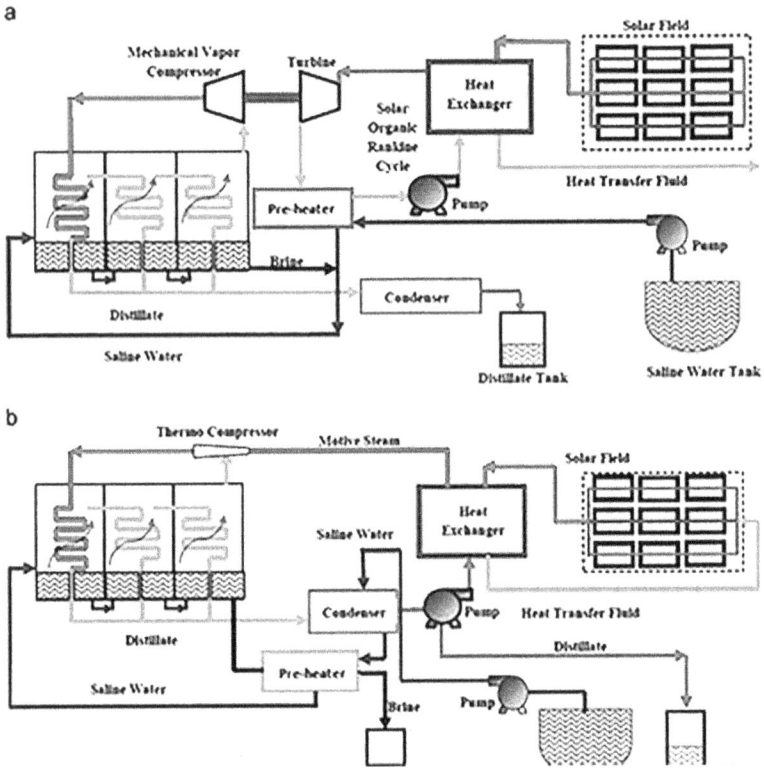

FIGURE 7.9   (a) Solar-powered MVC distillation system; (b) solar-powered TVC distillation system.

Source: Sharon and Reddy (2015).

be superior to that of the other systems in the record. MVC requires a lower value of 0.8 kWh/m³ specific power consumption for 14 effects of forward feed, 42,000 ppm salt of feed water, and a 50% recovery rate. For a 14-effect system, the best total efficiency is around 7%. When the number of effects was increased from 4 to 14, just a 1% decrease in the solar collector area was observed. However, PR increased from 3.5 to about 9 due to a decrease in motive steam mass flow rate from 3.17 kg/s to 1.21 kg/s. It has a significant impact on the functioning of the MED subsystem. As the number of effects increases, so does the performance ratio. For four effects, the specific area of MED is 200 m²-s/kg; for 14 effects, it is 700 m²-s/kg. When the effects increase from 4 to 14, there is a reduction in particular power usage of roughly 250%. As the number of effects increases from 4 to 6, the system efficiency improves from 0.1% to 3.2%. However, increasing the number of effects to 14 results in lesser incremental benefits of 6.9% of system efficiency. In addition, lowering the condenser temperature would improve the cycle's efficiency (Almatrafi, 2017).

The maximum universal performance ratio (UPR) for thermal-based desalination processes and membrane-based RO processes is 105. The optimal thermodynamic limit, on the other hand, is 828. Except for RO plants, all other viable desalination technologies are inefficient, with only 10–13% of the thermodynamic limit (TL) reflecting approaches that are extremely energy-intensive and unsustainable for long-term desalination. Desalination methods should account for 25–30% of the TL in the future to ensure sustainability. The MED system is powered by the seawater thermocline (ST), employing thermocline energy with a low-temperature gradient of 20°C between 200 and 300 metres depth. At four distinct working temperatures of 30–5°C, 30–8°C, 30–10°C, and 30–13°C, and with the corresponding number of stages of 6, 5, 4, and 3, the system generates freshwater at rates of 12.6, 10.6, 8.6, and 6.6 LPM, respectively, with a UPR of 147, 158, 145, and 120. ST-MED may obtain the highest UPR of 158, which is approximately a two-fold improvement (18.8% of the TL) over existing approaches. Only four stages are required to effectively unitize available thermal energy for a temperature differential of 20°C (Shahzad, Burhan, Ghaffour, & Ng, 2018).

An innovative LT-MED device with mechanical vapour compression powered by the supercritical Rankine cycle and utilizing a low-grade solar heat source via an ETC was investigated. High salt/brine concentrated water with a salinity of more than 100,000 ppm or power plant effluent streams can potentially be desalinated with greater efficiency and lower energy consumption than previously available systems. Increased motive steam temperature, supercritical ORC pressure, salt concentration, and a number of effects on the solar collector area, specific power consumption, and system efficiency were discussed. The MVC would require 4 kWh/m^3 with a saltwater salinity of 1,00,000 ppm with 14 effects and a 50% recovery rate. The system's overall efficiency was 14%. While the number of effects increased from 4 to 16, the MVC work requirement decreased because the last effect's vapour mass flow rate decreased, resulting in a 75% reduction in the solar collector area. As the temperature differential between the vapour created in the effect and the motive steam decreases, the specific area of the MED system increases from 165 m^2-s/kg to 1200 m^2-s/kg for 4 and 16 effects, respectively. While the number of effects increases from 4 to 16, the specific power consumption lowers by around 75%. Because the reduced mass flow rate of the vapour generated in the previous effect caused the MVC to perform less work, it is approximately 3 kWh/m^3 for 16 effects. The ORC pressure and cycle efficiency had no impact on MVC and MED's specific power. Salinity, on the other hand, has only a local impact on the MED system. It has been discovered that there is a 25% rise in salinity of the feed water as the boiling point elevation (BPE) goes from 50,000–120,000 ppm. Salinity had little effect on particular power usage. The RO system, on the other hand, is extremely sensitive to feed water salinity. When the salinity of the input water reaches 100,000 ppm, the RO system requires 250% more specific power. The motive steam temperature affected the MED specific area, solar field, and specific power consumption for the LT-MED system with an operating temperature of 90°C. The specific area of the MED system reduces by nearly 60% as the temperature difference of the effect increases. However, because of the increased compression ratio, MVC's specific power consumption, solar field area was increased by almost 80% (Almatrafi, 2018).

Using the international mathematical and statistical library (IMSL), a theoretical model of an ST-MED system was constructed and simulated in FORTRAN (IMSL). Four different working settings with cold seawater temperatures ranging from 5 to 13°C and MED stages ranging from 3 to 6. This proposed system with six stages may produce 12.5 LPM of freshwater. A four-stage MED system operating at 30°C hot and 10°C cold seawater temperatures can reach the maximum UPR of 158, which is 18.8% of the TL. To date, the lowest capital and greatest UPR have been reported in the literature (Ng & Shahzad, 2018).

At the University of Almeria (Spain), the unique integration of a thermal desalination system employing vacuum MED technology and a solar thermal field based on static collectors was evaluated experimentally and optimized the distillation unit performance under various environmental circumstances such as 60, 70, and 80°C inlet temperature of the membrane module. According to a quasi-dynamic model-based simulation analysis, 41.7, 68.4, and 70.5 m³ of distilled water are generated, respectively. The flat plate collector solar thermal field of 17 kWth total power at 90°C coupled to the buffer thermal storage tank allows the overall system to run more consistently even when there are solar radiation perturbations present throughout the day. Because of the significant loss in production using heat at 60°C, it is unable to modify the operational expenses. However, using the heat at 70°C was a superior alternative. Even if there is a slight increase in output at 80°C, reaching this temperature is challenging because it is only feasible for 2.6% of the running period. However, as compared to the 70°C working temperature, it adds less to operational expenditure and so makes it worthwhile (Andrés-Mañas et al., 2020).

A theoretical assessment of geothermal energy was conducted, as well as the thermal desalination potential of the PTCs. The results indicated that both solar and geothermal resources attain an annual time of 76% at a 490 m well depth and a temperature of 41.8°C climatic zones. On the other hand, a geothermal source of energy may power a MED plant for an entire year. Additionally, twofold effect absorption can improve 30% of annual time results. According to this analysis, 790 m of depth is sufficient to produce the 70°C operating temperature, since a thermal gradient of 8.870 can be achieved per 100 m of depth. By combining the two resources, CO_2 emissions of 510,387,920 kg/year can be minimized, and existing plants can be amortized in six years (Colmenar-Santos, Palomo-Torrejón, Mur-Pérez, & Rosales-Asensio, 2020).

The effect of saltwater temperature, salinity, and sun thermal energy on the cost of freshwater has been investigated. Some researches focused on specific performance parameters, such as the unit cost of freshwater, the rate of water production, and the total thermal efficiency. The integration of the MED with low-temperature solar collectors (60°C–95°C), medium-temperature solar collectors (165°C–200°C), and high-temperature solar collectors (370°C–530°C) results in water production costs of 2\$/m³–3.6\$/m³, 1.4\$/m³–3.1\$/m³, and 1.8\$/m³–2.2\$/m³, respectively, with a payback period of 4 to 16 years. Additionally, information was provided to assist in the selection of the most optimal MED/solar technology pairings for both small- and large-scale applications (Baniasad Askari & Ameri, 2021).

A novel spray evaporation (SE)-MED system with three-stage heat recovery was developed to achieve zero liquid discharge (ZLD) and significantly reduce waste-water output. The SE technique is used to recover salts and freshwater from saline concentrate disposal (Guo, Li, Wang, & Li, 2021).

The MED unit, when combined with a solar pond of 30,000–40,000 m², can produce 100,000 tonnes of pure water in a year at a cost equivalent to traditional desalination technologies. The cost of desalinating water is mostly determined by the price of salt, which gets cheaper as the area of the solar pond increases (Tsilingiris, 1995).

To increase the efficiency of thermal desalination plants such as MSF and MVCs, a detailed micro-thermal energy and exergy analysis was performed. The minimum work required for separation (i.e. work required to separate the unit potable water from the feed salty water) was evaluated in this study. MSF separation requires 0.729 kWh/m³ of work. The MSF system's overall specific exergy loss was roughly 63 kJ/kg. Because the second law efficiencies of MSF and MVC are only approximately 4% and 7%, respectively, there is scope for enhancing the performance of both MSF and MVC systems by lowering the maximum exergy destruction in the systems. In the stages, 78% of total exergy was destroyed (distillers). Enabling a co-generation plant by connecting the desalination and power plants can improve system performance. Thermal desalination can be more efficient when a low-grade heat source is used (El-Feky, 2015).

An in-depth examination of the various configurations of flash evaporation desalination systems, including single and multi-stage flash evaporation, flat plate collector integrated single-stage flash evaporator, HDH with flash evaporation, and solar still with flash evaporation, along with the factors affecting the performance of the system, including non-equilibrium fraction (NEF), operating pressure and temperature, nozzle diameter, degree of superheat, and spray velocity, has been carried out and summarized that the evaporation rate is greatly influenced by the operating pressure and degree of superheat (Raj, Murugavel, Rajaseenivasan, & Srithar, 2015).

7.6 EXISTING LARGE-SCALE SOLAR DESALINATION: A CASE STUDY

7.6.1 Solar-Powered Tomato Farm, Port Augusta

The world's first CSP plant, measuring 20 hectares in size, was completed and commenced operating in 2016 in Port Augusta of the south Australian desert. The plant is composed of a 51,500-square-metre solar field with over 23,000 heliostats that will provide heat, power, and freshwater for the annual production of 15 million kilogrammes of tomatoes as shown in Figure 7.10 (a) and (b). Danish renewables created this technique at a cost of approximately $200 million US dollars. Additionally, the Sundrop tomato farm includes a greenhouse that will produce over 450,000 m³ of freshwater annually, resulting in a 2 million litre reduction in annual diesel consumption (Vorrath, 2016).

FIGURE 7.10a   Concentrated solar power (CSP) tower plant, Port Augusta.

Source: Vorrath (2016).

FIGURE 7.10b   Tomato production utilizing solar energy.

Source: Amelinckx (2016).

FIGURE 7.11   Solar-powered desalination plant, Kiunga.

Source: Colagrossi (2019).

7.6.2 Solar-Powered Desalination Plant, Kiunga

A desalination plant, Figure 7.11, powered by a solar panel array that produces 75,000 litres of freshwater per day (19,800 gallons/day) and continues to serve 25,000 people per day, was built in July 2018 for a total cost of around 500,000 dollars after the Kiunga Village in Kenya started to notice serious issues like children's kidney damage due to polluted water (Colagrossi, 2019).

7.7 ENVIRONMENTAL CONSEQUENCES OF DESALINATION

Despite the fact that desalination is critical for generating freshwater for people and industries all over the world, it has some serious environmental consequences (Lior, 2017; Shahabi, McHugh, Anda, & Ho, 2015). Pipeline leakage, chemical disposal after pre-treatment of seawater, and brine disposal after freshwater production all have an impact on aquifers and marine ecosystems because they change the salinity, temperature, and density of seawater, which may not be a suitable environment for marine dependent species (Enríquez-de-Salamanca, Díaz-Sierra, Martín-Aranda, & Santos, 2017). Furthermore, the chlorine in the feed water may induce corrosion and fouling of heat exchangers (Cherchi, Badruzzaman, Becker, & Jacangelo, 2017; Hoepner & Lattemann, 2003). Also, heavy metals in feed water react negatively with desalination plant equipment (Cherif, Champenois, & Belhadj, 2016). Additionally, owing to the high energy requirement for their operation, conventional desalination technologies emit a significant quantity of greenhouse gas (GHG).

   To address environmental challenges associated with desalination technologies, the following guidelines and approaches are employed:

- By adding air filters, NO_x burners, and particle separators to reduce air pollutants such as CO, CO_2, NO, and SO_2 generated by desalination plants.
- Implementing a Zero Liquid Discharge (ZLD) brine discharge management plan to economically remove waste liquid in the inland disposal region for concentrates and to utilize waste water in an effective manner (Tong & Elimelech, 2016).
- The standard of post-desalination by-products must be maintained.
- Prior to the actual release of brine concentrates, the desalination plant might be paired with mineral extraction techniques.

Many researchers have studied the environmental impact of desalination plants. M. Darwish et al. developed the seawater RO (SWRO) method in 2012 to replace the MSF desalination process with 1.2 Mm^3/day capacity in order to satisfy Qatar's freshwater need to a great extent (Darwish, Hassabou, & Shomar, 2013). It has been demonstrated that the planned SWRO plant requires three times less feed water (3.6 Mm^3/day) than the comparable 8.4 Mm^3/day for MSF, hence decreasing the marine environment problem. Furthermore, this facility outputs just 2.4 Mm^3/day of concentrated brine and cooling water, compared to 7.2 Mm^3/day for MSF. Furthermore, the residuals of chlorine and thermal effluent were significantly decreased. Additionally, the SWRO plant cuts yearly CO_2 emissions from 3.564 Mtonnes to 0.891 Mtonnes by conserving up to 75% of the required energy.

7.8 CONCLUSIONS AND DEVELOPMENT DIRECTIONS

In this chapter, the basic desalination processes, specifically large-scale thermal-based desalination processes such as MSF evaporation, MED, MVC distillation, TVC distillation, and LTTD, have been discussed. In addition, the feasibility of a solar energy integrated desalination system as a long-term solution to the problem of water scarcity as well as a case study of existing large-scale solar energy-based desalination has been discussed in detail. Furthermore, the benefits and drawbacks of large-scale thermal desalination, as well as the numerous criteria that influence its performance, have been explored. Besides that, there were also examples of various renewable energy sources powering thermal desalination as well as co-generation systems such as the ORC paired with MED and MVC systems for obtaining freshwater while also generating electricity, among others. It has been shown that combining MSF, MED, and VCD among themselves as well as with the reverse osmosis (RO) process leads to better thermal efficiency and performance ratio. It is also noted that, when used in conjunction with other phase change technologies, such as MSF and TVC, MED provides increased thermal efficiency and performance ratio for the intended application. More importantly, much like with MED-TVC, it has been seen that the use of RO increases heat recovery while concurrently lowering energy consumption, brine flow, and freshwater salinity.

Solar ponds, PTC, and parabolic dish collector systems are all used to provide the thermal energy required for heating the saline feed water required for thermal desalination operations. However, the corrosion of solar thermal collectors, owing to the handling of saline water, is a major concern with this system, and this can be solved by utilizing special polymers instead of typical corrosion-prone copper collectors and cleaning the desalination plant with corrosion inhibitors from time to time. With increasing attention being paid these days to solar PV modules, a significant reduction in the cost of water production could be achieved by lowering their cost while simultaneously increasing the desalination system efficiency.

Among the desalination methods described, LTTD stands out as one of the eco-friendly, high-capacity desalination approaches with the lowest carbon footprint and global warming potential (GWP). However, it is necessary to explore the effect of dissolved non-condensable gases in seawater on the operation of LTTD plants, and also developing a cost-effective hybrid system with an LTTD plant is regarded as advantageous for the future. The stand-alone system can be customized for small-scale applications, such as those involving remote residents. In addition, the LTTD system can be coupled to the concentrating collectors to produce freshwater during off-hours using thermal energy storage.

REFERENCES

Abdallahan, W. (1991). Design and performance desalination systems of solar MSF. *Desalination*, *82*, 175–185.

Abraham, R. (2007). Experimental studies on a desalination plant using ocean temperature difference. *International Journal of Nuclear Desalination*, *2*(4), 383–392.

Abraham, R., & Robert Singh, T. (2006). Thermocline-driven desalination: The technology and its potential. *International Journal of Nuclear Desalination*, *2*(2), 109–116. https://doi.org/10.1504/IJND.2006.012513.

Abutayeh, M., Humood, M., Alsheghri, A. A., Al Hammadi, A. J., & Farraj, A. R. (2013). Experimental study of a solar thermal desalination unit. *ASME International Mechanical Engineering Congress and Exposition, Proceedings (IMECE)*, *6 B*, 1–9. https://doi.org/10.1115/IMECE2013-66174.

Aguilar-Jiménez, J. A., Velázquez, N., López-Zavala, R., Beltrán, R., Hernández-Callejo, L., González-Uribe, L. A., & Alonso-Gómez, V. (2020). Low-temperature multiple-effect desalination/organic Rankine cycle system with a novel integration for fresh water and electrical energy production. *Desalination*, *477*(November 2019), 114269. https://doi.org/10.1016/j.desal.2019.114269.

Almatrafi, E., Moloney, F., & Goswami, D. (2017). Multi effects desalination-mechanical vapor compression powered by low temperature supercritical organic rankine cycle. In *Proceedings of the ASME 2017 International Mechanical Engineering Congress and Exposition*. Volume 6: Energy, 1–9. Tampa, Florida, USA. November 3–9, 2017. V006T08A020. ASME.https://doi.org/https://doi.org/10.1115/IMECE2017-72230.

Almatrafi, E., Moloney, F., & Goswami, D. (2018). Performance analysis of solar thermal powered supercritical organic rankine cycle assisted low-temperature multi effect desalination coupled with mechanical vapor compression. *Proceedings of the ASME 2018 Power Conference Collocated with the ASME 2018 12th International Conference on Energy Sustainability and the ASME 2018 Nuclear Forum*. Volume 2: Heat Exchanger Technologies; Plant Performance; Thermal Hydraulics and Computational Fluid Dynamics; Water Management for Power Systems; Student Competition. Lake Buena Vista, Florida, USA. June 24–28, 2018. V002T11A002. ASME. https://doi.org/10.1115/POWER2018-7307.

Al-Othman, A., Tawalbeh, M., El Haj Assad, M., Alkayyali, T., & Eisa, A. (2018). Novel multi-stage flash (MSF) desalination plant driven by parabolic trough collectors and a solar pond: A simulation study in UAE. *Desalination*, *443*(April), 237–244. https://doi.org/10.1016/j.desal.2018.06.005

Alsehli, M., Choi, J. K., & Aljuhan, M. (2017). A novel design for a solar powered multistage flash desalination. *Solar Energy, 153*, 348–359. https://doi.org/10.1016/j.solener.2017.05.082.

Amelinckx, A. (2016). *This Farm Uses Only Sun and Seawater to Grow Food*. Retrieved from https://modernfarmer.com/2016/10/sundrop-farms/.

Andrés-Mañas, J. A., Roca, L., Ruiz-Aguirre, A., Acién, F. G., Gil, J. D., & Zaragoza, G. (2020). Application of solar energy to seawater desalination in a pilot system based on vacuum multi-effect membrane distillation. *Applied Energy, 258*(August 2019), 114068. https://doi.org/10.1016/j.apenergy.2019.114068.

Baig, H., Antar, M. A., & Zubair, S. M. (2011). Performance evaluation of a once-through multi-stage flash distillation system: Impact of brine heater fouling. *Energy Conversion and Management, 52*(2), 1414–1425. https://doi.org/10.1016/j.enconman.2010.10.004.

Balaji, D. (2016). Experimental study on the effect of feed water nozzles on non-equilibrium temperature difference and flash evaporation in a single-stage evaporator and an investigation of effect of process parameters on the liquid flashing in a LTTD desalination process. *Desalination and Water Treatment, 3994*(May), 1–17. https://doi.org/10.1080/19443994.2016.1172511.

Baniasad Askari, I., & Ameri, M. (2021). A techno-economic review of multi effect desalination systems integrated with different solar thermal sources. *Applied Thermal Engineering, 185*, 116323. https://doi.org/10.1016/j.applthermaleng.2020.116323.

Bendig, M., Maréchal, F., & Favrat, D. (2013). Defining "waste heat" for industrial processes. *Applied Thermal Engineering, 61*, 134–142. https://doi.org/10.1016/j.applthermaleng.2013.03.020.

Chaibi, M. T. (2000). An overview of solar desalination for domestic and agriculture water needs in remote arid areas. *Desalination, 127*(2), 119–133. https://doi.org/10.1016/S0011-9164(99)00197-6.

Chandrakanth, B., Venkatesan, G., Prakash Kumar, L. S. S., Jalihal, P., & Iniyan, S. (2018). Thermal design, rating and second law analysis of shell and tube condensers based on Taguchi optimization for waste heat recovery based thermal desalination plants. *Heat and Mass Transfer/Waerme- Und Stoffuebertragung, 54*(9), 2885–2897. https://doi.org/10.1007/s00231-018-2326-2.

Chandrasekharam, D., Lashin, A., Al Arifi, N., Al-Bassam, A. M., & Chandrasekhar, V. (2020). Geothermal energy for sustainable water resources management. *International Journal of Green Energy, 17*(1), 1–12. https://doi.org/10.1080/15435075.2019.1685998.

Cherchi, C., Badruzzaman, M., Becker, L., & Jacangelo, J. G. (2017). Natural gas and grid electricity for seawater desalination: An economic and environmental life-cycle comparison. *Desalination, 414*, 89–97. https://doi.org/10.1016/j.desal.2017.03.028.

Cherif, H., Champenois, G., & Belhadj, J. (2016). Environmental life cycle analysis of a water pumping and desalination process powered by intermittent renewable energy sources. *Renewable and Sustainable Energy Reviews, 59*, 1504–1513. https://doi.org/10.1016/j.rser.2016.01.094.

Chik, M. A. T., Othman, N. A., Sarip, S., Ikegami, Y., My, A., Othman, N., . . . Izzuan, H. (2015). Design optimization of power generation and desalination application in Malaysia utilizing ocean thermal energy. *Jurnal Teknologi, 77*(1), 177–185. https://doi.org/10.11113/jt.v77.4144.

Colagrossi, M. (2019). *Solar-powered desalination plant in Kenya gives fresh water to 25,000 people a day*. Retrieved from https://bigthink.com/the-present/solar-power-desalination/.

Colmenar-Santos, A., Palomo-Torrejón, E., Mur-Pérez, F., & Rosales-Asensio, E. (2020). Thermal desalination potential with parabolic trough collectors and geothermal energy in the Spanish southeast. *Applied Energy, 262*(August 2019), 114433. https://doi.org/10.1016/j.apenergy.2019.114433.

Darwish, M., Hassabou, A. H., & Shomar, B. (2013). Using Seawater Reverse Osmosis (SWRO) desalting system for less environmental impacts in Qatar. *Desalination*, *309*, 113–124. https://doi.org/10.1016/j.desal.2012.09.026.

El-Feky, A. K. (2015). A comprehensive micro-thermal analysis of thermal desalination plants for improving their efficiency. *International Journal of Environmental Protection and Policy*, *2*, 16–25. https://doi.org/10.11648/j.ijepp.s.2014020601.13.

Eltawil, M. A., Zhengming, Z., & Yuan, L. (2009). A review of renewable energy technologies integrated with desalination systems. *Renewable and Sustainable Energy Reviews*, *13*(9), 2245–2262. https://doi.org/10.1016/j.rser.2009.06.011.

Enríquez-de-Salamanca, Á., Díaz-Sierra, R., Martín-Aranda, R. M., & Santos, M. J. (2017). Environmental impacts of climate change adaptation. *Environmental Impact Assessment Review*, *64*, 87–96. https://doi.org/10.1016/j.eiar.2017.03.005.

Ettouney, H. (2002). Performance of the once-through multistage flash desalination process. *Proceedings of the Institution of Mechanical Engineers, Part A: Journal of Power and Energy*, *216*, 229–240. https://doi.org/10.1243/095765002320183559.

Fath, H., Abbas, Z., & Khaled, A. (2011). Techno-economic assessment and environmental impacts of desalination technologies. *Desalination*, *266*, 263–273. https://doi.org/10.1016/j.desal.2010.08.035.

Feria-Díaz, J. J., López-Méndez, M. C., Rodríguez-Miranda, J. P., Sandoval-Herazo, L. C., & Correa-Mahecha, F. (2021). Commercial thermal technologies for desalination of water from renewable energies: A state of the art review. *Processes*, *9*(2), 1–22. https://doi.org/10.3390/pr9020262.

García-Rodríguez, L. (2003). Renewable energy applications in desalination: State of the art. *Solar Energy*, *75*(5), 381–393. https://doi.org/10.1016/j.solener.2003.08.005.

García-Rodríguez, L., Palmero-Marrero, A. I., & Gómez-Camacho, C. (2002). Comparison of solar thermal technologies for applications in seawater desalination. *Desalination*, *142*(2), 135–142. https://doi.org/10.1016/S0011-9164(01)00432-5.

Gnaneswar, V., Nirmalakhandan, N., & Deng, S. (2010). Renewable and sustainable approaches for desalination. *Renewable and Sustainable Energy Reviews*, *14*(9), 2641–2654. https://doi.org/10.1016/j.rser.2010.06.008.

Guo, P., Li, T., Wang, Y., & Li, J. (2021). Energy and exergy analysis of a spray-evaporation multi-effect distillation desalination system. *Desalination*, *500*, 114890. https://doi.org/10.1016/j.desal.2020.114890.

Hoepner, T., & Lattemann, S. (2003). Chemical impacts from seawater desalination plants—A case study of the northern Red Sea. *Desalination*, *152*(1–3), 133–140. https://doi.org/10.1016/S0011-9164(02)01056-1.

Hou, J., Cheng, H., Wang, D., Gao, X., & Gao, C. (2010). Experimental investigation of low temperature distillation coupled with spray evaporation. *Desalination*, *258*(1–3), 5–11. https://doi.org/10.1016/j.desal.2010.03.030.

Hou, S., Zhang, Z., Huang, Z., & Xie, A. (2008). Performance optimization of solar multistage flash desalination process using Pinch technology. *Desalination*, *220*, 524–530. https://doi.org/10.1016/j.desal.2007.01.052.

Ibrahim, A. G. M., Rashad, A. M., & Dincer, I. (2017). Exergoeconomic analysis for cost optimization of a solar distillation system. *Solar Energy*, *151*, 22–32. https://doi.org/10.1016/j.solener.2017.05.020.

Jin, Z., & Wang, H. (2015). Modelling and experiments on ocean thermal energy for desalination. *International Journal of Sustainable Energy*, *34*(2), 103–112. https://doi.org/10.1080/14786451.2013.820187.

Kalogirou, S. A. (2005). Seawater desalination using renewable energy sources. *Progress in Energy and Combustion Science*, *31*(3), 242–281. https://doi.org/10.1016/j.pecs.2005.03.001.

Kucera, J. (2014). Introduction to desalination. In *Desalination: Water from Water*. Hoboken, NJ: John Wiley & Sons, Ltd., 1–37. https://doi.org/10.1002/9781118904855.ch1.

Latteman, S. (2010). *Development of an Environmental Impact Assessment and Decision Support System for Seawater Desalination Plants*. 1st Edition. London: CRC Press. https://doi.org/10.1201/b10829.

Lior, N. (2017). Sustainability as the quantitative norm for water desalination impacts. *Desalination*, *401*, 99–111. https://doi.org/10.1016/j.desal.2016.08.008

Low, S. C. (1991). Vacuum desalination using waste heat from a steam turbine. *Desalination*, *81*, 321–331.

Luo, T., Young, R., & Reig., P. (2015). *Aqueduct Projected Water Stress Country Rankings. Technical Note*. Washington, D.C.: World Resources Institute. (August), 1–16.

Luqman, M., Ghiat, I., Maroof, M., Lahlou, F. Z., Bicer, Y., & Al-Ansari, T. (2020). Application of the concept of a renewable energy based-polygeneration system for sustainable thermal desalination process—A thermodynamics' perspective. *International Journal of Energy Research*, *44*(15), 12344–12362. https://doi.org/10.1002/er.5161.

Manjarrez, R., & Galván, M. (1979). Solar multistage flash evaporation (SMSF) as a solar energy application on desalination processes. Description of one demonstration project. *Desalination*, *31*(1–3), 545–554. https://doi.org/10.1016/S0011-9164(00)88557-4.

Manju, S., & Sagar, N. (2017a). Progressing towards the development of sustainable energy: A critical review on the current status, applications, developmental barriers and prospects of solar photovoltaic systems in India. *Renewable and Sustainable Energy Reviews*, *70*(May 2016), 298–313. https://doi.org/10.1016/j.rser.2016.11.226.

Manju, S., & Sagar, N. (2017b). Renewable energy integrated desalination: A sustainable solution to overcome future fresh-water scarcity in India. *Renewable and Sustainable Energy Reviews*, *73*(January), 594–609. https://doi.org/10.1016/j.rser.2017.01.164

Moustafa, S. M. A., & Jarrar, D. I. (1985). Performance of a self—regulating solar multistage flash desalination system. *Solar Energy*, *35*(4), 333–340.

Mutair, S., & Ikegami, Y. (2014). Design optimization of shore-based low temperature thermal desalination system utilizing the ocean thermal energy. *Journal of Solar Energy Engineering, Transactions of the ASME*, *136*(4), 1–8. https://doi.org/10.1115/1.4027575

Muthunayagam, A. E., Ramamurthi, K., & Paden, J. R. (2005). Low temperature flash vaporization for desalination. *Desalination*, *180*, 25–32. https://doi.org/10.1016/j.desal.2004.12.028

Natarajan, S. K., Suraparaju, S. K., Elavarasan, R. M., Pugazhendhi, R., & Hossain, E. (2022). An experimental study on eco-friendly and cost-effective natural materials for productivity enhancement of single slope solar still. *Environmental Science and Pollution Research*, *29*(2), 1917–1936. https://doi.org/10.1007/s11356-021-15764-8.

Ng, K. C., & Shahzad, M. W. (2018). Sustainable desalination using ocean thermocline energy. *Renewable and Sustainable Energy Reviews*, *82*(June 2017), 240–246. https://doi.org/10.1016/j.rser.2017.08.087.

Qi, C., Lv, H., Feng, H., Lv, Q., & Xing, Y. (2017). Performance and economic analysis of the distilled seawater desalination process using low-temperature waste hot water. *Applied Thermal Engineering*, *122*, 712–722. https://doi.org/10.1016/j.applthermaleng.2017.05.064

Rabiee, H., Khalilpour, K. R., Betts, J. M., & Tapper, N. (2018). Energy-water nexus: Renewable-integrated hybridized desalination systems. In *Polygeneration with Polystorage: For Chemical and Energy Hubs*. Cambridge, MA: Academic Press, 409–458. https://doi.org/10.1016/B978-0-12-813306-4.00013-6.

Raj, M. M. A., Murugavel, K. K., Rajaseenivasan, T., & Srithar, K. (2015). A review on flash evaporation desalination. *Desalination and Water Treatment*, *57*, 1–10. https://doi.org/10.1080/19443994.2015.1070283.

Saari, R. (1978). Desalination by very low-temperature nuclear heat. *Nuclear Technology*, *38*(2), 209–214. https://doi.org/10.13182/NT78-A32014.

Safi, M. J. (1998). Performance of a flash desalination unit intended to be coupled to a solar pond. *Renewable Energy*, *14*(1–4), 339–343. https://doi.org/10.1016/S0960-1481(98)00087-1.

Sahu, S. K., Arjun Singh, K., & Natarajan, S. K. (2020). Design and development of a low-cost solar parabolic dish concentrator system with manual dual-axis tracking. *International Journal of Energy Research*, *45*(October), 1–11. https://doi.org/10.1002/er.6164.

Sahu, S. K., Arjun Singh, K., & Natarajan, S. K. (2021). Impact of double trumpet-shaped secondary reflector on flat receiver of a solar parabolic dish collector system. *Energy Sources, Part A: Recovery, Utilization, and Environmental Effects*, *41*, 1–19. https://doi.org/10.1080/15567036.2021.1918803.

Saidur, R., Elcevvadi, E. T., Mekhilef, S., Safari, A., & Mohammed, H. A. (2011). An overview of different distillation methods for small scale applications. *Renewable and Sustainable Energy Reviews*, *15*(9), 4756–4764. https://doi.org/10.1016/j.rser.2011.07.077.

Sampathkumar, A., & Natarajan, S. K. (2021a). Experimental investigation of single slope solar still with Eucheuma (agar-agar) fiber for augmentation of freshwater yield: Thermo-economic analysis. *Environmental Progress and Sustainable Energy*, *41*(August), 2–9. https://doi.org/10.1002/ep.13750.

Sampathkumar, A., & Natarajan, S. K. (2021b). Experimental investigation on productivity enhancement in single slope solar still using Borassus Flabellifer micro-sized particles. *Materials Letters*, *299*, 130097. https://doi.org/10.1016/j.matlet.2021.130097

Sampathkumar, K., Arjunan, T. V., Pitchandi, P., & Senthilkumar, P. (2010). Active solar distillation-A detailed review. *Renewable and Sustainable Energy Reviews*, *14*(6), 1503–1526. https://doi.org/10.1016/j.rser.2010.01.023.

Senthil Kumar, R., Mani, A., & Kumaraswamy, S. (2007). Experimental studies on desalination system for ocean thermal energy utilisation. *Desalination*, *207*(1–3), 1–8. https://doi.org/10.1016/j.desal.2006.08.001.

Shahabi, M. P., McHugh, A., Anda, M., & Ho, G. (2015). Comparative economic and environmental assessments of centralised and decentralised seawater desalination options. *Desalination*, *376*, 25–34. https://doi.org/10.1016/j.desal.2015.08.012

Shahzad, M. W., Burhan, M., Ghaffour, N., & Ng, K. C. (2018). A multi evaporator desalination system operated with thermocline energy for future sustainability. *Desalination*, *435*(January 2017), 268–277. https://doi.org/10.1016/j.desal.2017.04.013.

Sharon, H., & Reddy, K. S. (2015). A review of solar energy driven desalination technologies. *Renewable and Sustainable Energy Reviews*, *41*, 1080–1118. https://doi.org/10.1016/j.rser.2014.09.002.

Shatat, M., & Riffat, S. B. (2014). Water desalination technologies utilizing conventional and renewable energy sources. *International Journal of Low-Carbon Technologies*, *9*(1), 1–19. https://doi.org/10.1093/ijlct/cts025.

Signh, D., & Sharma, S. K. (1989). Performance ratio, area economy and economic return for an integrated solar energy/multi-stage flash desalination plant. *Desalination*, *73*(C), 191–195. https://doi.org/10.1016/0011-9164(89)87013-4.

Sistla, P. V. S., Venkatesan, G., Jalihal, P., & Kathiroli, S. (2009). Low temperature thermal desalination plants. In *Proceedings of the ISOPE Ocean Mining Symposium*. Chennai, India, CA: International Society of Offshore and Polar Engineers (ISOPE), 59–63.

Subramani, A., Badruzzaman, M., Oppenheimer, J., & Jacangelo, J. G. (2011). Energy minimization strategies and renewable energy utilization for desalination: A review. *Water Research*, *45*(5), 1907–1920. https://doi.org/10.1016/j.watres.2010.12.032.

Suraparaju, S. K., Dhanusuraman, R., & Natarajan, S. K. (2021). Performance evaluation of single slope solar still with novel pond fibres. *Process Safety and Environmental Protection*, *154*, 142–154. https://doi.org/10.1016/j.psep.2021.08.011.

Suraparaju, S. K., & Natarajan, S. K. (2020). Performance analysis of single slope solar desalination setup with natural fiber. *Desalination and Water Treatment*, *193*(February), 64–71. https://doi.org/10.5004/dwt.2020.25679.

Suraparaju, S. K., & Natarajan, S. K. (2021a). Augmentation of freshwater productivity in single slope solar still using Luffa acutangula fibres. *Water Science and Technology*, *84*(10–11), 2943–2957. https://doi.org/10.2166/wst.2021.298.

Suraparaju, S. K., & Natarajan, S. K. (2021b). Augmentation of freshwater productivity in single slope solar still using Luffa acutangula fibres. *Water Science and Technology*, *84*(10–11), 2943–2957. https://doi.org/10.2166/wst.2021.298.

Suraparaju, S. K., & Natarajan, S. K. (2021c). Experimental investigation of single-basin solar still using solid staggered fins inserted in paraffin wax PCM bed for enhancing productivity. *Environmental Science and Pollution Research*, *28*, 20330–20343. https://doi.org/10.1007/s11356-020-11980-w.

Suraparaju, S. K., & Natarajan, S. K. (2021d). Productivity enhancement of single-slope solar still with novel bottom finned absorber basin inserted in phase change material (PCM): Techno-economic and enviro-economic analysis. *Environmental Science and Pollution Research*, *28*, 45985–46006. https://doi.org/10.1007/s11356-021-13495-4

Suri, R. K., Al-Marafie, A. M. R., Al-Homoud, A. A., & Maheshwari, G. P. (1989). Cost-effectiveness of solar water production. *Desalination*, *71*(2), 165–175. https://doi.org/10.1016/0011-9164(89)80007-4.

Swaminathan, J., Nayar, K. G., & Lienhard V, J. H. (2016). Mechanical vapor compression—Membrane distillation hybrids for reduced specific energy consumption. *Desalination and Water Treatment*, *57*(55), 26507–26517. https://doi.org/10.1080/19443994.2016.1168579.

Szacsvay, T., Hofer-Noser, P., & Posnansky, M. (1999). Technical and economic aspects of small-scale solar-pond- powered seawater desalination systems. *Desalination*, *122*, 185–193.

Tay, J. H., Low, S. C., & Jeyaseelan, S. (1996). Vacuum desalination for water purification using waste heat. *Desalination*, *106*(1–3), 131–135. https://doi.org/10.1016/S0011-9164(96)00104-X.

Thimmaraju, M., Sreepada, D., Babu, G. S., Dasari, B. K., Velpula, S. K., & Vallepu, N. (2018). Desalination of water. In *Desalination and Water Treatment*. London, UK: IntechOpen. https://doi.org/10.5772/intechopen.78659.

Tiwari, G. N., Singh, H. N., & Tripathi, R. (2003). Present status of solar distillation. *Solar Energy*, *75*, 367–373. https://doi.org/10.1016/j.solener.2003.07.005.

Tong, T., & Elimelech, M. (2016). The global rise of zero liquid discharge for wastewater management: Drivers, technologies, and future directions. *Environmental Science and Technology*, *50*(13), 6846–6855. https://doi.org/10.1021/acs.est.6b01000.

Toth, A. J. (2020). Modelling and optimisation of multi-stage flash distillation and reverse osmosis for desalination of saline process wastewater sources. *Membranes*, *10*(10), 1–18. https://doi.org/10.3390/membranes10100265.

Tsilingiris, P. T. (1995). The analysis and performance of large-scale stand-alone solar desalination plants. *Desalination*, *103*, 249–255.

Ullah, I., & Rasul, M. G. (2019). Recent developments in solar thermal desalination technologies: A review. *Energies*, *12*(1). https://doi.org/10.3390/en12010119

Vassilis, B., Soteris, K., & Emmy, D. (2016). *Thermal Solar Desalination Methods and Systems*. London, UK: Academic Press, 1–19. https://doi.org/10.1016/c2015-0-05735-5.

Venkatesan, G., Iniyan, S., & Jalihal, P. (2014). A theoretical and experimental study of a small-scale barometric sealed flash evaporative desalination system using low grade thermal energy. *Applied Thermal Engineering*, *73*(1), 629–640. https://doi.org/10.1016/j.applthermaleng.2014.07.059.

Venkatesan, G., Iniyan, S., & Jalihal, P. (2015). A desalination method utilising low-grade waste heat energy. *Desalination and Water Treatment*, *3994*(September). https://doi.org/10.1080/19443994.2014.960459.

Vorrath, S. (2016). World-first solar tower powered tomato farm opens in Port Augusta. Retrieved from https://reneweconomy.com.au/world-first-solar-tower-powered-tomato-farm-opens-port-augusta-41643/.

Wakil, M., Choon, K., Thu, K., & Baran, B. (2014). Multi effect desalination and adsorption desalination (MEDAD): A hybrid desalination method. *Applied Thermal Engineering*, *72*, 289–297. https://doi.org/10.1016/j.applthermaleng.2014.03.064.

Wang, X., Christ, A., Regenauer-Lieb, K., Hooman, K., & Chua, H. T. (2011). Low grade heat driven multi-effect distillation technology. *International Journal of Heat and Mass Transfer*, *54*(25–26), 5497–5503. https://doi.org/10.1016/j.ijheatmasstransfer.2011.07.041.

Zheng, H. (2017). Solar desalination system combined with conventional technologies. In *Solar Energy Desalination Technology*. Beijing, China: Elsevier, 537–622. https://doi.org/10.1016/b978-0-12-805411-6.00007-5.

8 Solar Thermal Technologies for Process Heat Applications in Oil and Gas Industries

N.V.V. Krishna Chaitanya, Ram Kumar Pal, and K. Ravi Kumar

CONTENTS

DOI: 10.1201/9781003263326-10

8.1 INTRODUCTION

The requirement of the oil and gas industries is increasing due to the continuous increment in the global energy demand. The increase in oil and gas production increases the energy requirement in its various production stages. A depiction of projected oil requirements in various regions of the world to meet energy demand for the year 2030 is shown in Figure 8.1 [1]. In the United States (US) alone, the energy consumption by petroleum and other liquid fuels is estimated to be up to 39 quadrillion British thermal units by the year 2050 [2]. An outline of year-wise energy consumption by petroleum and other liquid fuels in the US is shown in Figure 8.2. Around 10% of the total oil produced was consumed in the production and processing of oil, primarily consumed for heat generation.

The use of conventional methods such as fossil fuels to produce energy has implications on economics and the carbon footprint that has a serious role in the environment. Further, the process complexities and the energy required to extract the oil increase with the depth of the oil well. The variation in the drilling depth to extract oil in the Gulf of Mexico is shown in Figure 8.3. Renewable energy technologies could be employed in oil and gas industries to generate the required energy for oil production and processing

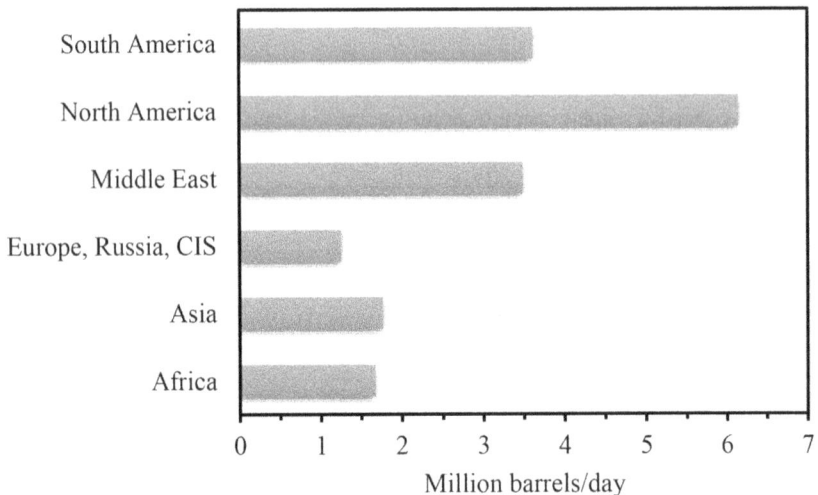

FIGURE 8.1   Projected crude oil requirement in the various regions of the world by the year 2030 (1 barrel petroleum = 0.15 m^3) [1].

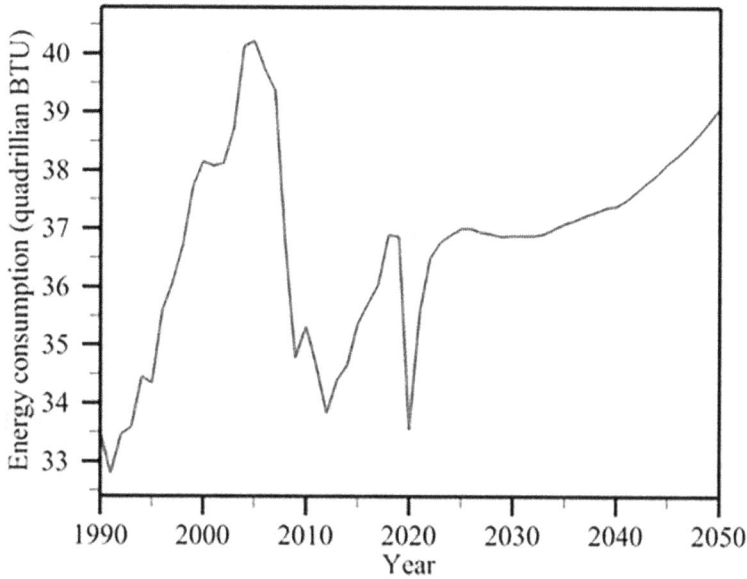

FIGURE 8.2 Energy consumption by petroleum and other liquid fuels in the US [2].

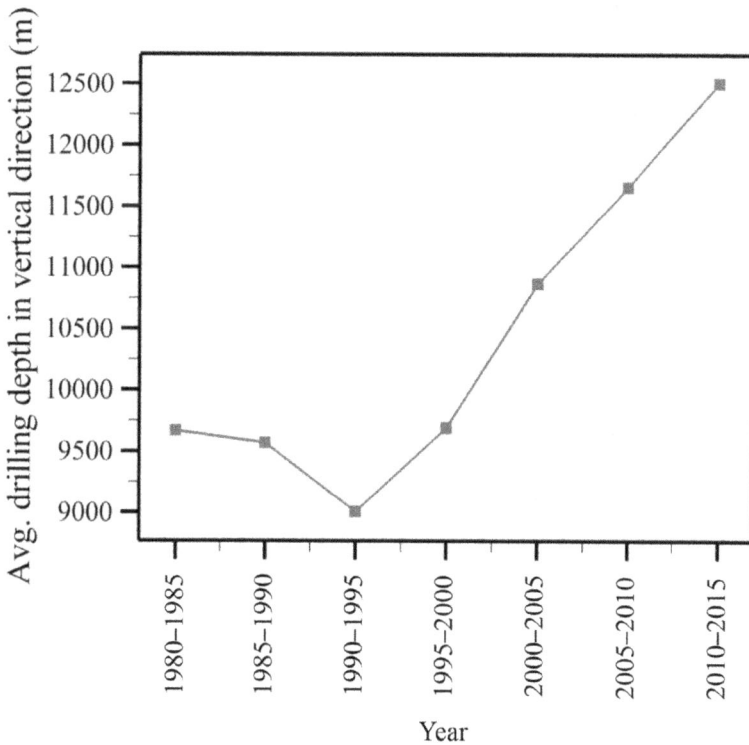

FIGURE 8.3 Drilling depth in the Gulf of Mexico [4].

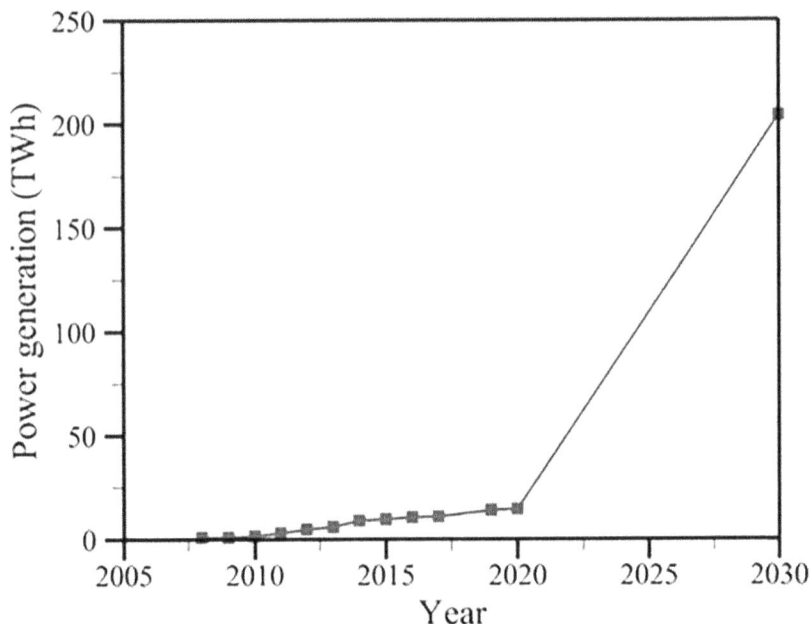

FIGURE 8.4   Past and projected CSP power generation [5].

in a sustainable way. Renewable sources such as solar, wind, biomass, hydropower, etc., can generate clean energy. Among them, solar thermal energy technologies are developing faster in energy generation and utilization, particularly in countries like India, with an installed capacity of more than 48 GW as of mid-2021 [3].

In addition, the CSP generation around the world increased to 14.5 TWh in 2020, with a predicted growth of 204 TWh by the year 2030 (as shown in Figure 8.4) and net zero-emission by the year 2050 [5]. However, to achieve net zero-emission, much more effort is required in research and development, energy storage, cost reduction, and important industry support in converting to sustainable generation from conventional ways.

The oil and gas industry chain is mainly classified into upstream, midstream, and downstream industries, as shown in Figure 8.5. In the first segment, the upstream industry comprises of identification of an oil field (offshore/onshore), exploration, drilling of the oil field, production, and extraction. This phase plays a vital role in finding the oil field resources and extracting the oil using various techniques such as primary, secondary, and tertiary. The enhanced oil recovery is one of the tertiary processes that require temperatures up to 300 °C. The infrastructures include the level of the wellhead, well platform, production platform, and crude oil terminal [6]. The upstream industry is also known as exploration and production (E&P) due to its activities related to oil searching, extraction, and production.

The midstream involves activities that are related to transportation and storage. In most cases, the oil fields are located far away from refineries and at various geographical

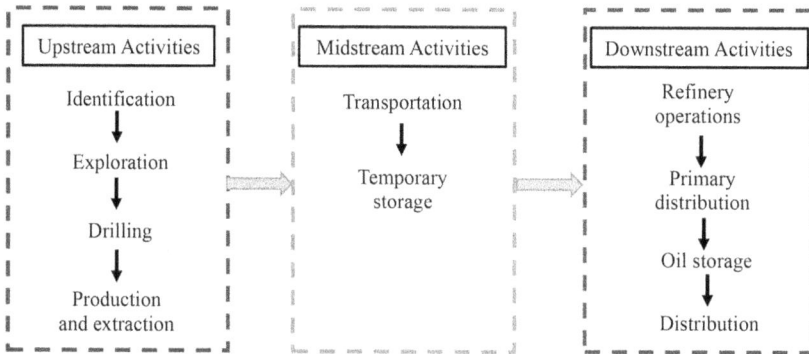

FIGURE 8.5   Various activities involved in the oil and gas industry supply chain.

locations. Hence, it is important to safely transport the oil to the refineries and store the oil temporarily for transportation. Activities such as refining, storing, and distribution occur in the third phase, known as the downstream industry. The oil refinery includes various processes such as stripping (160–180 °C), fractionation (230–260 °C), quenching (135–150 °C), mechanical drive (300–350 °C), and power generation (~500 °C) that requires various temperature ranges depending on the applications. More details on the upstream and downstream processes are discussed in Section 8.4.

Integrating solar thermal technology with process heat applications in oil and gas industries is considered feasible in terms of clean energy generation compared with conventional methods. Studies on the application of solar energy in oil and gas industries started in the 70s by the world's leading oil and gas companies such as Exxon, Chevron, Shell, and BP. However, their progress in the utilization of solar energy was slow. Recently, the international restrictions on greenhouse gas emissions [7] motivated the oil and gas industries to study the scope of renewable energy. The oil and gas companies have been competing to move to sustainable heat production [8]. The oil field regions generally have high solar insolation and utilization of solar energy to generate heat and save fossil resources. The replacement of fossil fuel-based steam generation systems with solar steam generation is feasible and cost-effective, especially for regions with high solar radiation. Solar energy can be converted to heat or electricity depending on the application and the type of technology used. For example, solar photovoltaic (PV) can convert solar radiation to electricity. Solar thermal energy technologies can generate heat that can be used for process heat applications or generate electricity by using steam/gas turbines. In many applications, heat is required to meet the energy demand of the various processes in the oil and gas industries. Hence, solar thermal technologies are considered suitable technology compared with solar PV. Solar thermal energy technologies for process heating have tremendous opportunities in oil and gas industries for their upstream and downstream processes. Solar thermal energy technologies have the potential to ensure the process energy demand, protect the environment, and promote the industry's economic development [9, 10].

Over the past decades, many researchers have investigated the use of solar thermal energy for process heating in oil and gas industries [2]. In the oil extraction process (thermal enhanced oil recovery), the temperature requirement is in the range of 250–300 °C [11]. Solar thermal energy technologies can provide the required amount of energy. In addition, integrating solar thermal energy systems with oil and gas industries save the CO_2 release up to 24.6 kton/yr depending on the oil field characteristics [12]. Furthermore, solar thermal technologies can also be used for heat requirements in the hydrogen generation process in oil and gas industries to replace conventional methods [13].

The solar heating system can also be used in piping transport to avoid oil freezing [14]. Since the raw fuel is highly viscous under ambient conditions, the fuel is typically heated up to 50 °C. It is relatively expensive to produce the required heat with conventional fuels. Therefore, using renewable energy resources, particularly solar energy, is considered a sustainable way to generate energy for piping transport. In recent years, the utilization of solar energy for heating the oil in oil transportation has been gaining attention and has also been commercialized [15–17]. It is also estimated that with the use of solar heating systems, around 30% of the daily conventional fuel consumption can be saved [14, 18]. An outline of the solar heating system in piping application is shown in Figure 8.6. Initially, the heat is transferred from the solar collectors to the heat transfer fluid. Later, the heat is transferred from the heat transfer fluid to the oil through a heat exchanger. The oil is then transported to the water mantle oven, where an automatic ignition system maintains the required oil temperature. The solar heat can be used during the day, and backup storage can be used at night and when there are fluctuations in solar radiation.

This chapter discusses important aspects of upstream and downstream oil and gas industries and their integration with solar thermal energy technologies. In addition, suitable concentrated solar thermal (CST) energy technologies are also discussed in detail, along with the integration with the processes and challenges involved in the utilization of solar energy in the oil and gas industries.

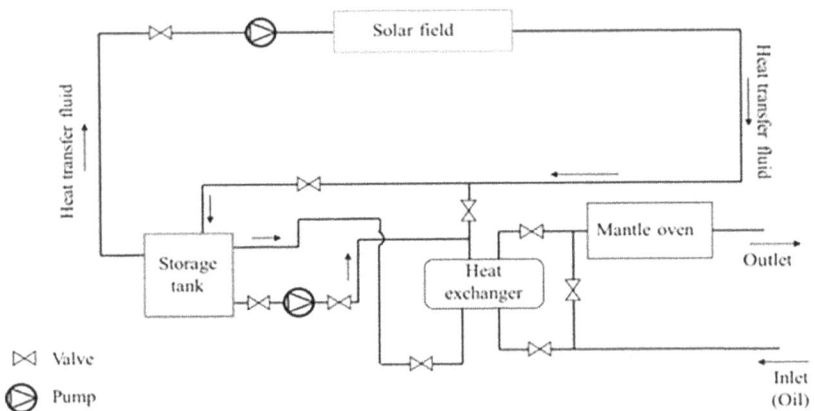

FIGURE 8.6 A schematic of a typical solar industrial process heating system [14].

8.2 RELEVANCE OF HARNESSING SOLAR THERMAL ENERGY FOR OIL AND GAS INDUSTRIES

As discussed earlier, the heat requirement in various processes of oil and gas industries is largely fulfilled by fossil fuels and has a substantial footprint in greenhouse gas emissions. Therefore, it is necessary to minimize fossil fuels consumption while meeting the increasing energy demand. There are two ways to minimize fossil fuels consumption: (i) improving the efficiency of various processes utilizing fuel and electricity and (ii) substituting the fossil fuels by harnessing renewable energy sources. Solar energy is environmentally clean and generally accepted as a sustainable source of energy. Solar thermal technologies are promising alternatives to harness solar energy and have been successfully implemented for domestic and industrial heating applications as well as power generation [19]. The solar to thermal conversion efficiency of CST is much higher (up to 70%); solar to electric conversion for solar thermal power generation systems is 17–25%, while for photovoltaics it is 15–20% [20]. As discussed in the introduction section, the heat required for the various processes of the oil and gas industries exists in the temperature range of 150–500 °C and can be successfully provided using existing solar thermal energy technologies.

Solar thermal energy technologies are mature, technically proven, and commercially available. Many successful small- to large-scale commercial solar power plants are running across the world [21]. Further, these technologies have been successfully implemented to fulfill the heat demand for various process heating applications in dairy, food, chemical, paper and pulp, plastics, and bricks industries [20]. There is a strong possibility of hybridizing the existing conventional fossil fuel-based energy systems with CST systems. The integration of thermal energy storage systems improves energy dispatchability.

Following are the advantages of integration of CST technologies with oil and gas industries:

Lower operating and maintenance costs: The CST systems do not need any fossil fuels that substantially reduce operating and maintenance costs [22].

Long life span: The CST systems operate in a moderate temperature range (150–500 °C) for process heating applications. Hence, the CST components have a longer life span (~25 years).

Low emissions to the environment during the system's life cycle: Based on the life cycle assessment of CST systems, it is understood that emission to the environment during the life cycle (from the manufacturing of components to their final deposition) of CST systems is lower than that from the technologies using fossil fuels [23]. However, there are some disadvantages of CST systems which are discussed in Section 8.7.

8.3 SELECTION OF SOLAR THERMAL TECHNOLOGIES FOR PROCESS HEATING IN OIL AND GAS INDUSTRIES

There are various technologies to convert solar energy into useful energy in the form of heat (solar thermal energy technologies) or electrical energy (photovoltaics). The

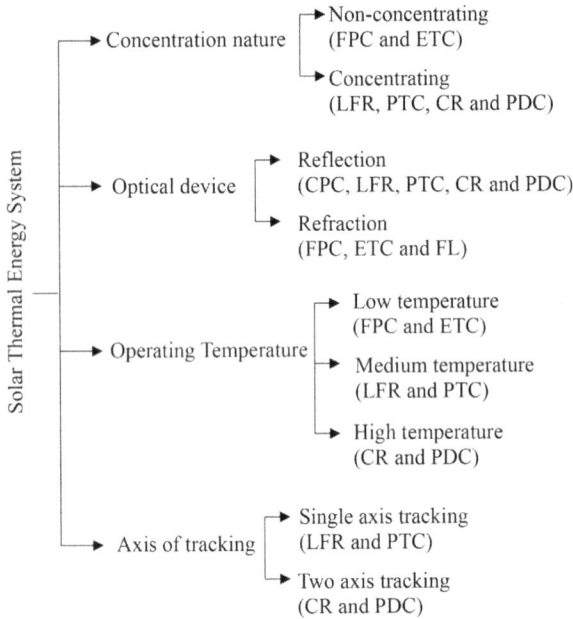

FIGURE 8.7   Classification of solar energy technologies [24].

solar thermal energy technologies are most suitable for process heating applications. There are various available solar thermal technologies that can be classified based on various parameters such as operating temperature range, tracking requirement, concentration achieved on the focal plane, and optical principle. The classifications of solar thermal technologies based on these parameters are presented in Figure 8.7. Based on the operating temperature range, these technologies can be classified as low-temperature (40–120 °C), medium-temperature (100–550 °C), and high-temperature collectors (250–1500 °C). Based on the tracking principle, these technologies can be classified as single-axis tracking or two-axis tracking collectors. Further, based on the concentration nature, the solar collectors can be classified as non-concentrating collectors and concentrating collectors. The non-concentrating collectors have an aperture area equivalent to the absorber area and could deliver low-temperature heat. The concentrating collectors collect direct normal irradiance (DNI) from a large area and further concentrate in a small area (called the receiver area). These collectors are suitable for medium- and high-temperature heat. The ratio of the opening area of the concentrator (called aperture area) and the receiver surface area is called the concentration ratio. The concentrated solar thermal technologies have a concentration ratio of more than one. Further, the solar collectors could be classified as reflection and refraction systems based on the optical principle. The solar collectors using mirrors are reflection-based systems, and collectors using transparent glass or lenses are refraction-based systems.

Solar thermal technologies can generate heat in the temperature range of 50–1200 °C. The non-concentrated solar thermal technologies are mainly suitable for low-temperature (< 120 °C) applications, whereas the concentrated solar thermal technologies are suitable for medium- to high-temperature applications. As per the temperature requirements in the various process of oil and gas industries, the CST technologies are suitable. There are four types of commercially available concentrated solar thermal technologies: parabolic trough collector (PTC), linear Fresnel reflector (LFR), central receiver (CR), and parabolic dish collector (PDC).

8.3.1 PARABOLIC TROUGH COLLECTOR

Parabolic trough collector (PTC) technology is suitable to generate heat in the range of 150–400 °C. The PTC consists of a parabolic-shaped mirror (shown in Figure 8.8) which reflects the DNI on the linear receiver placed along the focal line of the parabola. The parabolic-shaped mirror is called a concentrator. The receiver consists of a metallic tube called an absorber and is enclosed by a glass cover. The heat transfer fluid (HTF) flows through the absorber and is heated by concentrated radiation [25]. The PTCs can deliver temperatures in the range of 150–400 °C, which is the temperature range where many industrial processes are carried out.

The concentrated solar flux on the receiver can be calculated by the following equation [24]:

$$S = I_b \rho_{conc} \gamma \left(\tau \alpha \right)_{rec} K \left(\theta \right) \tag{1}$$

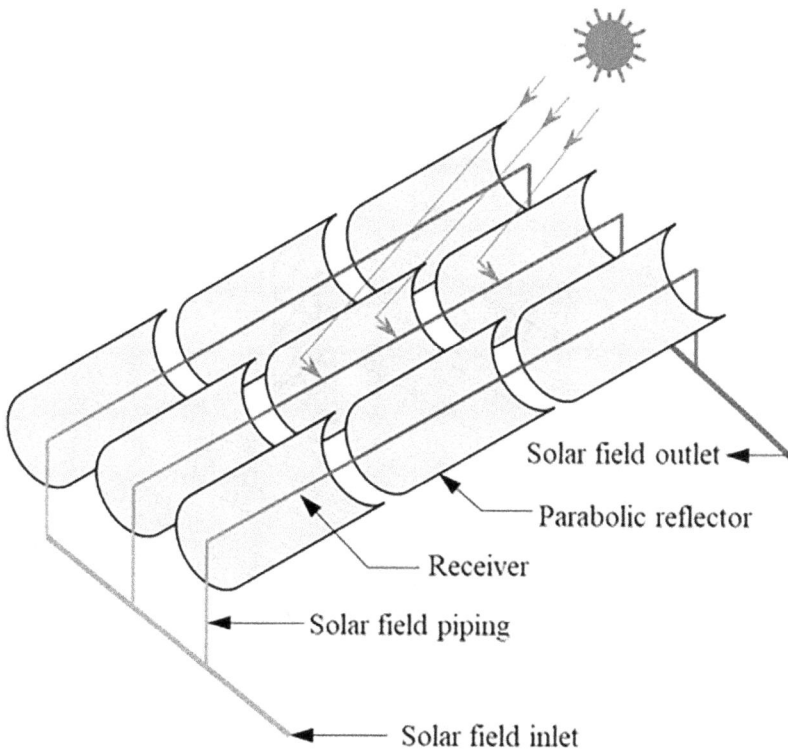

FIGURE 8.8 Schematic of parabolic trough collector system.

Here, I_b, ρ_{conc}, $(\tau\alpha)_{rec}$, $K(\theta)$ are DNI, the reflectivity of the concentrator, transmissivity-absorptivity product of receiver, and incidence angle modifier, respectively.

The useful heat collected from the PTC can be calculated by the following formula [24]:

$$Q_u = \dot{m}C_p\left(T_{fo} - T_{fi}\right) = F_R\left(W - D_o\right)L_{coll}\left[S - \frac{U_L}{CR}\left(T_{fi} - T_{amb}\right)\right] \qquad (2)$$

Here, T_{fo} and T_{fi} are fluid inlet and outlet temperature, respectively, and T_{amb} is the ambient temperature. The terms F_R, W, and D_o are heat removal factor, aperture width, and absorber's outer diameter, respectively.

The efficiency of the PTC can be expressed as [26]:

$$\eta = \frac{Q_u}{SM.I_b.L_{coll}.W} \qquad (3)$$

Here, SM and L_{coll} are solar multiple and collector length, respectively.

The PTC rows can be arranged in a north-south or east-west direction [27]. The HTFs used in PTC are thermal oils, molten salts, or pressurized water.

8.3.2 LINEAR FRESNEL REFLECTOR

Linear Fresnel reflector (LFR) technology is based on the line focus technique similar to PTC technology. LFR consists of concentrators placed close to the ground and a fixed

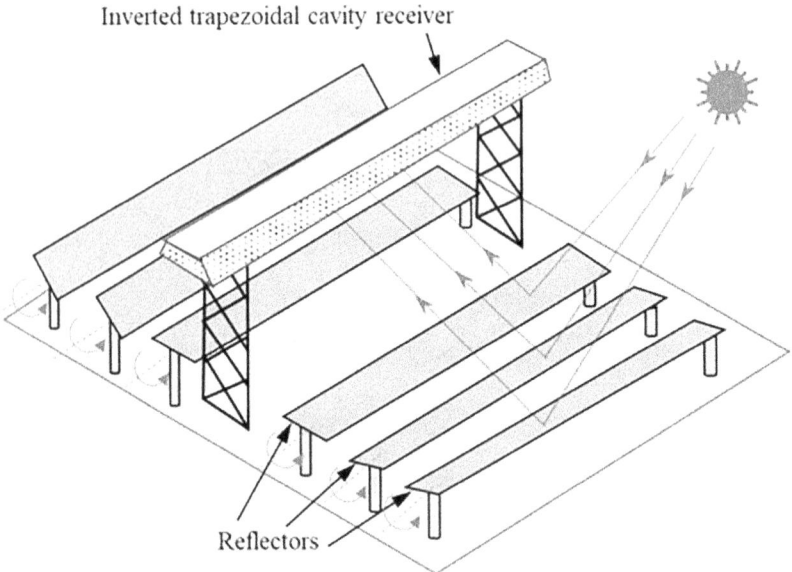

FIGURE 8.9   Linear Fresnel reflector solar collector system.

inverted trapezoidal linear receiver placed on the top (shown in Figure 8.9). The individual concentrator is made of 10 flat ground-mounted mirrors, and each rotates to track the sun's apparent motion and reflect the DNI onto the linear receivers. The receiver consists of a selective coated absorber tube through which HTF flows and is heated by concentrated solar radiation. The advantage of the fixed receiver is it does not require high-pressure flexible piping systems, as in the case of PTC. Water is the most commonly used HTF in the LFR systems. However, there is more flexibility in selecting HTF because the fixed receiver and other HTFs such as therminol oil and molten salts can easily be employed.

The thermal energy collected by the absorber tube of LFR can be expressed as follows [28]:

$$Q_{abs} = 0.7 \cdot \alpha \cdot \rho \cdot \gamma \cdot A_c \cdot I_b \sqrt{1 - \cos^2(\delta)\sin^2(\psi)} \tag{4}$$

Here, α, ρ, γ, A_c, δ, and ψ are the absorber's absorptivity, mirror reflectance, intercept factor, effective reflector surface area, declination angle, and solar altitude angle, respectively.

The overall heat loss coefficient (U_L) between the absorber and ambient air could be calculated using the following formula [28]:

$$U_L = \varepsilon_a \sigma \left(T_a^2 + T_{amb}^2\right)\left(T_a + T_{amb}\right) \tag{5}$$

Here, T_a and T_{amb} are the temperatures of the absorber's surface and ambient, respectively.

The thermal efficiency of the LFR can be expressed as [24]:

$$\eta = \eta_{opt} - \frac{U_L A_a \left(T_a - T_{amb}\right)}{I_b \times A_{a,coll}} \tag{6}$$

Here, η_{opt} is the optical efficiency of LFR. U_L, $A_{a,out}$, $A_{a,coll}$ are overall heat loss coefficient, absorber's external surface area, and collector aperture area, respectively.

The capital cost of LFR is lower than PTC as it requires light structural support and a fixed receiver without any flexible and movable joints [27]. However, the optical efficiency is lower as compared to PTC. The advanced Compact Linear Fresnel Reflector (CLFR) is more promising in terms of cost-saving and land use. In the CLFR system, multiple receivers are employed to minimize energy loss and make the solar field compact. The consecutive mirrors of the CLFR solar field redirect the incoming solar radiation on two different receivers installed on each side of the solar field.

8.3.3 CENTRAL RECEIVER

The central receiver (CR) is based on the point focus technique and requires two-axis tracking systems. The concentrator (also called heliostat) is made of a number of flat (or slightly curved) mirrors and equipped with a two-axis tracking system to track

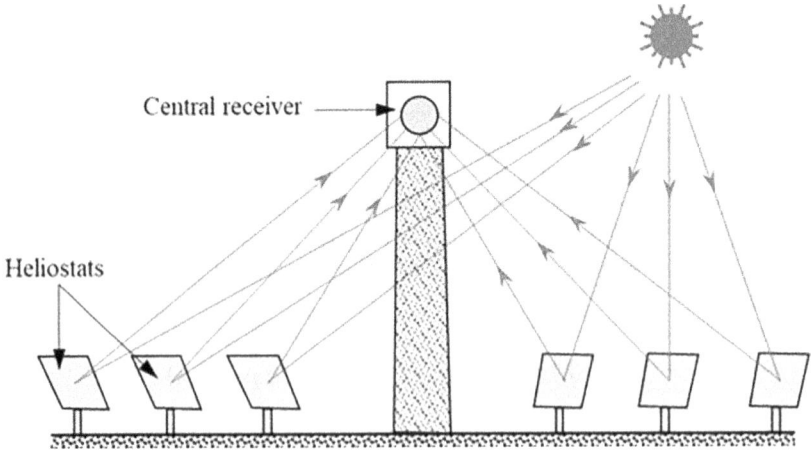

FIGURE 8.10   Schematic of a central receiver system.

the apparent motion of the sun (shown in Figure 8.10). The heliostats concentrate the DNI on the receiver placed at the top of the tower.

The central receiver system employs two types of receivers: (1) surface receiver and (2) cavity receiver. The arrangement of heliostats and receivers is designed to minimize the cost of the solar field and the energy loss. In the case of the surface receiver, the cylindrical surface receiver is mounted on the tower, and heliostats are placed around the tower. In the case of the cavity receiver, the receiver is placed on the tower with a heat transfer surface facing to the north, and the heliostats are placed on the northern side of the tower (for the northern hemisphere).

The solar to the thermal conversion efficiency of the CR solar collector system can be defined as [24]:

$$\eta = \eta_{opt} \times \eta_{rec} \tag{7}$$

Here, η_{opt} and η_{rec} represent optical and receiver efficiency, respectively. The optical efficiency can be expressed as follows [29]:

$$\eta_{opt} = \eta_{cos} \times \eta_{att} \times \eta_{int} \times \eta_{s\&b} \times \eta_{ref} \times \eta_{abs} \tag{8}$$

Here, η_{cos}, η_{att}, η_{int}, $\eta_{s\&b}$, η_{ref}, and η_{abs} represent cosine efficiency, atmospheric attenuation efficiency, interception efficiency, shading and blocking efficiency, and absorption efficiency, respectively.

The receiver efficiency can be calculated as [24]:

$$\eta_{rec} = \frac{(\alpha Q_{inc} - Q_{loss})}{Q_{inc}} = \alpha\alpha - \frac{Q_{loss}}{Q_{inc}} \tag{9}$$

Here, Q_{inc} and Q_{loss} are incident concentrated heat on the receiver and heat losses from the receiver.

8.3.4 PARABOLIC DISH COLLECTOR

The parabolic dish collector consists of a parabolic dish-shaped concentrator that concentrates DNI on the focal point where the receiver is placed (shown in Figure 8.11). The concentrator is equipped with a two-axis tracking system to track the sun's apparent motion to maintain the aperture normal to the beam radiation. The concentration ratio of a parabolic dish collector (PDC) system can be up to 10,000, and it could generate heat up to the temperature of 1,200 °C. The PDC system has the highest solar to thermal conversion efficiency as compared to the other solar thermal technologies.

The solar energy collected by the concentrator can be calculated as follows [30]:

$$Q_s = A_c \times I_b \tag{10}$$

Here, A_c is the effective aperture area of the dish. The useful heat can be calculated as [24]:

$$Q_u = \dot{m} \times C_p \times \left(T_{fo} - T_{fi}\right) \tag{11}$$

The thermal efficiency can be given as follows:

$$\eta_{th} = \frac{Q_u}{Q_s} \tag{12}$$

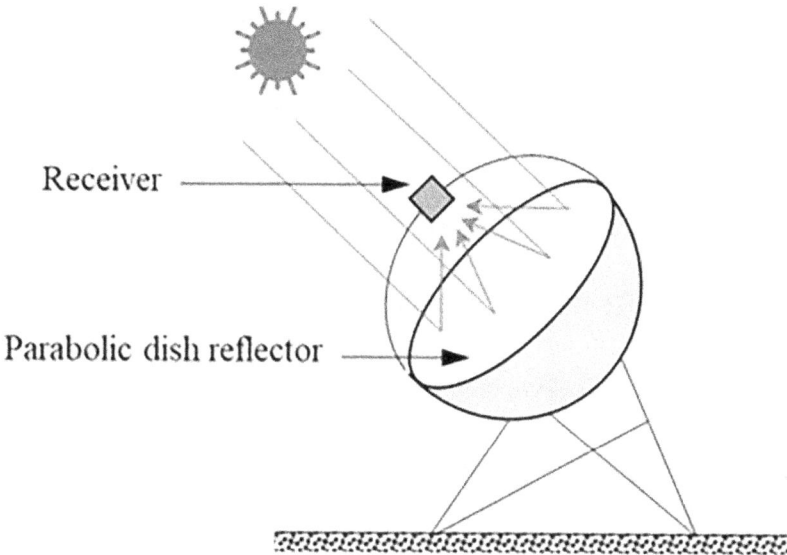

FIGURE 8.11   Schematic of a parabolic dish collector system.

TABLE 8.1
Comparison of CST Technologies [24, 31]

Parameters	PTC	LFR	CR	PDC
Operating principle	Concentrates DNI on linear tube receiver placed along the focal line	Concentrates DNI on the elevated inverted linear fixed receiver	Concentrates DNI on receiver placed at the top of a tower	Concentrates DNI on receiver placed at the focus point
Tracking system	Single-axis	Single-axis	Two-axis	Two-axis
Maturity	Commercially proven	Recent commercial projects	Commercially proven	Demonstration projects
Operating temperature (°C)	350–550	250–350	250–1200	250–1500
Concentration ratio	10–80	>60	>1000	Up to 10000
Receiver/Absorber	Movable receiver attached with collector; complex design	Fixed receiver, no evacuation, secondary reflector	Fixed, external surface or cavity receiver	Movable receiver attached with collector
Storage system	Direct or Indirect storage	Short-term pressurized steam storage	Direct two-tank storage	No proven storage
Storage with molten salt	Commercially available	Possible, but not proven	Commercially available	Possible, but not proven
Land occupancy	Large	Medium	Medium	Small
Cost ($/kW)	4700–7300	5000	6400–10700	12578
Cost ($/m²)	270	–	–	300–600

TABLE 8.2
Thermal Efficiency of CST Technologies [20]

CST Technology	Efficiency (%)				
	100 °C	200 °C	300 °C	400 °C	500 °C
Parabolic trough collector	79.14	77.24	73.58	66.73	54.68
Linear Fresnel reflector	65.94	63.85	60.12	53.48	42.14
Central receiver	69.11	69.11	65.96	62.34	58.10
Parabolic dish collector	83.38	79.88	76.00	71.56	66.35

The summary of CST technologies is presented in Table 8.1. Further, the thermal efficiency of the CST technologies in various operating temperature ranges is presented in Table 8.2. The selection of an appropriate solar thermal technology depends on mainly five factors: (i) operating temperature, (ii) solar collector's efficiency, (iii) annual energy yield, (iv) cost, and (v) available area.

8.4 INTEGRATION OF SOLAR THERMAL TECHNOLOGIES WITH PROCESSES OF OIL AND GAS INDUSTRIES

8.4.1 UPSTREAM OIL AND GAS INDUSTRY

The upstream oil and gas industries require considerable heat for oil extraction. An outline of the upstream and downstream flowchart is shown in Figure 8.12. The term upstream is used for operations that involve exploration and extraction. Hence, it is also called E&P (exploration and production). A series of processes are involved in oil production, such as (i) primary recovery, (ii) secondary recovery, and (iii) tertiary recovery. In the primary recovery stage, the pressure is sufficient for the extraction of oil and gas. In addition, methods such as artificial lift use electric submersible pumps for oil extraction to enhance the flow rate. It is estimated that around 4–15% of the oil reserves could be extracted in the primary recovery stage [32]. However, the precise percentage of oil extraction depends on the geological characteristics of the oil well.

There is a need for external pressure in the secondary recovery stage due to a fall in the well pressure. Sources such as fluids or gas (e.g., water, hydrocarbons, and carbon dioxide) are injected to increase the well pressure for advancing the oil to the surface. Around 11–20% of the reserves can be extracted through the secondary recovery stage. In the tertiary recovery stage, there are a number of methods to stimulate production. This recovery process is also known as enhanced oil recovery (EOR). The EOR process is classified into three main categories, namely (i) gas injection, (ii) thermal injection, and (iii) chemical injection.

1. Gas injection

   The gas injection, which is also called miscible flooding, is an injecting technique that uses miscible fluids as working fluid to inject into the oil reservoir, as shown in Figure 8.13. The miscible fluid maintains the pressure in the reservoir and improves the oil flow rate. Some of the most commonly used gases include carbon dioxide (CO_2) and nitrogen which reduce oil viscosity and are considered less expensive than LPG (liquefied petroleum

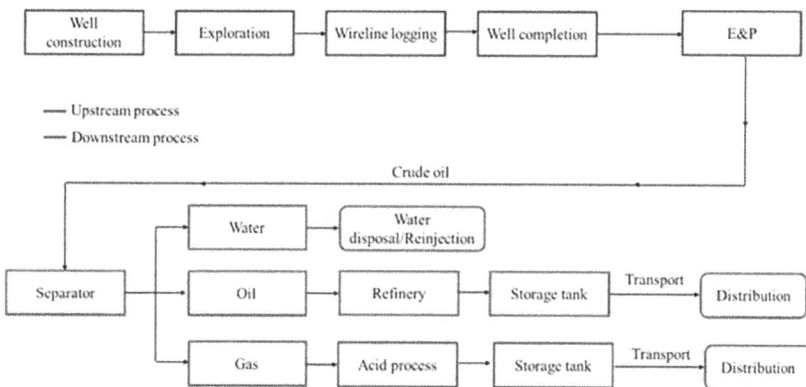

FIGURE 8.12   An outline of the upstream and downstream processes in oil and gas industries [33].

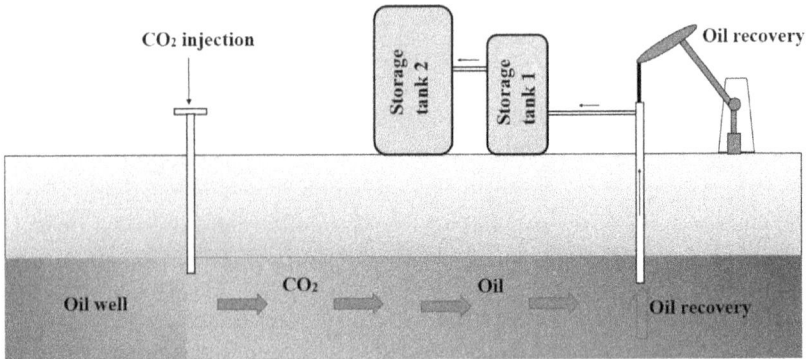

FIGURE 8.13   Illustration of CO_2 flooding in oil extraction.

gas). 60% of the enhanced oil recovery projects in the United States use 2 billion cubic feet of CO_2 gas to extract 280,000 barrels of oil per day with more than 114 ongoing plants [34]. 20% of the total initial reserves may be extracted in this stage. The type of recovery process depends on the characteristics of the oil field, including geological structure, fluid viscosity, reservoir pressure, gas solubility, and so on. [35]

2. Thermal injection

    The thermal injection technique, also known as thermal enhanced oil recovery (TEOR), involves the injection of heated fluid/gas such as steam that lowers the viscosity of heavy viscous oil. The optimum operating temperature is in the range of 200–300 °C [36]. Some TEOR methods include steam flooding, cyclic steam injection, and in-situ combustion. An outline of the steam flooding technique is shown in Figure 8.14. The steam that is injected is produced by burning fossil fuels.

The thermal EOR technique is considered to be commercially successful, with around 67% usage around the world, as shown in Figure 8.15.

3. Chemical injection

    The chemical injection technique uses chemicals such as alkaline or caustic solutions and involves polymers to increase the solubility of heavily depleted oil formations [39–41]. The chemical formulations can reduce the surface tension and ease the flow. However, each chemical mixture has different effects on the oil [42]. For example, the water-soluble dilute solution polymer can increase oil recovery. The use of alkaline solutions as working fluid creates fatty acids such as soap and increases the interfacial tension, further increasing the production. Chemically enhanced oil recovery (EOCR) has improved significantly with the developments of the new chemicals. However, the limitations of EOCR include high chemical costs, sensitivity, and incompatibility of polymers/chemicals to the reservoirs.

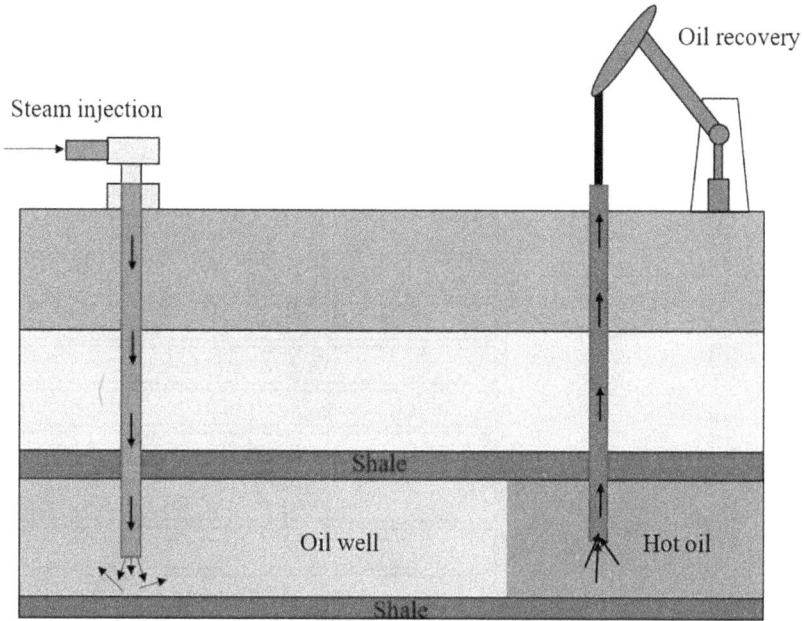

FIGURE 8.14   Schematic of steam flooding technique in the oil recovery process.

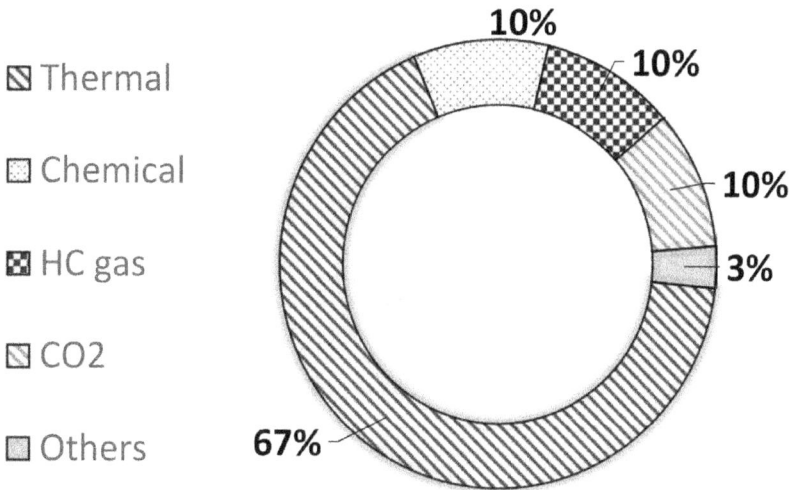

FIGURE 8.15   The proportion of EOR methods used in projects worldwide [37, 38].

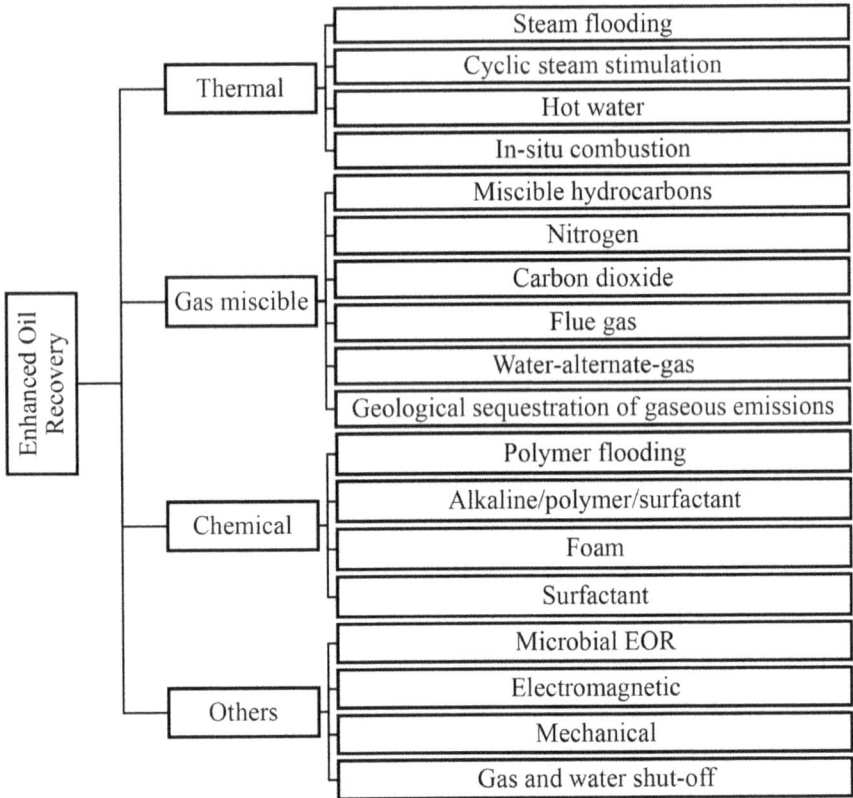

FIGURE 8.16   Classification of secondary and EOR processes.

This technique accounts for 1% of the total EOR production in the US [34]. Classifications in the secondary and enhanced oil recovery process are shown in Figure 8.16.

Among the EOR techniques, thermal EOR (steam flooding and cyclic steam injection) accounts for more than two-thirds of the projects in the United States. EOR process implementation is on a fast track in oil extraction and is expected to continue for the next two decades. The steam used in TEOR is produced by burning fossil fuels that increase carbon footprint and substantially affect the environment. Hence, using renewable energy such as solar to produce steam may reduce the effects on the environment and lower the costs for steam generation [43]. In a feasibility study, Sandler et al. [44] concluded that solar-generated steam could be considered as an alternative for conventional fossil fuels where sufficient solar insolation is available. In oil fields, the concentrated solar thermal energy technologies can replace the conventional boilers and achieve temperatures ranging from 240–300 °C with a 70–100 bar pressure.

8.4.1.1 Integration of Solar Energy Systems with Enhanced Oil Recovery System

Solar thermal technologies such as LFR or PTC can generate steam replacing conventional boilers that use fossil fuels. Depending on the quality of the oil field, it requires 2 to 4 barrels of steam to produce a barrel of oil [45]. Integrating solar thermal technology with existing TEOR systems may reduce the use of fossil fuels by up to 80%. The solar thermal systems could be integrated with the existing TEOR projects in multiple ways, and modifications to the facility could be required according to the applications. Solar energy can be utilized in the following manner to minimize conventional fuel consumption in the oil field.

1. Feedwater preheat: The feed water could be preheated during the day and stored in the insulated tanks and can be used at night or when there are fluctuations in solar energy. This technique requires fewer changes to the existing facility and can reduce fossil oil consumption by 10%.
2. Continuous steaming: The steam injection into the oil wells should be carried out on a continuous basis. Around 25% of the oil consumption can be reduced by directly generating steam using CST during the daytime and using the conventional methods during the night and disruptions due to cloudy days.
3. Variable-rate steaming: Petroleum Development Oman (PDO) and Shell conducted a study on injecting more steam during the day and less steam during the night. It was concluded that the rate of oil extracted with the use of continuous steaming is the same as variable-rate steaming.

Figure 8.17 shows the cumulative production of oil for four scenarios of continuous and cyclic steam injection (STB—stock tank barrel per day) for a period of 10 years. The scenarios include [46]: (i) continuous fossil fuel-based steam generation, (ii) cyclic 100% solar-based steam generation, (iii) cyclic 50% solar and 50% fossil-fuel-based steam generation, (iv) cyclic 100% fossil-fuel-based steam generation. In a continuous stream generation, the steam is injected continuously, whereas, in cyclic steam generation, the steam is injected in the defined interval. It is observed that when continuous fossil fuel and cyclic 100% scenarios are compared, the difference is around 15000 STB which is negligible in view of total oil production for a period of 10 years. It is also important to mention that in cyclic 100% solar steam, the steam injection rate is around 1000 STB/day and 30 STB/day during daytime and night time respectively.

Solar-assisted EOR with 100% solar fraction has good economic impacts, such as low operational costs, and is less sensitive to cost fluctuations [47]. At the same time, there may be cost fluctuations in the case of conventional fuel systems. A patent on a hybrid system that uses solar energy for steam during the daytime and a conventional method at night (oxy-fuel combustion) was granted by Mokheimer and Habib in the year 2017 [38, 48].

8.4.2 DOWNSTREAM OIL AND GAS INDUSTRY

The downstream process in the oil and gas industries includes refining crude oils extracted in the upstream process. Heat is used to separate the impurities by breaking

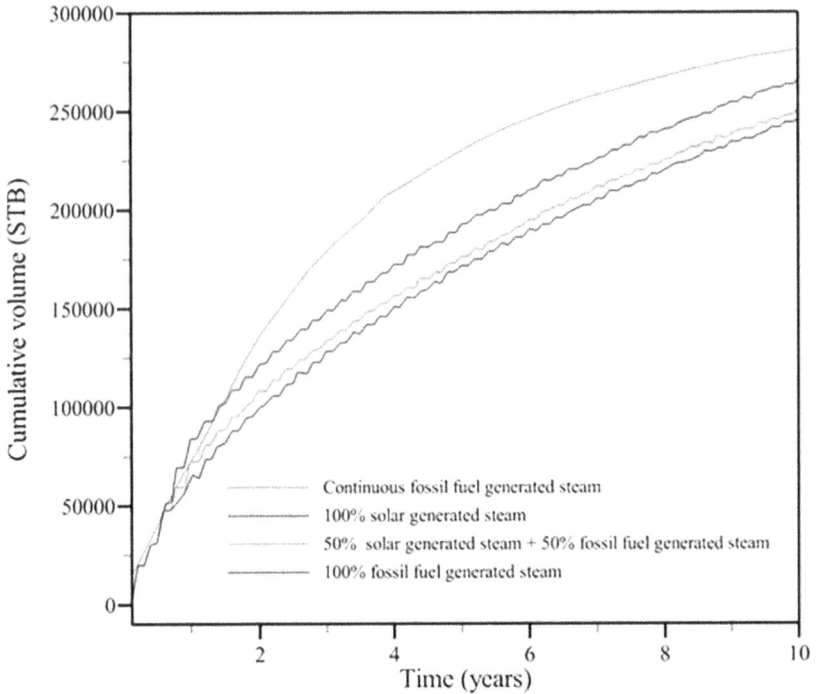

FIGURE 8.17 Comparison of cumulative oil production at various cyclic steam injection scenarios for a period of 10 years (STB: stock tank barrel per day) [46].

carbon bonds. Products such as kerosene, aviation fuel, gasoline, petrol, diesel, etc., are extracted from crude oil. The temperature and pressure play an important role in the refining process to produce the required outcome. Close to 90% of the energy is used for oil refining for direct heating and steam generation [49]. Steam plays an important role in around 11 refining processes that include (i) isopentane/iso-hexane, (ii) isobutane, (iii) alkylation, (iv) catalytic reforming, (v) catalytic hydrocracking, (vi) catalytic hydrotreating, (vii) coking operation, (viii) visbreaking, (ix) vacuum distillation, (x) atmospheric distillation, and (xi) fluid catalytic cracking. It is estimated that running these operations requires 900 trillion BTUs of energy and is generated using conventional energy systems in the existing industries [50]. The energy consumption in various processes is presented in Table 8.3.

The steam is used for processes such as (i) stripping, (ii) fractionation, (iii) mechanical drive, (iv) quenching, (v) process heating, and (vi) captive power generation. The oil and gas downstream industry consumes more fuel due to energy requirements at high temperatures up to 550 °C depending on the crude type and distillate demand. The US petroleum industry spends $9 billion to produce energy for the refining requirements [4, 51]. This increases the operational costs and has much impact on the environment due to the use of fuel generators [2, 52]. Hence, it is important to incorporate renewable sources for its heating operations in petroleum refinery industries.

TABLE 8.3
Energy Consumption in the Downstream Refinery Industry [2]

Refinery Process	Energy Requirement (MJ/barrel)	Temperature Requirement (°C)
Atmospheric crude distillation	86–196	340–350
Vacuum distillation	53–119	370–380
Fluid catalytic cracking	168–339	480–550
Delayed coking	120–242	485–505
Catalytic hydrotreating	64–173	250–320
Hydrogen production	66–167	200–300

8.4.2.1 Integration of Solar Energy Systems with Downstream Oil and Gas Industries

An outline of the solar thermal energy technology integrated with the refinery processes is shown in Figure 8.18. It is important to mention that CST technologies are more suitable to integrate with the process due to their capability to generate the heat at required temperatures ranging from 135 °C to 290 °C and 500 °C for captive power generation. Various processes require different heat treatments in the form of direct heat or through steam (e.g., a turbine). In such cases, the steam can be extracted at any location of the solar field. Thermal energy storage may generate steam continuously and reduce the fluctuations and supply at a constant rate.

Hydrogen plays a vital role in removing the nitrogen and sulphur impurities in the diesel through a process known as hydrotreatment. Due to its high demand in the refining process, the hydrogen requirement increased by 50% between 2008 and 2014 [4]. Hydrogen is produced through the steam-reforming process of natural gas

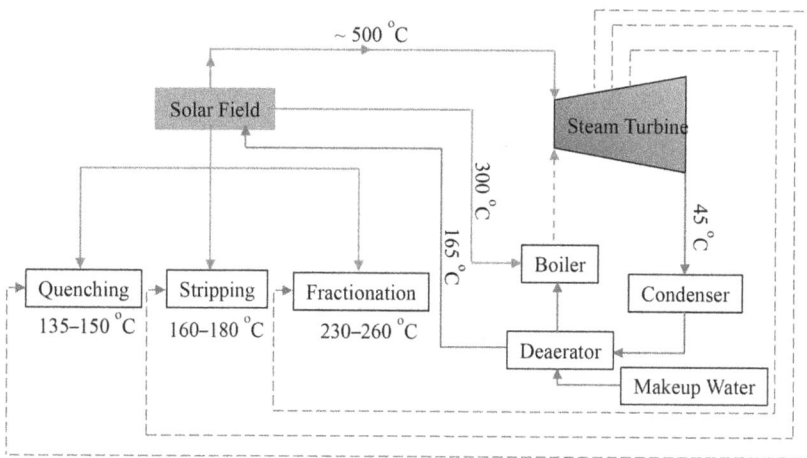

FIGURE 8.18 Integration of solar thermal technology with the downstream process in oil and gas industries [24].

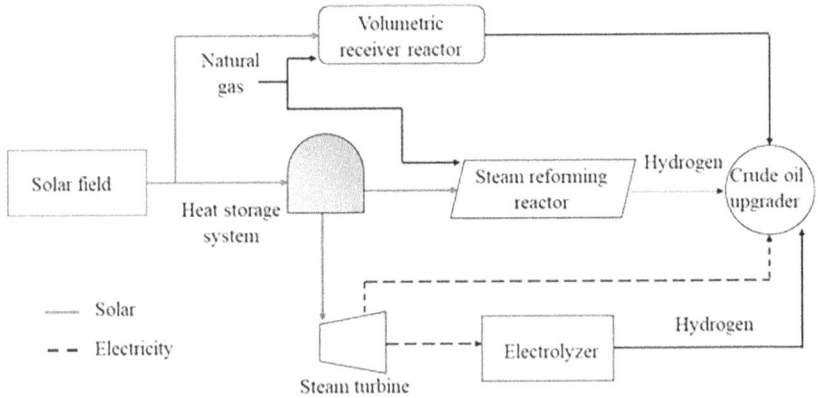

FIGURE 8.19   Integrating the solar thermal system with the hydrogen production process [53].

[53]. This process requires a large amount of heat which can be supplied using concentrated solar technology. Likkasit et al. [53] investigated the process of hydrogen production using solar energy, as shown in Figure 8.19.

8.5 CURRENT STATUS OF SOLAR PROCESS HEATING IN OIL AND GAS INDUSTRIES

In the present scenario, the solar integrated process heating applications are gaining attention, particularly in the commercial sector, due to their reliability and low maintenance costs when compared to the conventional methods. Several projects in the oil and gas industries have integrated solar thermal technology to generate energy as per their requirement. The following are some of the existing oil fields that have already integrated/in-process solar energy for TEOR.

8.5.1 McKittrick, California

This solar-based EOR project was commissioned in 2011 and is one of the existing EOR oil fields in McKittrick, California. The pilot project partnership between Glass-Point and Berry Petroleum aims to produce steam and preheated water for the oil field using solar steam generators. The key specialty of this pilot project is its glass enclosure around mirrors that protect from dust and sand covering the glass panels. The setup is also fitted with automated washing tools that eliminate manual cleaning, further reducing maintenance costs. The project reduces the use of natural gas for steam production by up to 80% and produces 1 million BTUs per hour of solar heat [54].

8.5.2 Coalinga, California

A 29 MW_{th} steam facility was installed at Coalinga in partnership with Chevron Technology Ventures and BrightSource. The project is considered to be the world's largest solar TEOR. It is reported that with the integration of solar thermal technology, the oil field can extract around 70–90% of the original oil, whereas the conventional technology extracts around 25%. The facility is installed on 100 acres of land, having central receiver solar thermal technology with 3,822 heliostats and 7644 mirrors. The tracking mirrors concentrate the sunlight onto the boiler placed on the top of the tower. The steam produced by the solar system is pumped into the oil reservoir to increase the well pressure [55].

8.5.3 Petroleum Development Oman

GlassPoint Solar, in partnership with Petroleum Development Oman (PDO), commissioned a 7 MW_{th} solar enhanced oil recovery project in 2013. The facility can generate 50 tons of steam per day and be pumped directly into the existing TEOR project at Amal West oil field. The facility is fitted with automatic cleaning technology to remove dust in all weather conditions. Another project named Miraah of 1 GW_{th} solar field by PDO and GlassPoint Solar started in the year 2015. The first block of the Miraah project was completed in 2017, and as of 2018, the Miraah project is in operation with a 100 MW_{th} capacity. The project is aimed to save 5.5 trillion BTUs of natural gas per year [56].

8.5.4 Belridge Solar

Belridge Oil Plant is considered to be one of the largest oil fields in the United States, producing more than 70,000 barrels of oil a day. The solar integrated EOR project can generate 12 million barrels of steam per year, and a thermal power plant installed in the same facility can generate 26.5 MW of electricity. The project was announced in the year 2017 with the partnership by Shell, ExxonMobil, and Aera Energy. The present usage of natural gas to generate steam is around 4.5 bcf (billion cubic feet) a year which can be saved if integrated with solar thermal technology. The project can further reduce the carbon footprint with a predicted 376,000 metric tons of carbon emission per year [57].

8.5.5 Mukhaizna Oil Field

A solar integrated enhanced recovery facility is an upcoming project in partnership with GlassPoint Solar and Occidental of Oman at Mukhaizna oil field. The steam facility can generate 100,000 barrels of solar steam per day, which can reduce 800,000 tons of carbon emission per year [58].

Solar integration in downstream oil and gas industries can reduce carbon emissions, and the risk of climate change could be controlled. At present, fossil fuels are used to refine oils, and many industries are making plans to integrate renewable energies, particularly solar. Shell is considering powering their oil refinery using solar photovoltaics with an expected 3,000 MWh of electricity production per year [59].

8.5.5.1 Case Study

McKittrick Solar EOR Oil Field

The world's first commercial solar EOR project was constructed by GlassPoint Solar in partnership with Berry Petroleum at McKittrick, California, in 2011. The solar field is specifically designed for a rough oil field environment with a unique technology that can generate low-cost steam at the industrial level. The greenhouse is constructed on a half acre of barren land. The interior consists of curved aluminum sheets that are used as flimsy mirrors. Furthermore, motors are arranged for the movement of the mirrors to track the sun [60]. A photograph of solar collectors placed in a enclosure is shown in Figure 8.20. The system can generate one million BTUs per hour of solar heat [61]. Other specifications of the McKittrick solar project are shown in Table 9.4.

FIGURE 8.20 Solar collectors placed in a enclosure at McKittrick oil field, California [61].

TABLE 8.4
Specifications of McKittrick Solar EOR Oil Field at McKittrick, California [62]

Specification	Value
Peak output	0.3 MW
Energy output	15.5 GWh/yr
Daily steam output	2 tons

Specification	Value
Solar field area	640 m²
Total project area	1925 m²
Technology	GlassPoint enclosed trough
Greenhouse blocks	1
Construction start	2010
First steam	2011
Gas savings	2 billion BTUs per year
CO_2 emission saved	104 tons per year

Miraah Solar Thermal Project

The Miraah project is a 1,021 MW solar thermal facility that works on parabolic trough-based CST technology located in South Oman. GlassPoint Solar in partnership with Petroleum Development Oman (PDO) constructed the plant with operations starting in 2017. The facility can generate 6000 tons of steam per day for oil production. The steam is supplied to TEOR operations at Amal oil field in addition to 7 MW pilot plant that was commissioned in 2013. The total project area including all infrastructure, 36 glasshouse modules, span an equivalent of 640 acres of land. The facility is expected to lower their CO_2 emissions by 300,000 tons annually. The glasshouse modules consisting of sun tracking curved mirrors that focus sunlight on the receiver are placed in glasshouse modules that save mirrors from dust and rain storms. The specifications of the project are given in Table 8.5.

TABLE 8.5
Specifications of Miraah Solar Thermal Project, Oman [56, 63]

Specification	Value
Peak output	28 MW
Daily steam output	6,000 t
Total project area	1925 m²
Technology	GlassPoint enclosed trough
Greenhouse blocks	36
Construction start	2015
First steam	2018
Gas savings	5.6 trillion British thermal units per year
CO_2 emission saved	300,000 tons per year

8.6. BARRIERS TO THE LARGE-SCALE DEPLOYMENT OF SOLAR PROCESS HEATING SYSTEMS

Solar thermal energy has a wide range of applications, such as TEOR, fractionation, stripping, and quenching, and can generate electricity with steam turbines in oil upstream and downstream industries. Integrating CST technologies also improves the environment and controls the carbon footprint. However, there are challenges in implementing the solar thermal systems for process heat applications, and some of them include:

- High capital cost.
- Need for a backup system.
- Space constraints.
- Environmental and safety issues.

8.6.1 HIGH CAPITAL COST

The oil and gas industries require large-scale solar energy systems, and hence it can be fulfilled by the extensive installation of solar thermal energy technology. The initial investment for solar facilities is relatively high and depends on the type of solar thermal energy technology. Apart from equipment purchase costs, there is a need to modify existing machinery to make them compatible with solar integration, further increasing costs. Hence, cost-effective materials and technologies should be developed to bring down the initial costs. The use of modern equipment such as 3D printing can be used to reduce manufacturing costs further.

8.6.2 NEED OF A BACKUP SYSTEM

The solar thermal system operates during the daytime when there is solar radiation. The processes in oil and gas industries need heat at all times. However, solar energy cannot be generated at nighttime and sometimes in the daytime due to clouds. In such cases, there is a need of an energy storage/backup system. The energy can be stored in the daytime using thermal energy storage systems and utilized when required. Thermal energy storage systems are divided into three types, namely (i) sensible heat storage, (ii) latent heat storage, and (iii) thermochemical heat storage. A sensible heat storage system stores the energy by increasing the storage media temperature without changing its phase. It is considered to be one of the low-cost and long-lasting methods to store heat. Latent heat storage systems store heat by changing the phase of the storage media. In thermochemical heat storage, an endothermic reaction is performed to store the heat, and when the energy needs to be released, an exothermic reaction is performed. The use of the specific type of heat storage depends on the type of investment and temperature requirement. In addition, a precise mechanism to predict the weather conditions in combination with solar radiation sensors such as pyranometer and pyrheliometer and solar radiation forecasting techniques for predicting the meteorological data should also be installed to make sure that there is a continuous supply of energy [64, 65].

8.6.3 Environmental and Safety Issues

Solar thermal systems have a significant positive impact on the environment and can reduce the use of fossil fuels which further reduces the carbon footprint. However, there are a few environmental issues while installing the solar thermal systems, such as building the structures for the facility causing air and noise pollution. Incorrect handling may lead to high-temperature working fluids leakage that damages the ecosystems. The solar facility needs a large amount of land that sometimes requires clearing the forest/agricultural land, further disturbing flora and fauna. The water to generate steam for process heating may not be sufficiently available in desert regions. Hence, water wastage should be controlled, and new techniques for reusing the condensed water should be employed.

8.6.4 Space Constraints

Solar thermal energy systems need a large space based on the technology to install the solar collectors and to integrate with oil and gas industries. It requires approximately 1 km^2 of land or more to install the solar technology that produces 60 MW of electricity [66]. The upstream oil and gas wells are mostly found offshore (in the middle of oceans), and it is difficult to install large-scale solar thermal systems, particularly on the water surface. In a few cases, non-concentrating technologies such as flat plates or evacuated tube collectors can be installed on the roof of a building or premises rather than requiring space in the ground. The oil extraction and refinery facilities installed onshore can use brownfields or abandoned lands to install solar systems. In addition, rooftop technology can also be used to generate the required amount of energy from solar.

8.7 FUTURE SCOPE OF SOLAR THERMAL TECHNOLOGIES IN OIL AND GAS INDUSTRIES

In the present scenario, the transformation of energy production from conventional ways to renewable energy is a topic of interest for researchers as well as industries. The contribution of solar energy to the process heat applications is huge, and some of them are discussed in the previous sections. Recently, oil and gas companies have shown their interest in integrating solar thermal energy technologies with their processes to meet energy demand. However, energy production is locked up to some extent. For example, solar systems cannot work at night and completely depend on solar radiation availability. Although there are backup systems available for the continuous supply of energy on a limited basis, extensive research and development should be performed to meet the demand when there is no availability of solar energy.

Most of the existing oil and gas companies are established for a long period with their conventional way of producing energy. In some cases, the integration of solar technology may not be possible due to the compatibility of equipment with the latest ones. Hence, the manufacturer should construct products that are compliant with the existing facilities.

One of the important challenges faced when installing solar thermal systems on a large scale is the availability of land (as mentioned in the previous section). Solar thermal energy systems such as CR and PTC need a large amount of land to generate the required energy. However, the oil and gas industries located offshore or in urban areas have limited land, and it is difficult to install solar systems. In such cases, barren lands may be used for installations. If the land is far away from the industry, there are possibilities for efficiency loss in the energy, which should also be addressed. Another futuristic area is a floating solar energy system where the solar panels float on the water surface. This technique may offer solutions for the land demand for solar energy system installation. However, there are limitations to the floating technology, such as a change in the life of the structure, water quality, ecological damage, etc. Since it is in the initial stages of development, more research effort should be performed to examine the technology's limitations and implementation.

Solar energy is abundantly available, and with the right use, solar energy can meet global demand. Considering the drawbacks of solar thermal energy systems, there are two important points to be addressed (i) competitive costs of various components of the system and (ii) diurnal variation of solar radiation. The levelized costs of the energy produced from the solar thermal energy systems are in competition with other renewable and conventional energy technologies. Hence, effort should be performed in view of reducing manufacturing costs. On the other hand, the variation of solar energy is not consistent due to the earth's rotation and changes in the dynamics of the atmosphere. At the same time, most of the processes require a continuous energy supply. Hence, better control techniques should be developed to capture the radiation and predict the weather. The temperature requirement varies with process applications. Although temperature control techniques are available to achieve the energy demand for process heating, the focus should be made on the required energy at a given temperature to be withdrawn from the solar field itself to minimize exergy loss. In other words, a single solar thermal energy system can provide the required heat at a given temperature for multiple processes in the industry. This is challenging due to various temperature requirements for different process heating applications and can be accomplished using control techniques.

8.8 SUMMARY

Solar energy is considered to be one of the favorable renewable sources to produce clean energy. In industries such as oil and gas, high-quality steam is required to run the process heat applications. Solar thermal energy technologies can be integrated with conventional methods to achieve the required temperature range and meet the demand. Depending on the solar energy technologies, the solar thermal energy system can generate heat up to ~1000 °C and also can be used to generate electricity through steam turbines. The thermal EOR process in the upstream oil and gas industry requires high-temperature steam (250–300 °C) in order to extract the crude oil. Several oil and gas companies have sustainably shifted their steam production. The downstream oil and gas industries require steam for various process heating applications such as stripping, fractionation, quenching, etc. The heat required for these processes in different temperature ranges (150–550 °C) can be fulfilled by solar thermal energy technologies.

This chapter provides a comprehensive overview of energy demand in oil and gas industries with more inclination towards the integration of solar thermal energy technologies with process heat applications in both upstream and downstream industries. In addition, the barriers to the deployment of solar energy technologies on a large scale are also discussed in detail. This chapter is useful for researchers and manufacturers to fill the gap that limits the growth of renewable technologies in industries. Finally, more research and development work should be performed to provide solutions in a more sustainable way that controls the emission of greenhouse gases into the environment and helps move forward toward a greener world.

NOMENCLATURE

Acronyms

CLFR   Compact Linear Fresnel Reflector
CPC   Compound Parabolic Collector
CR   Central Receiver
CSP   Concentrated Solar Power
CST   Concentrated Solar Thermal
DNI   Direct Normal Irradiance
EOCR   Chemically Enhanced Oil Recovery
EOR   Enhanced Oil Recovery
ETC   Evacuated Tube Collector
FL   Fresnel Lens
FPC   Flat Plate Collector
HTF   Heat Transfer Fluid
LFR   Linear Fresnel Reflector
LPG   Liquefied Petroleum Gas
PDC   Parabolic Dish Collector
PDO   Petroleum Development Oman
PTC   Parabolic Trough Collector
PV   Photovoltaics
TEOR   Thermal Enhanced Oil Recovery
US   United States

Symbols

$A_{a, coll}$ Collector aperture area (m^2)
A_c Effective reflector surface area (m^2)
C_p Specific heat (J/kg-K)
CR Concentrator ratio
D_o Absorber outer diameter (m)
F_r Heat removal factor
I_b Beam radiation (W/m^2)
K Incident angle modifier (Degree)
L_{coll} Collector length (m)

$\dot{m}$	Mass flow rate (kg/s)
Q_{inc}	Incident concentrated heat (W)
Q_{loss}	Heat loss from the receiver (W)
Q_s	Heat energy collected by concentrator (W)
Q_u	Useful heat (W)
S	Concentrated solar flux on receiver surface (W/m^2)
SM	Solar multiple
T	Temperature (K)
U_L	Overall heat loss coefficient
W	Aperture diameter (m)

Greek letters

α	Absorptivity
δ	Declination angle (Degree)
ε	Emissivity
γ	Intercept factor
η	Efficiency
τ	Transmissivity
ρ	Reflectivity of the concentrator
ψ	Solar altitude angle (Degree)
σ	Stefan-Boltzmann constant (Wm^{-2}K^{-4})

Subscripts

a	Absorber
amb	Ambient
b	Beam radiation
$conc$	Concentrator
f	Fluid
i	Inlet
o	Outlet
opt	Optical
rec	Receiver

REFERENCES

[1] "Annual Energy Outlook 2021." www.eia.gov/outlooks/aeo/ (accessed Feb. 08, 2022).
[2] M. Absi Halabi, A. Al-Qattan, and A. Al-Otaibi, "Application of solar energy in the oil industry—Current status and future prospects," *Renew. Sustain. Energy Rev.*, vol. 43, pp. 296–314, 2015, doi: 10.1016/j.rser.2014.11.030.
[3] "MNRE ‖ Physical Progress," "Physical progress," *Ministry of New and Renewable Energy*, 2021.
[4] S. Ericson, J. Engel Cox, and D. Arent, "Approaches for Integrating Renewable Energy Technologies in Oil and Gas Operations," Technical Report (NREL/TP-6A50-72842), United States, pp. 1–26, 2019, doi: 10.2172/1491378, 2019.
[5] "Concentrated Solar Power (CSP)—Analysis," *IEA*. www.iea.org/reports/concentrated-solar-power-csp (accessed Dec. 26, 2021).

[6]   H. Sahebi, S. Nickel, and J. Ashayeri, "Strategic and tactical mathematical programming models within the crude oil supply chain context-A review," *Comput. Chem. Eng.*, vol. 68, pp. 56–77, 2014, doi: 10.1016/j.compchemeng.2014.05.008.

[7]   "What is the Kyoto Protocol?" *UNFCCC*. https://unfccc.int/kyoto_protocol (accessed Feb. 09, 2022).

[8]   C. Temizel *et al.*, "Technical and economical aspects of use of solar energy in oil & gas industry in the Middle East," *Soc. Pet. Eng.—SPE Int. Heavy Oil Conf. Exhib. 2018 HOCE 2018*, 2018, doi: 10.2118/193768-MS.

[9]   B. Bierman *et al.*, "Performance of an enclosed trough EOR system in South Oman," *Energy Procedia*, vol. 49, pp. 1269–1278, 2014, doi: 10.1016/j.egypro.2014.03.136.

[10]   B. Bierman, J. O'Donnell, R. Burke, M. McCormick, and W. Lindsay, "Construction of an Enclosed Trough EOR System in South Oman," *Energy Procedia*, vol. 49, pp. 1756–1765, Jan. 2014, doi: 10.1016/j.egypro.2014.03.186.

[11]   A. Askarova *et al.*, "Thermal enhanced oil recovery in deep heavy oil carbonates: Experimental and numerical study on a hot water injection performance," *J. Pet. Sci. Eng.*, vol. 194, p. 107456, Nov. 2020, doi: 10.1016/j.petrol.2020.107456.

[12]   S. Gupta, R. Guédez, and B. Laumert, "Market potential of solar thermal enhanced oil recovery-a techno-economic model for Issaran oil field in Egypt," *AIP Conf. Proc.*, vol. 1850, 2017, doi: 10.1063/1.4984573.

[13]   C. Likkasit, A. Maroufmashat, A. Elkamel, and H. Ku, "Integration of renewable energy into oil & gas industries: Solar-aided hydrogen production," *Proc. 2016 Int. Conf. Ind. Eng. Oper. Manage.*, Detroit, Michigan, USA, pp. 897–904, 2016.

[14]   Z. He, "Application of solar heating system for raw petroleum during its piping transport," *Energy Procedia*, vol. 48, pp. 1173–1180, 2014, doi: 10.1016/j.egypro.2014.02.132.

[15]   "Solar-heating system studied for heavy-oil pipelines," *Oil & Gas Journal*, Mar. 29, 1999. www.ogj.com/home/article/17230768/solarheating-system-studied-for-heavyoil-pipelines (accessed Feb. 08, 2022).

[16]   A. Sharma, "A comprehensive study of solar power in India and World," *Renew. Sustain. Energy Rev.*, vol. 15, no. 4, pp. 1767–1776, May 2011, doi: 10.1016/j.rser.2010.12.017.

[17]   M. Abdibattayeva, K. Bissenov, G. Askarova, N. Togyzbayeva, and G. Assanova, "Transport of heavy oil by applying of solar energy," *Environ. Clim. Technol.*, vol. 25, no. 1, pp. 879–893, Jan. 2021, doi: 10.2478/rtuect-2021-0066.

[18]   A. E. Rahman AM, N. As, and H. Mhm, "Application of solar energy heating system in some oil industry units and its economy," *J. Fundam. Renew. Energy Appl.*, vol. 07, no. 04, 2017, doi: 10.4172/2090-4541.1000233.

[19]   R. K. Pal, and K. R. Kumar, "Two-fluid modeling of direct steam generation in the receiver of parabolic trough solar collector with non-uniform heat flux," *Energy*, vol. 226, p. 120308, 2021, doi: https://doi.org/10.1016/j.energy.2021.120308.

[20]   A. K. Sharma, C. Sharma, S. C. Mullick, and T. C. Kandpal, "Solar industrial process heating: A review," *Renew. Sustain. Energy Rev.*, vol. 78, no. May, pp. 124–137, 2017, doi: 10.1016/j.rser.2017.04.079.

[21]   R. K. Pal, and K. R. Kumar, "Thermo-hydrodynamic modeling of flow boiling through the horizontal tube using Eulerian two-fluid modeling approach," *Int. J. Heat Mass Transf.*, vol. 168, p. 120794, 2021, doi: 10.1016/j.ijheatmasstransfer.2020.120794.

[22]   P. D. Tagle-Salazar, K. D. P. Nigam, and C. I. Rivera-Solorio, "Parabolic trough solar collectors: A general overview of technology, industrial applications, energy market, modeling, and standards," *Green Process. Synth.*, vol. 9, no. 1, pp. 595–649, 2020, doi: 10.1515/gps-2020-0059.

[23]   S. J. W. Klein, and E. S. Rubin, "Life cycle assessment of greenhouse gas emissions, water and land use for concentrated solar power plants with different energy backup systems," *Energy Policy*, vol. 63, pp. 935–950, 2013, doi: 10.1016/j.enpol.2013.08.057.

[24] K. Ravi Kumar, N. V. V. K. Chaitanya, and N. Sendhil Kumar, "Solar thermal energy technologies and its applications for process heating and power generation—A review," *J. Clean. Prod.*, vol. 282, p. 125296, Dec. 2020, doi: 10.1016/j.jclepro.2020.125296.

[25] R. K. Pal, and K. R. Kumar, "Effect of transient concentrated solar flux profile on the absorber surface for Direct Steam Generation in the parabolic trough solar collector," *Renew. Energy*, vol. 186, pp. 226–249, 2021, doi: 10.1016/j.renene.2021.12.105.

[26] K. S. Reddy, and K. R. Kumar, "Solar collector field design and viability analysis of stand-alone parabolic trough power plants for Indian conditions," *Energy Sustain. Dev.*, vol. 16, no. 4, pp. 456–470, 2012, doi: 10.1016/j.esd.2012.09.003.

[27] D. A. Baharoon, H. A. Rahman, W. Z. W. Omar, and S. O. Fadhl, "Historical development of concentrating solar power technologies to generate clean electricity efficiently—A review," *Renew. Sustain. Energy Rev.*, vol. 41, pp. 996–1027, 2015, doi: 10.1016/j.rser.2014.09.008.

[28] G. Mokhtar, B. Boussad, and S. Noureddine, "A linear Fresnel reflector as a solar system for heating water: Theoretical and experimental study," *Case Stud. Therm. Eng.*, vol. 8, no. August 2010, pp. 176–186, 2016, doi: 10.1016/j.csite.2016.06.006.

[29] S. M. Besarati and D. Yogi Goswami, "A computationally efficient method for the design of the heliostat field for solar power tower plant," *Renew. Energy*, vol. 69, pp. 226–232, 2014, doi: 10.1016/j.renene.2014.03.043.

[30] V. P. Stefanovic, S. R. Pavlovic, E. Bellos, and C. Tzivanidis, "A detailed parametric analysis of a solar dish collector," *Sustain. Energy Technol. Assess.*, vol. 25, no. December 2017, pp. 99–110, 2018, doi: 10.1016/j.seta.2017.12.005.

[31] R. K. Pal, and K. R. Kumar, "Investigations of Thermo-Hydrodynamics, Structural Stability, and Thermal Energy Storage for Direct Steam Generation in Parabolic Trough Solar Collector: A Comprehensive Review," *J. Clean. Prod.*, vol. 311, no. March, p. 127550, 2021, doi: 10.1016/j.jclepro.2021.127550.

[32] A. V Abramova, V. O. Abramov, S. P. Kuleshov, and E. O. Timashev, "Analysis of the modern methods for enhanced oil recovery," *Energy Sci. Technol.*, vol. 3, no. January, pp. 118–148, 2015, doi: 10.13140/2.1.2709.4726.

[33] S. S. Mardhika, "Determine environment impacts in upstream processes of oil and gas industries," *E3S Web Conf.*, vol. 73, pp. 8–11, 2018, doi: 10.1051/e3sconf/20187305008.

[34] "Enhanced oil recovery," *Fossil Energy and Carbon Management*, 2021. https://www.energy.gov/fecm/science-innovation/oil-gas-research/enhanced-oil-recovery (accessed Feb. 09, 2022).

[35] W. C. Lyons, "Mechanisms & recovery of hydrocarbons by natural means (Chapter 3)," *Working Guide to Reservoir Engineering*, Gulf Professional Publishing, pp. 233–239, 2010, doi: 10.1016/b978-1-85617-824-2.00003-4, ISBN: 978-1-85617-824-2.

[36] K. A. Lawal, and O. Olamigoke, "On the optimum operating temperature for steam floods," *SN Appl. Sci.*, vol. 3, no. 1, p. 9, 2021, doi: 10.1007/s42452-020-04082-2.

[37] S. Kokal, and A. Al-Kaabi, "Enhanced Oil Recovery: Challenges and Opportunities," *World Petroleum Council*, pp. 64–69, 2010. http://www.world-petroleum.org/docs/docs/publications/2010yearbook/P64-69_Kokal-Al_Kaabi.pdf.

[38] E. M. A. Mokheimer, M. Hamdy, Z. Abubakar, M. R. Shakeel, M. A. Habib, and M. Mahmoud, "A comprehensive review of thermal enhanced oil recovery: Techniques evaluation," *J. Energy Resour. Technol. Trans. ASME*, vol. 141, no. 3, 2019, doi: 10.1115/1.4041096.

[39] D. A. Z. Wever, F. Picchioni, and A. A. Broekhuis, "Polymers for enhanced oil recovery: A paradigm for structure-property relationship in aqueous solution," *Prog. Polym. Sci. Oxf.*, vol. 36, no. 11, pp. 1558–1628, 2011, doi: 10.1016/j.progpolymsci.2011.05.006.

[40] A. Al Adasani and B. Bai, "Analysis of EOR projects and updated screening criteria," *J. Pet. Sci. Eng.*, vol. 79, no. 1–2, pp. 10–24, 2011, doi: 10.1016/j.petrol.2011.07.005.

[41] M. Baviere, *Basic Concepts in Enhanced Oil Recovery Processes*. United Kingdom, 1991.

[42] J. D. Shosa, and L. L. Schramm, "Surfactants: Fundamentals and applications in the petroleum industry," *Palaios*, vol. 16, no. 6, p. 614, 2001, doi: 10.2307/3515635.

[43] D. Kraemer, A. Bajpayee, A. Muto, V. Berube, and M. Chiesa, "Solar assisted method for recovery of bitumen from oil sand," *Appl. Energy*, vol. 86, no. 9, pp. 1437–1441, 2009, doi: 10.1016/j.apenergy.2008.12.003.

[44] J. Sandler, G. Fowler, K. Cheng, and A. R. Kovscek, "Solar-generated steam for oil recovery: Reservoir simulation, economic analysis, and life cycle assessment," *Energy Convers. Manag.*, vol. 77, pp. 721–732, 2014, doi: 10.1016/j.enconman.2013.10.026.

[45] A. Amarnath, "Enhanced Oil Recovery Scoping Study," An Electric Power Research Institute (EPRI) Report, 1999. https://www.adv-res.com/pdf/electrotech_opps_tr113836.pdf (accessed Apr. 10, 2022).

[46] M. M. Yegane, F. Bashtani, A. Tahmasebi, S. Ayatollahi, and Y. M. Al-Wahaibi, "Comparing different scenarios for thermal enhanced oil recovery in fractured reservoirs using hybrid (solar-gas) steam generators, a simulation study," *All Days*, Vienna, Austria, May 2016, p. SPE-180101-MS, doi: 10.2118/180101-MS.

[47] A. P. G. Van Heel, J. N. M. Van Wunnik, S. Bentouati, and R. Terres, "The impact of daily and seasonal cycles in solar-generated steam on oil recovery," *SPE EOR Conf. Oil Gas West Asia*, Muscat, Oman, April 11–13, 2010, pp. 347–360, 2010, doi: 10.2118/129225-ms, ISBN: 978-1-55563-285-4.

[48] E. M. A. Mokheimer and M. A.-A. M. Habib, "Hybrid solar thermal enhanced oil recovery system with oxy-fuel combustor," US9845667B2, Dec. 19, 2017 [Online]. Available: https://patents.google.com/patent/US9845667B2/en?oq=US9845667B2. (accessed Mar. 18, 2022).

[49] "U.S. Manufacturing Energy Use and Greenhouse Gas Emissions Analysis," *Energy Efficiency & Renewable Energy*, 2012. https://www.energy.gov/eere/amo/downloads/us-manufacturing-energy-use-and-greenhouse-gas-emissions-analysis (accessed Feb. 10, 2022).

[50] "Steam System Opportunity Assessment for the Pulp and Paper, Chemical Manufacturing, and Petroleum Refining Industries," Technical Report, Office of Energy Efficiency and Renewable Energy, U.S. Department of Energy, 2002. https://www.nrel.gov/docs/fy03osti/32806.pdf (accessed Mar. 15, 2022).

[51] "Energy Efficiency Improvement and Cost Saving Opportunities for Petroleum Refineries," Technical Report (Document Number: 430-R-15-002), Energy Star, 2016.

[52] "Inventory of U.S. Greenhouse Gas Emissions and Sinks," Technical Report (Document Number: 430-R-22-003), U.S. Environmental Protection Agency, 2015.

[53] C. Likkasit, A. Maroufmashat, A. Elkamel, H. M. Ku, and M. Fowler, "Solar-aided hydrogen production methods for the integration of renewable energies into oil and gas industries," *Energy Convers. Manag.*, vol. 168, no. January, pp. 395–406, 2018, doi: 10.1016/j.enconman.2018.04.057.

[54] "GlassPoint Unveils First Commercial Solar Enhanced Oil Recovery Project," 2011. https://www.businesswire.com/news/home/20110224006666/en/GlassPoint-Unveils-First-Commercial-Solar-Enhanced-Oil-Recovery-Project (accessed Apr. 10, 2022).

[55] "Brightsource Energy Delivers World's Largest Solar-to-Steam Facility for Enhanced Oil Recovery to Chevron," 2011. http://www.brightsourceenergy.com/brightsource-energy-delivers-worlds-largest-solar-to-steam-facility-for-enhanced-oil-recovery-to-chevron#.YcS6mmhByUk (accessed Jan. 15, 2022).

[56] "Miraah Solar Thermal Project," 2021. https://www.power-technology.com/projects/miraah-solar-thermal-project/ (accessed Apr. 15, 2022).

[57] "Belridge Solar Thermal Power Plant, California," 2018. https://www.power-technology.com/projects/belridge-solar-thermal-power-plant-california/ (accessed Apr. 18, 2022).

[58] "GlassPoint and Occidental of Oman Sign Agreement to Cooperate on Project to Facilitate Oil Production in Oman," 2018. https://www.businesswire.com/news/home/20181113005392/en/GlassPoint-and-Occidental-of-Oman-Sign-Agreement-to-Cooperate-on-Project-to-Facilitate-Oil-Production-in-Oman (accessed Apr. 10, 2022).

[59] "Shell Considering Solar Power to Run Its Largest Oil Refinery," 2019. http://ieefa.org/shell-considering-solar-power-to-run-its-largest-oil-refinery/ (accessed Feb. 10, 2022).

[60] D. Llamas, "GlassPoint Solar unveiled the world's first commercial solar enhanced oil recovery," *HELIOSCSP*. https://helioscsp.com/glasspoint-solar-unveiled-the-worlds-first-commercial-solar-enhanced-oil-recovery/ (accessed Apr. 10, 2022).

[61] "Solar Thermal Enhanced Oil Recovery," Mar. 14, 2022. https://en.wikipedia.org/w/index.php?title=Solar_thermal_enhanced_oil_recovery&oldid=1077048062 (accessed May 10, 2022).

[62] "McKittrick," *GlassPoint*. www.glasspoint.com/projects/mckittrick (accessed Apr. 10, 2022).

[63] "Miraah Solar Project," *Petroleum Development Oman*. www.pdo.co.om/en/technical-expertise/solar-project-miraah/Pages/default.aspx (accessed May 10, 2022).

[64] M. Demirtas, M. Yesilbudak, S. Sagiroglu, and I. Colak, "Prediction of solar radiation using meteorological data," *2012 Int. Conf. Renew. Energy Res. Appl. (ICRERA 2012)*, Nagasaki, Japan, Nov. 11–14, 2012, pp. 2–5, 2012, doi: 10.1109/ICRERA.2012.6477329.

[65] Y. Dong, and H. Jiang, "Global Solar Radiation Forecasting Using Square Root Regularization-Based Ensemble," *Math. Probl. Eng.*, vol. 2019, 2019, doi: 10.1155/2019/9620945.

[66] S. P. Srivastava, and S. P. Srivastava, "Solar energy and its future role in Indian economy," *Int. J. Environ. Sci. Dev. Monit*, vol. 4, pp. 81–88, 2013.

Part III

Sustainability Assessment for Solar Industrial Process Heating

9 Effective Integration and Technical Feasibility Analysis of Solar Thermal Networks for Industrial Process Heating Applications

M.M. Matheswaran

CONTENTS

9.1 INTRODUCTION

The global industrial sector is responsible for a significant amount of the energy that is consumed today. More than half of industrial heating needs occur between 60 to 250°C, and process heat accounts for 74% of all industrial energy use [1]. As per the IEA report 2019, only 9% of the demand for heat used in industrial processes is met by renewable energy sources. There are 741 industrial units that effectively implemented the solar-based energy production plants, which encompasses an area

DOI: 10.1201/9781003263326-12

of 662,648 m^2 and an installed capacity that amounts to 567MWth [2]. The annual rise in the need for heat in the industry is 1.7% and by the year 2030, the required heat energy will be 2.2 EJ, and the required collector area will be 1300 million m^2 [3]. From an economic point of view, the high cost of capital is one of the biggest reasons why solar thermal technologies aren't used more often. From a technical point of view, it is also a big challenge to make sure there is a steady supply of heat even though the source of energy (solar radiation) isn't steady [4]. Another important technical barrier is the amount of space and sunlight available, which varies a lot depending on where the solar plant is located. Incorporation of solar thermal energy into engineering processes necessitates research into and familiarity with a variety of pertinent topics, including but not limited to the following: solar potential; the existing level of technology; methodologies and extensions of solar thermal heat addition; performance assessment of solar thermal equipment; environmental and economic assessment; obstacles to larger implementation. While integrating the solar thermal system in process heat industries the following parameters are simultaneously considered for effective decision making [5].

- Solar fraction—the fraction of energy supplied by a solar collector network.
- Installation area of solar collector network.
- Payback period.
- The maximum period of supply time.
- Ability to adopt various situations based on process requirements by the industry.

Solar thermal systems can be used in industrial processes in three ways. The solar collectors' efficiency should be maximized by designing this integration to operate at the lowest possible temperature (always within the range required by industry) [6].

- Preparing water for use in a particular technique.
- Supply-side integration.
- Heating a circulating fluid by direct connection to a process.

While integrating the system, temperatures should be kept as low as possible from the point of view of capture efficiency in order to maximize the amount of solar energy that can be collected and the overall effectiveness of the solar system. The second reason for lowering the temperature at which the capture occurs is that doing so reduces the temperature difference between the ambient environment and the collector, which in turn maximizes the capture efficiency by minimizing the amount of heat lost through the casing to the surrounding environment [7]. When it comes to heating a process, on the other hand, the temperature of a hot utility should be sufficiently high so that it can accommodate the temperature level of the heating demands in addition to the minimum temperature difference that is permitted. In order to make the effective trade-off explained previously, Pinch Methodology was effectively expanded to meet regulatory or environmental standards, costs totals, and reduction or exclusion of GHG emissions, among other things [8].

9.1.1 Effective Integration of Solar Thermal System with Process Heat Demand

The integration of a solar thermal system into a typical industrial process is depicted in more basic terms in Figure 9.1 [9].

The primary components of this approach are the a priori aims, the intrinsic variables of the method's components, and the explanation, which provides the ultimate design that achieves the a priori objectives. When it comes to developing the integration of solar heat systems in processes, there are a few different techniques that might be used. These procedures are carried out in order to realize the following ideas on integration [10]:

- An understanding of the system as a whole as well as its individual bottlenecks.
- A range of options for improving material and/or energy efficiency.
- A contribution to the planning of changes to the process.
- Ensuring that the design of a profitable energy mix for the supply of renewable energy is optimized, and that heat recovery is also taken into account.

The Pinch Analysis is a prominent tool for process integration, and it is also often used for solar thermal system integration. The integration methods are discussed further later. This methodology can be used to analyze the potential for heat recovery in an industrial process or location, and it can also be used to design the heat recovery or utility mix based on a fixed utility system. The steps to be followed for the effective integration of solar thermal system as per the requirement of industrial heat demand are explained in the following subsections.

Collecting and extracting data: The requirements for heating and cooling are developed for the industrial processes, and each of these streams is defined based on their mass flow, specific heat, and inlet temperature, as well as their goal temperature (Figure 9.2).

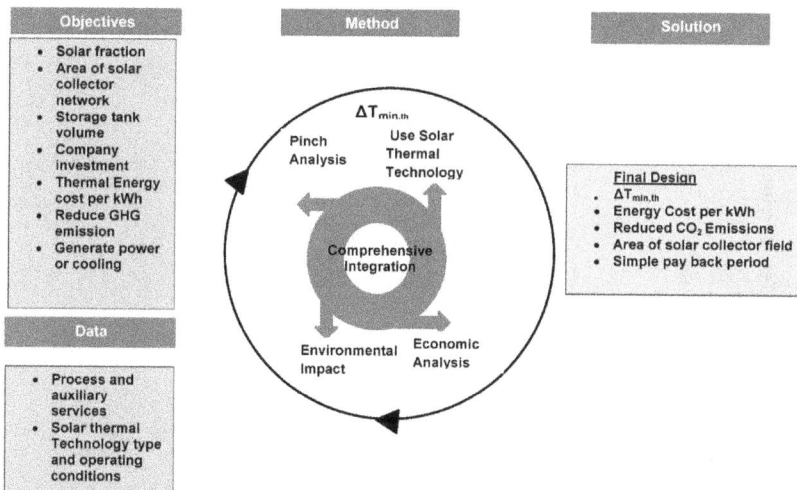

FIGURE 9.1 The design process of a solar thermal system.

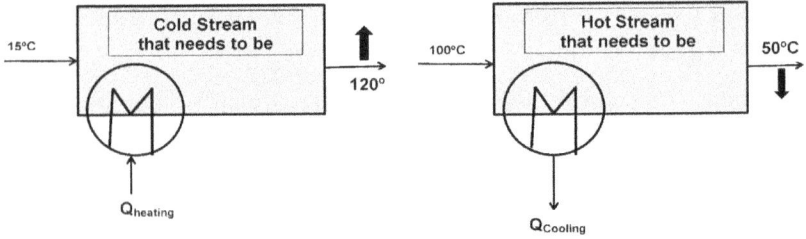

FIGURE 9.2   Identification of hot and cold workflows for Pinch Analysis.

9.1.2 Targeting

Maximizing heat recovery and determining the most cost-effective energy supply are the two main goals of this investigation. Here, we'll make use of two separate diagrams: the first objective may be seen by looking at the Composite Curves (CCs), while the second can be seen in the Grand Composite Curve (GCC). The Composite Curves are a graph of the temperature and enthalpy of all of the hot streams and all of the cold streams when they are constituted together. Figure 9.3 provides an illustration of a sample plot. The top curve, which can be found in the figure, is referred to as the hot composite curve. This curve depicts the aggregated hot streams of a heat integration problem (HCC).

The cold streams are represented by the Cold Composite Curve (CCC), which can be found at the bottom of Figure 9.3. The amount of heat that can be recovered from warm to cool streams is represented by the shadowed area's projection (indicated with arrows) along the ΔH axis. This is also known as heat recovery during the

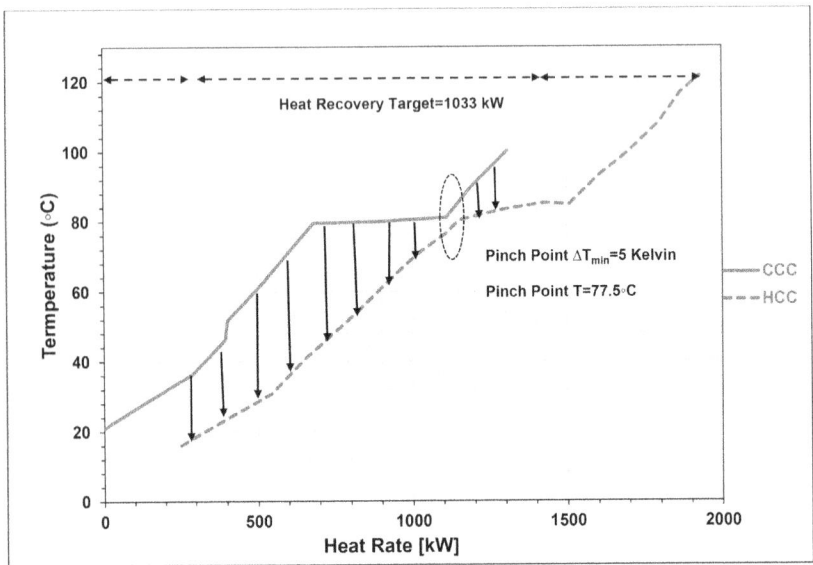

FIGURE 9.3   Composites curves based on cold and hot streams.

process. The term "pinch point" refers to the point at which the two CCs are equivalent to ΔT_{min} in terms of the degree of temperature approach that they have toward one another. The theoretical maximum heat recovery that can be achieved was determined by the horizontal overlap of the curves. The top portion of the figure contains a graphical representation of the minimum cooling and heating requirements for processes (hot and cold utilities). The CCs provide us with information regarding the possible integration of SHIP by defining the following three limitations:

- There must not be any external heating lower than the pinch temperature.
- There is to be no external cooling above the temperature of the pinch.
- Avoid transferring heat from heat sources located above the pinch to heat sinks located below it.

9.1.2.1 Grand Composite Curve (GCC)

This curve was made to make energy supply more efficient and find out things like which process has the lowest temperature heat demand (a relevant parameter to integrate solar systems). Figure 9.4 shows how this curve is made from the hot and cold CCs. First, the two CCs are moved up and down so that they touch in the middle of the ΔT_{min} (top left of figure). Then, the GCC is made by taking the difference in heat rates between hot and cold CCs. The minimum cold and hot utilities are shown at the end of the curve. Now, the x-axis shows the absolute heat rates.

The GCC shows the solar thermal system's maximum heating rate contribution based on its operating temperatures. Figure 9.5 shows that half of GCC's hot utility is required below 90°C, a desirable heat temperature for solar systems. Two important ideas must be analyzed for SHIP graph integration [11]:

- Maximum solar heat at minimum temperature (see GCC): Due to techno-economic feasibility, it may not be possible to cover all GCC processes in the

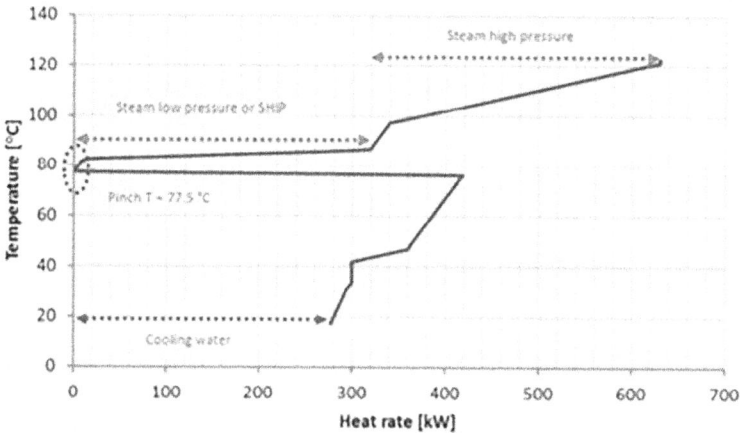

FIGURE 9.4 Grand Composite Curves.

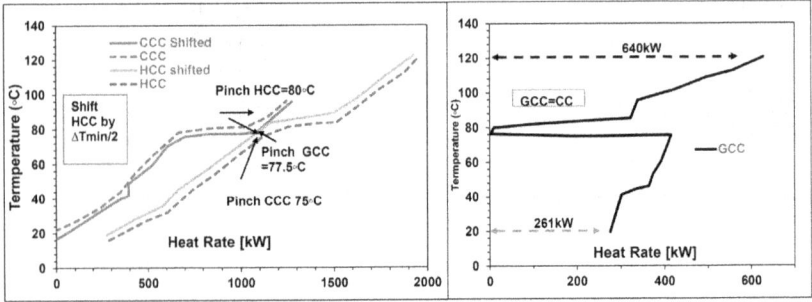

FIGURE 9.4A   Grand Composite Curves.

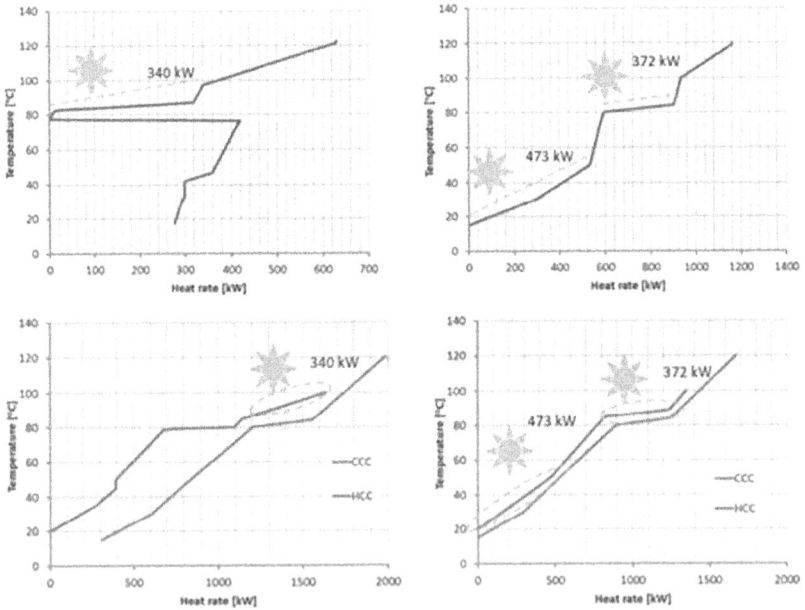

FIGURE 9.5   Integration of solar heat using GCC.

solar plant design. Different supply temperatures may be needed, therefore selecting one or two solar processes is feasible.

• List SHIP integration points (see CCs): The GCC solar heat rate is increased to hot CCs. This action evaluates potential integration points for continuous processes, identifying cold process stream candidates and their heat rates.

Figure 9.5 illustrates the many solar integrations that could be designed within the curves. On the right, the graphs illustrate the scenario in which there is no heat waste recovery; hence, solar heat is included in any region of the curve. On the left, it is only applicable to the curve that is located on the right-hand side of the pinch point.

The integration of solar heat with the Pinch Analysis presents a number of issues, including the following:

- Variable process streams are challenging to consider.
- The solar resource's sensitivity to time.
- Until the technological principles and temperature characteristics of the solar design are determined, it is impossible to know in full the solar heat gains.

The heat exchanger position is analyzed after utilities have been re-integrated into the composite curve during the design phase. The most efficient and cost-effective location for a heat exchanger can be determined using optimization methods. After considering the solar system, a feasible heat storage system and re-design of the stream network may be necessary.

9.2 INTEGRATION CONCEPTS

The categorization of industrial heat consumers for solar heat integration in industrial applications is shown in Table 9.1 [9].

9.2.1 SUPPLY AND PROCESS LEVEL INTEGRATION

Supply-level integration shown in Figure 9.6 entails producing steam or hot water for a central distribution system. This system transfers fluids to heat-consuming industry processes. However, process-level integration (shown in Figure 9.7) must meet the demands of each individual application [12]. Depending on the amount of heat needed, the process level can be segmented into "preheating of working fluids" and energizing and sustaining the temperature of the reservoir of machines.

TABLE 9.1
Concepts of Solar Heating for Industrial Processing

Level of Integration	Heat Transfer Medium/Equipment	Integration Procedure
Supply level	Steam	Parallel integration
		Heating feed water
		Heating makeup water
	Liquid	Parallel integration
		Return flow boost
		Heating of storage
Process level	External heat exchanger	Heating process medium
	Internal heat exchanger	Heating intermediate hot water circuit
		Heating bath
	Steam supply	Vacuum steam
		Low pressure steam

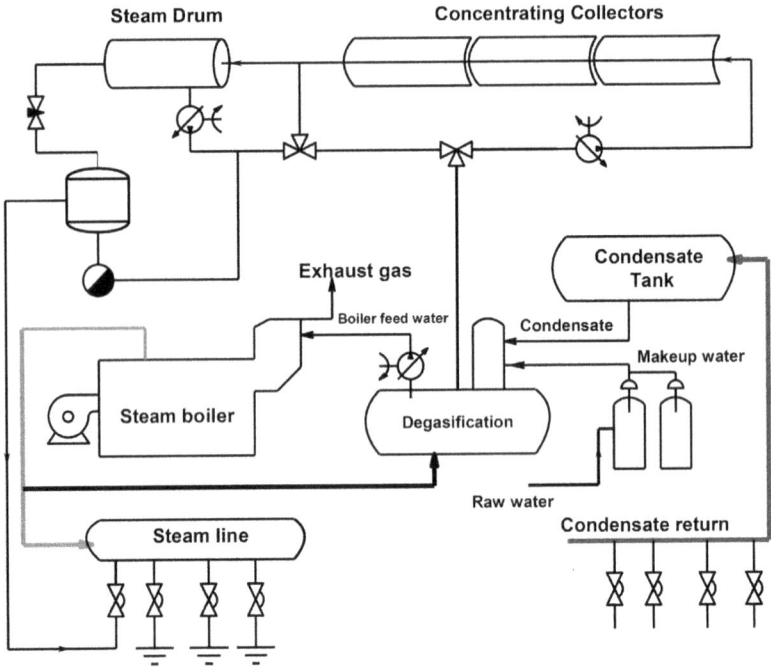

FIGURE 9.6 Supply level integration.

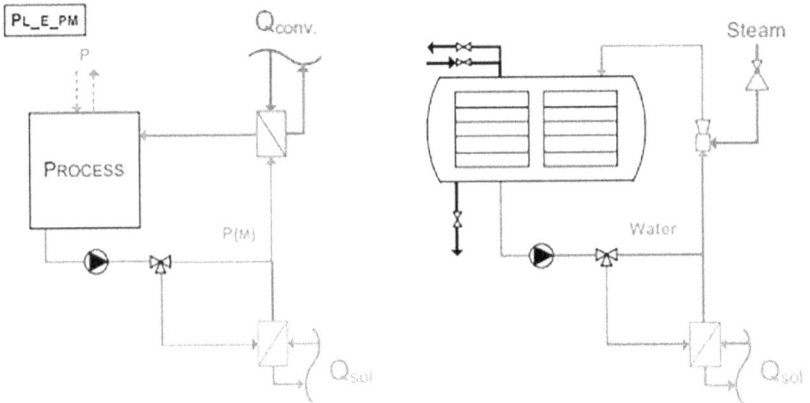

FIGURE 9.7 Process level integration.

At the supply level, the heat transfer medium that is being used (steam or liquid) in conjunction with the integration point that has been determined leads directly to an integration idea. Many different methods and tools may be used to implement a single integration notion at the process level. In situations like these, the actual process boundary conditions are the only factors that can determine which integration system is the most suitable. For instance, in the process of "heating and maintaining temperature of reservoirs, machines, or tanks" using direct steam injection, one could apply any one of the following four integration concepts:

- External bath heating process.
- Preheating at inlet.
- Internal HX.
- Low pressure steam supply.

When it comes to choosing an integration idea, the one that functions with a reduced temperature demand will be the most crucial criterion to take into consideration. In the event that the scalding bath receives a significant amount of freshwater as input, the most effective course of action would be to warm up this source of water [13]. If for any technical reason the bath needs to be filled with cold water (for example, to cool down the product), then this idea is not viable, and the subsequent ones with requirements for a lower temperature should be examined alternatively. In this case, the availability of space to heat outside (external bath heating) or internally (internal heat exchanger) determines the integration choice. Low pressure steam is easy to integrate because it doesn't interfere with the existing process, but it increases the solar system's temperature demand and affects collector design and performance. If the water input stream is discontinuous, it can be heated directly when provided. No freshwater input requires an external heat exchanger.

9.2.2 Assessment of Solar Collector Network Areas

The Composite Curves are utilized to calculate the area and auxiliary systems of the solar energy network. Prior to creating the network, the network's surface area is estimated. Equation 1 is utilized to construct a system by assuming "vertical" heat exchange [9, 14].

$$Q_T = UA\Delta T_{LMTD} \tag{1}$$

In this equation Q represents the total thermal energy requirement in kW, U indicates the overall heat transfer coefficient in W/m^2 K, and ΔT_{LMTD} shows the logarithmic mean temperature difference between the fluids. By suitably considering the convective heat transfer coefficients, equation 1 can be used to evaluate the area required for a solar collector network.

$$A_N = \frac{1}{\Delta T_{LMTD}} \left(\sum_i^{\text{Hot fluids}} \frac{Q_i}{h_i} + \sum_j^{\text{Cold fluids}} \frac{Q_j}{h_j} \right) \tag{2}$$

In equation 2, Q_i and Q_j represent hot and cold thermal load requirements in the industry in kW, h_i and h_j indicate the convective heat transfer coefficient between the fluids (W/m^{2}K). The evaluated area (A_N) is the total area of the solar thermal network. It can be used to design the thermal network with an array of collectors.

The configuration of the solar collector network depends on the thermal model that is proposed to provide the thermal load at the target temperature, as shown in

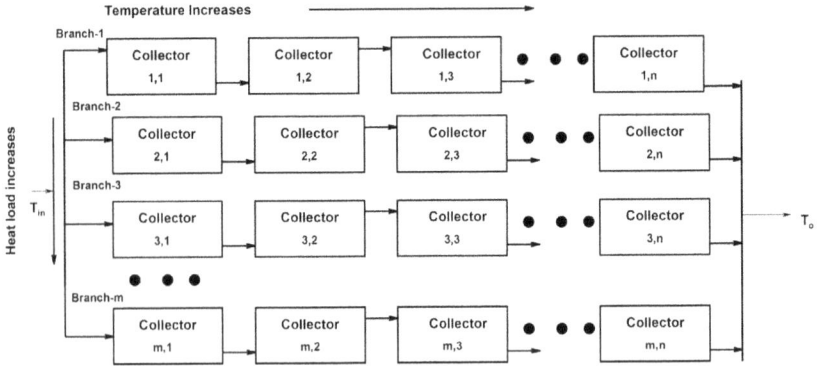

FIGURE 9.8 Solar collector network.

Figure 9.8. More specifically, the number of collectors connected in series deter-mines the temperature at which the load is to be maintained, while the quantity of collectors connected in parallel determines the amount of heat that must be main-tained at that temperature.

The physical measurements and features of the thermal system components, the characteristics of the heat transfer fluids, the operational parameters of the system, and the environment conditions are the input parameters of the collector system. Other design variables include the operating parameters of the process. When the difference between the exit temperature (T_e^n) and the inlet temperature (T_i^{n-1}) of the solar collector is taken into consideration, it is possible to calculate the bare mini-mum number of collectors that should be connected in series, denoted by N_s.

$$Q_T = UA\Delta T_{\text{LMTD}} \tag{3}$$

The solar collector must have a differential in the temperature of at least one degree Celsius. By using the thermodynamic basics it is found that the number of collector's series to be connected in parallel (N_P) is evaluated by using equation 4.

$$N_p = \frac{Q_T}{Q_{se}} \tag{4}$$

In equation 4, Q_T and Q_{se} represent the total heat requirement and thermal load to be supplied by the number of collectors in series of each row. The formula used to calculate the number of solar collectors, N_c, that constitute the structure of the solar field is as follows:

$$N_c = N_p N_s \tag{5}$$

The total collector area for the solar thermal energy network is given by

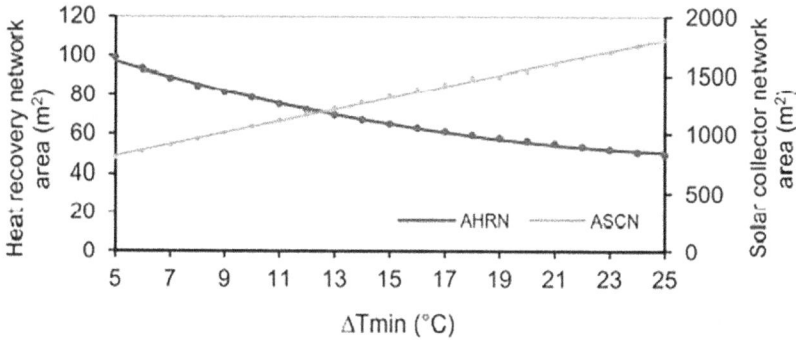

FIGURE 9.9 Influence of ΔT_{min}.

$$A_{SC} = WLN_c \tag{6}$$

In equation 6, L and W show the length and width of the thermal system of the collector. The calculated area A_{sc} is indirectly proportional to the ΔT_{min} evaluated by using Pinch Analysis. By maintaining the lower value of ΔT_{min} decreases the value of the required collector area. Figure 9.9 shows that the area of the SCN is 14 times greater as compared with the area of the thermal energy recovery system, while the cost of the heat recovery network is two and a half times greater than the cost of the SCN. This necessitates an analysis of the expenses and a discussion on the relationship between those economics and the worth of the areas, which includes the space that is existing for installation, the network area of the solar thermal system, and the available area that is dedicated to heat recovery.

9.2.2.1 Estimation of Expenditures for System Components

The incorporation of solar thermal technology will incur expenses no matter which of the examined scenarios is pursued. Following this, a description of the equations that were used to determine the expenses is provided (costs from thermal energy recovery network, secondary services, storage system, and SCN). Likewise, the pricing per kWh for the solar system and the price per kWh for the integrated system were calculated by taking into account the total cost of the system (regardless of whether it was solar or integrated) as well as the total quantity of energy that was generated by the system. The task of establishing whether or not the heat recovery network is cost-effective was carried out by equation (C_{HN}).

$$C_{HN} = N_e \left[a + b \left(\frac{A_{HN}}{N_e} \right)^c \right] \tag{7}$$

In the equation 7,
a = Number of Heat exchangers for energy recovery

b = Area of an energy recovery network

C = 266600 and 6500, taken based on the type of heat exchanger, construction material, and pressure loss.

The cost-utility of the system based on hot and cold streams is evaluated using the relation as follows:

$$C_{us} = Q_h C_{uh} + Q_c C_{uc} \qquad (8)$$

where the heating and cooling needs of the process, measured in kW, are referred to as Q_h and Qc, respectively, and C_{uh} and C_{uc} refer to the expenses associated with cooling water and steam, respectively. The expenses of heating, which were taken into account for this study, were 150 USD per kW year, and the costs of cooling services were 35 USD per kW year. The cost of solar network collectors is evaluated using the relation as follows

$$C_{SC} = N_c \left[\beta_o + \frac{A_t N_t}{\pi} \left(\beta_1 d + \beta_2 + \frac{\beta_3}{d} \right) + WL\beta_4 + \beta_{10} \frac{mL\mu}{\pi \rho d^4} \right] + \beta_5 \left(\frac{mH_b}{e_{ff}} \right) \qquad (9)$$

In equation 9, N_c, A_t, and N_t show the number of collectors, lateral area of solar collectors, and the number of tubes respectively. Further d, W, L, e_{ff}, m, and H_b are the diameter of the tube, the width of the collector, length of the collector, pump load, mass flow rate, and efficiency of the system respectively. The β_1 to β_{10} represent the setting constants. The sizing of the collector system and the amount of supply time that is required by the process both have an effect on the cost of the storage system, also known as the C_{TSS}. Following that, equation 10 will be used to calculate the cost of the thermal storage system.

$$C_{Tss} = a' + b' V_{Tss}{}^\wedge c' \qquad (10)$$

In equation 10, V_{Tss} is the volume of the tank in m³. The cost constant for storage tanks a' = 5800, b' = 1600, and c' = 0.7.

$$V_{Tss} = \frac{3600 \, Q_{Tss} t}{CP \Delta T_{Tss} \rho \eta_{Tss}} \qquad (11)$$

Q_{TSS}—daily heat load (kW), t—time (hrs), C_P—specific heat (J/kgK), ρ—density or working thermal fluid (kg/m³), ΔT_{Tss}—temperature variation, η_{Tss}—efficiency of the thermal storage system.

The total cost of the system is evaluated by using the relation

$$C_T = C_{HR} + C_{SH} + C_{Tss} + C_{us} \qquad (12)$$

It is also important to calculate the yearly cost ($/year) of our interest, which is the amount that is comparable to a cost per unit of time. The solution is found using equation 13.

$$C_{TA\,IS} = C_{TIS} \left[\frac{i(1+i)^n}{(1+i)^n - 1} \right] \tag{13}$$

The $C_{TA\,IS}$ allows for the correlation of all energy-integration-related variables. The amount of time it takes for an investment to get its money back is a key metric. Both the solar thermal system and the integrated system can be modelled using equation 14. This metric takes into account the yearly savings in fossil fuel costs and compares them to the total system cost in US dollars.

$$Payback = \frac{Total\ system\ cost}{S} \tag{14}$$

9.2.2.2 Solar Fraction

Although having a solar fraction of one might be the goal, there are limitations imposed by physical, financial, and thermal factors [15]. With a constant heat duty and varying values of ΔT_{min} and solar fraction, the STE cost (USD/kWh) is depicted in Figure 9.10. For a certain ΔT_{min}, the cost per kWh rises from $0.01/kWh to $0.05/kWh as solar penetration increases. When the solar portion is constant, the cost per kWh remains unchanged regardless of ΔT_{min}.

After determining the ΔT_{min} according to the desired outcome, GCC is implemented to provide the effectiveness of hot systems at two different temperatures and lengthen the source time of the first SCN. In this case, the surplus energy can be used to generate an additional service, such as power or cooling, or to shrink the SCN. To ensure the thermal load required by the process industries throughout the year, the SCN was constructed with lower solar power levels on year.

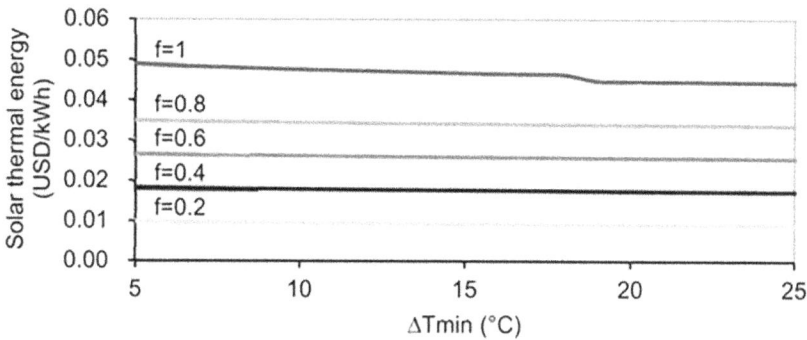

FIGURE 9.10   Influence of ΔT_{min} and solar fraction on cost of the systems.

9.3 CASE STUDIES

Case Study I: For Dairy Industry

The production of dairy encompasses a vast array of important nutrients and their byproducts [16]. This indicates that the process temperature levels correlate with the solar collector network temperature levels, and it may be able to deliver the entire heat load essential by the operation. These operations are carried out at temperatures below 100°C.

In this case study about dairy industries, the heating utility is needed for 5 hours, from 8:00 in the morning until 1:00 in the afternoon. This process is carried out in groups. Figure 9.11 shows a diagram of the main steps in the process of making dairy products. The process needs 4,401 kW of heat for 5 hours (8:00 a.m. to 1:00 p.m.).

The first step in the process of integration is to figure out what the goals are for the process data and the variables that will be used in the technique. Table 9.2 shows the information about the process streams that were used for the Pinch Analysis. T_{inlet} and T_{outlet} show the temperatures at the beginning and end of the stream, while C_p shows the heat capacity of the current in kilowatts per metre of current. Use the data from the streams to figure out how many minimum hot utilities are needed. In this case, the heat transfer coefficients that

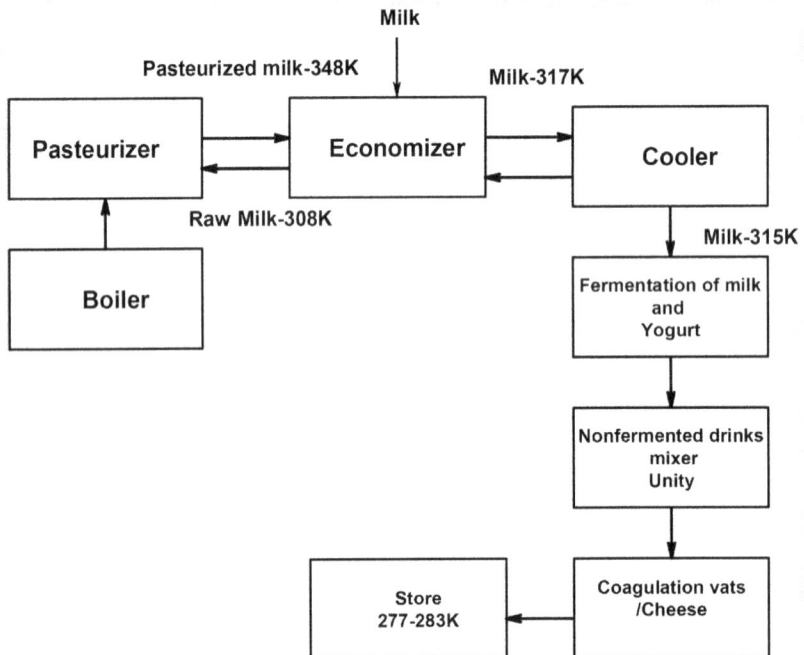

FIGURE 9.11   Dairy product process.

TABLE 9.2

Data for Dairy Processing

Streams	Parameter	T_i	T_{out}	C_P
Hot-1	Effluent-milk	75	44	4.38
Hot-2	Boiler-water	95	78	10.63
Hot-3	Milk-input	44	36	5.84
Hot-4	Water-storage tank	73	40	1.94
Hot-5	Water-rennet	40	38	3.89
Hot-6	Water-storage tank	62	25	9.3
Hot-7	Surplus water	62	25	12.9
Cold-1	Main water-boiler	12	95	10.6
Cold-2	Milk-input-raw	4	35	4.38
Cold-3	Milk-input	35	75	4.38
Cold-4	Water cooler	12	18	6.3
Cold-5	Main water	12	38	2.6
Cold-6	Water-rennet	34	35	3.89

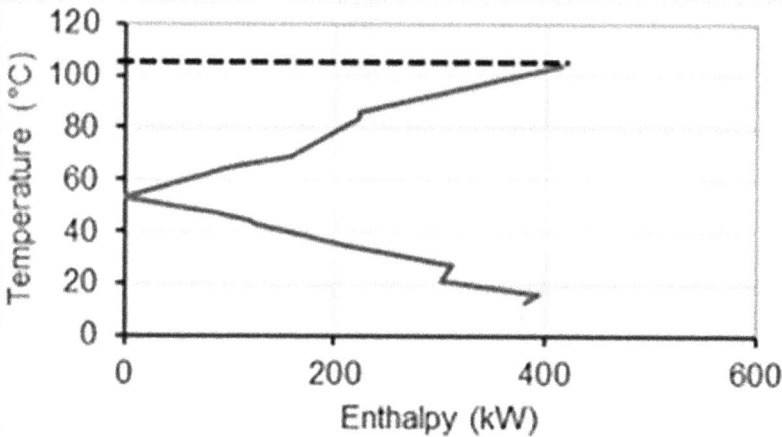

FIGURE 9.12 Grand Composite Curve (a) $\Delta T_{min} = 8°C$; (b) $\Delta T_{min} = 18°C$.

are taken into account are 0.8 kW/(m²°C) for water and mildly viscous liquids and 0.3 kW/(m²°C) for viscous substances.

During the process of developing the SCN for the dairy sector analysis, a required temperature of 100°C was chosen, and heat load levels were adjusted depending on the ΔT_{min}. For a ΔT_{min} of 14.5°C, 435 collectors are equivalent to 796 m². When ΔT_{min} increases, these numbers also go up, as does the thermal demand.

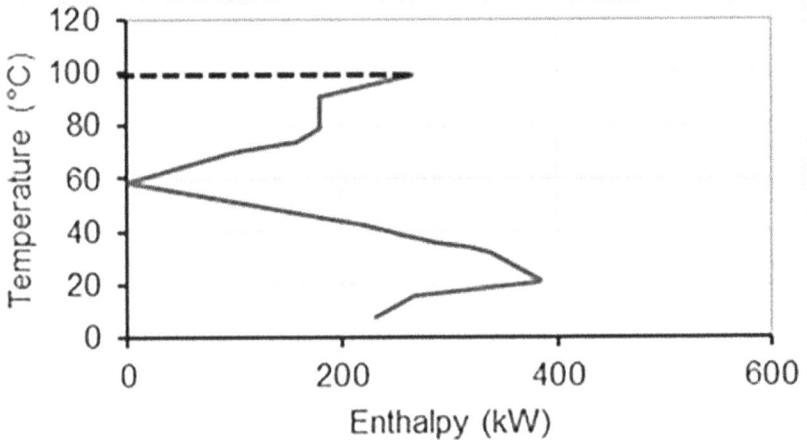

FIGURE 9.12B   Grand Composite Curve (a) $\Delta T_{min} = 8°C$; (b) $\Delta T_{min} = 18°C$.

Once the process energy needs are determined using Pinch Analysis, the heat recovery network and utility costs are estimated [17]. Also, the solar fraction that should be attained by incorporating STE through solar collectors may be determined. By determining temperatures below the maximum, it increases supply time. Figure 9.12 (a) and (b) demonstrate pinch point and thermal loads by adjusting ΔT_{min}. The heating service supply for GCC is 99°C for ΔT_{min} 8°C. When the data providing by both curves is compared, it can be seen that the supply temperature for Figure 9.12 (a) is 95°C, and the heat load for that figure is 265°C. On the other hand, the supply kW temperature for Figure 9.12 (b) is also 95°C; however, the thermal load for that figure is 415 kW.

In all scenarios, the temperature level for the energy need can be met by using solar collectors operating at low temperatures, and the fraction of solar energy is equal to one. Keeping with the process described in the earlier section shows the results of the thermal system (Q_h and Q_c), the area of the heat recovery network (AHRN), and the price for the dairy industry when ΔT_{min} is changed from 5°C to 25°C. The results show that the price of the dairy business goes down when ΔT_{min} goes from 5°C to 25°C. In the Pinch Analysis, the minimum hot utility has gone up from 220 kW to 522 kW, which is a 58% increase. However, the area covered by the heat recovery network has gone down by 60%.

The solar fraction is one at temperatures ranging from 5°C to 25°C ΔT_{min}. STE is able to fulfil the entire heat duty. The chosen scenario will be determined by the amount of space available for STE integration, the cost of the solar system, and the total cost of the integrated system.

With a rise in ΔT_{min}, there is a discernible shift in the annualized price of each module of the combined solar thermal system, as follows: when compared to the annualized expenditure of the heat recovery and solar collector network, the costs of ancillary services and the annualized expenditure of the thermal energy storing system are noticeably cheaper. These relationships can be utilized in a variety of ways to accomplish various design goals for the solar thermal system. It is conceivable to do away with the usage of fossil fuels entirely, which results in zero emissions of greenhouse gases being released into the atmosphere [11]. Despite the fact that this target is met with each kg of CO_2 produced, an extra objective is being pursued.

Case Study II

Case Study II examines the TSP curves and tools. It has a two-stream process, A, and s four-stream process, B. The difference in temperature between the process streams and the storage is $\Delta T_{pmin} = 10°C$, while the difference in temperature between the solar capture system and the storage is $\Delta T_{Cmin} = 5°C$. For the amount of heat that comes from solar sources, the data collected were the same as in the previous case study: solar collectors with an area of 1,950 m² and efficiencies ranging from 58 to 20%, operating at temperatures between 86 and 125°C. The amount of heat that comes from sun irradiation was obtained from a report published by the European Commission in 2008.

In order to construct a practicable heat recovery, it is necessary to take into account the differences in temperature shown in Figure 9.13. Equalization of the heat source 5 and the heat demand 6 is required in order to calculate the amount of heat that can be recovered (Figure 9.13). In conclusion, the construction of an minimal capture temperature curve as well as a captured solar energy curve is required in order to incorporate the solar thermal energy (7 and 8, Figure 9.13). The point where the captured solar energy curve and minimal capture temperature

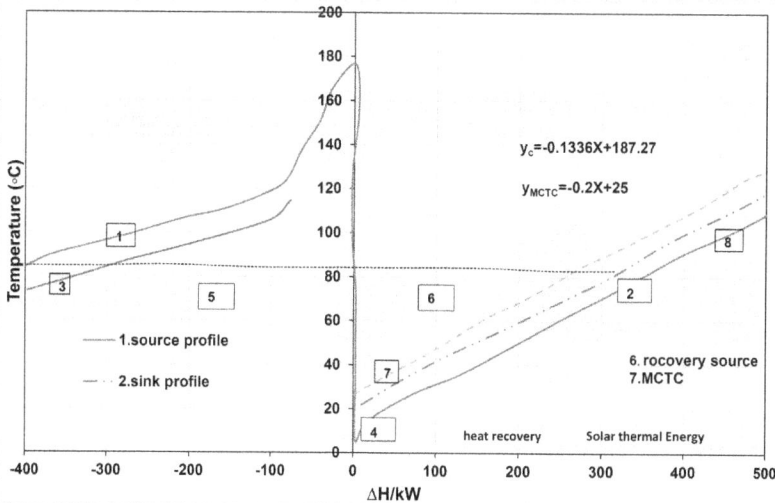

FIGURE 9.13   Grand Composite Curve for Case Study II.

curve intersect is located at a ΔH value of 486.4 kW and a TC value of 122°C (Figure 9.13). It reflects the total quantity of the heat demand that was satisfied by heat recovery in addition to the heat that was generated by solar thermal energy. When that quantity is subtracted from the heat recovery capacity (312 kW), the solar thermal energy integration capacity is found to be 174.4 kW. It is important to note that the total heat need was 1,050 kW (Figure 9.13), but that only 563.6 kW should now be supplied by a steady source of energy.

9.4 CONCLUSION

The method of design that was proposed, known as wide-ranging solar thermal amalgamation for process industries, made it possible to achieve multiple goals, including optimizing, in each of the possible cases, the utilization of STE, the amount of time the SCN was in operation, and the solar fraction. In both of the studies, the temperature measured at the pinch is lower than 100°C. In each of the case studies, many situations were analyzed, and the goals that were created varied from one case study to the next depending on the requirements of the processes. On the other hand, it takes a strategy that is founded on design purposes and aims to enhance the integration of STE in order to produce more ecological processes. A design space that allows for the fulfilment of goals to be reached simultaneously can be opened up with the assistance of ΔT_{min}. These goals include producing zero CO_2 emissions, increasing the working time of the SCN, producing excess energy that can be utilized in other systems, and adapting to partial installation surfaces, among other things.

REFERENCES

[1]. Nemet, A., Kravanja, Z., & Klemeš, J. J. (2012). Integration of solar thermal energy into processes with heat demand. Clean Technologies and Environmental Policy, 14(3), 453–463.

[2]. Fuentes-Silva, A. L., Lizárraga-Morazán, J. R., Picón-Núñez, M., & Martínez-Rodríguez, G. (2019). Incorporating the concept of flexible operation in the design of solar collector fields for industrial applications. Energies, 12(570), 1–20.

[3]. Eiholzer, T., Olsen, D., Hoffmann, S., Sturm, B., & Wellig, B. (2017). Integration of a solar thermal system in a medium-sized brewery using pinch analysis: methodology and case study. Applied Thermal Engineering, 113, 1558–1568.

[4]. Allouhi, A., Agrouaz, Y., Amine, M. B., Rehman, S., Buker, M. S., Kousksou, T., . . . & Benbassou, A. (2017). Design optimization of a multi-temperature solar thermal heating system for an industrial process. Applied Energy, 206, 382–392.

[5]. Abdelhady, F., Bamufleh, H., El-Halwagi, M. M., & Ponce-Ortega, J. M. (2015). Optimal design and integration of solar thermal collection, storage, and dispatch with process cogeneration systems. Chemical Engineering Science, 136, 158–167.

[6]. Baniassadi, A., Momen, M., & Amidpour, M. (2015). A new method for optimization of Solar Heat Integration and solar fraction targeting in low temperature process industries. Energy, 90, 1674–1681.

[7]. Martínez-Rodríguez, G., Baltazar, J. C., Fuentes-Silva, A. L., & García-Gutiérrez, R. (2022). Economic and environmental assessment using two renewable sources of energy to produce heat and power for industrial applications. Energies, 15(7), 2338.

[8]. Walmsley, T. G., Walmsley, M. R., Tarighaleslami, A. H., Atkins, M. J., & Neale, J. R. (2015). Integration options for solar thermal with low temperature industrial heat recovery loops. Energy, 90, 113–121.

[9]. Martínez-Rodríguez, G., Fuentes-Silva, A. L., Velázquez-Torres, D., & Picón-Núñez, M. (2022). Comprehensive solar thermal integration for industrial processes. Energy, 239, 122332.

[10]. Muster, B., Hassine, I. B., Helmke, A., Heß, S., Krummenacher, P., Schmitt, B., & Schnitzer, H. (2015). Solar Process heat for Production and Advanced Applications. Integration Guideline; IEA SHC Task, 49.

[11]. de Santos López, G. (2021). Techno-economic analysis and market potential study of solar heat in industrial processes: A Fresnel direct steam generation case study [Internet] [Dissertation] (TRITA-ITM-EX). Available from: http://urn.kb.se/resolve?urn=urn:nbn:se:kth:diva-293918

[12]. Brunner, C., Muster-Slawitsch, B., Meitz, S., & Frank, E. (2020). Solar heat integrations in industrial processes. IEA Solar Heating and Cooling Technology Collaboration Programme.

[13]. Atkins, M. J., Walmsley, M. R., & Morrison, A. S. (2010). Integration of solar thermal for improved energy efficiency in low-temperature-pinch industrial processes. Energy, 35(5), 1867–1873.

[14]. Ali, E. N., Liaquat, R., Ali, M., Waqas, A., & Shahzad, N. (2022). Techno-economic and GHG mitigation analyses based on regional and seasonal variations of non-concentrating solar thermal collectors in textile sector of Pakistan. Renewable Energy Focus, 42, 165–177.

[15]. Kumar, P., Sinha, K. K., Đurin, B., Gupta, M. K., Saxena, N., Banerjee, M. K., . . . & Kanga, S. (2022). Economics of implementing solar thermal heating systems in the textile industry. Energies, 15(12), 4277.

[16]. Tasmin, N., Farjana, S. H., Hossain, M. R., Golder, S., & Mahmud, M. P. (2022). Integration of solar process heat in industries: a review. Clean Technologies, 4(1), 97–131.

[17]. Mohammadi, K., Khanmohammadi, S., Immonen, J., & Powell, K. (2021). Techno-economic analysis and environmental benefits of solar industrial process heating based on parabolic trough collectors. Sustainable Energy Technologies and Assessments, 47, 101412.

10 Solar Thermal Energy Systems Life Cycle Assessment

Abolfazl Ahmadi and Amir Hosein Saedi

CONTENTS

DOI: 10.1201/9781003263326-13

10.1 INTRODUCTION

Solar thermal energy (STE), which is much more effective than photovoltaics, is a method of obtaining thermal energy from the sun. There are low-temperature flat-plate collectors, mid-temperature (typically flat-plate) collectors, and high-temperature (lens/mirror) collectors, which are used to heat air and water in commercial and residential settings, and generate electricity, respectively (Morley, David *et al.* 2014).

10.1.1 Low-Temperature Collectors

Basically utilized to warm swimming pools, this type of collector can also be used to heat space. By using water/air as a medium, heat is transferred to a destination Norton, Brian (2013).

10.1.1.1 Heating, Ventilation and Cooling

In America, heating, ventilation and air conditioning (HVAC) systems are responsible for more than 25% (4.75EJ) of commercial use and about 50% (10.1EJ) of residential use of energy. To tackle the problem, solar systems can be a good choice. Thermal mass materials, such as water, concrete and stone, save the sun's energy in the daytime and release it in cooler periods. Daylight, shading and climate conditions are important in deciding on the thermal mass proper placement and proportion, in a way that comfort temperatures are maintained and energy consumption is reduced. A thermal or solar chimney is a passive ventilation system for connecting a building's internal and external parts (Allen, J.A., Rogers, D.F 2014). When the system and the air inside are heated, an updraft results, thus pulling air through the building. With a history as long as the Roman emperor, these chimneys are still used in the Middle East (Alliger, G. 2011). In America and Canada, where most buildings are equipped with ventilation systems for cooling and heating, air collectors are preferred to liquid collectors for space heating (Ardente, F., Beccali, Atl 2003). Solar air panels are either unglazed or glazed. The latter is mainly used for heating purposes in residential buildings, in which air circulation is done via a solar air panel that heats the air and returns it to the building. These space heating devices need two or more house penetrations and work only if the solar collector has a higher air temperature than the building. The former is mainly used for preheating purposes in industrial, institutional and commercial places with high loads of ventilation, in which the building walls are transformed into efficient, inexpensive unglazed solar collectors. These transpired solar panels utilize a painted perforated metallic heat absorber, functioning as the external wall of the building. Heat is conducted from the absorber's surface to the 1-mm thick thermal boundary air layer outside absorber and to the passing air behind it. Via convection to the air outside, the boundary air layer is caught into an adjacent perforation, and subsequently the hot air is drawn from behind the absorption plate into the ventilator. As a ventilation and passive solar heating device, a Trombe wall involves an air conduit fit in between the window and the heat mass under solar irradiation. Sunlight reserves warmth in the mass and heats the air passage, causing a vent cycle and circulation at the wall's bottom and top.

Through heating, the Trombe wall emits the reserved heat. In the 1960s, Harold Hay developed a solar roof pond system (Harold R. Hay 1960), which is basically a roof-installed aqua bladder insulated with a portable shield, by which heat transfer between indoor and outdoor spaces is controlled. The bladder is uncovered in daylight to save heat for dark hours. For cooling purposes, the system is covered in the daytime to extract heat from the building interior while it is uncovered in the nighttime for heat radiation to the cooler ambience (Veenstra, A. *et al.* 2002). A rooftop basin prototype was developed by Skytherm in Atascadero, California for cooling and heating. Active solar cooling is attained by the cycles of absorption, refrigeration and desiccation under mechanical procedures. Paul Collins (2002) introduced solar cooling through ice making by means of a solar steam engine connected to a refrigerating apparatus (Ardente, F. *et al.* 2003). Cooling can also be provided using smart windows, thermal mass and shading approaches. Deciduous tree leaves create natural shade in summertime, whereas bare branches in the wintertime permit heat and light entrance into houses. Local temperatures are also moderated by the trees' water content. To supply hot water or space heating for nonresidential areas, solar process heating can be employed (Tsilingiridis, G. *et al.* 2004). The shallow evaporation ponds evaporate and concentrate dissolved solids. Evaporation pond applications range from the traditional process of salt extraction from seawater to more modern applications like concentrating saline solutions for leachate extraction and omitting dissolved solids from effluents. As a perforated wall exposed to sunlight, the Unglazed Transpired Collector (UTC) is intended for the ventilation air preheating as it can heat the inlet air to 22°C maximally and can supply an outlet temperature from 45 to 60°C. The short-time (3–12 year) payback of transpired collectors makes them an inexpensive choice in glazed systems. By 2009, more than 1,500 systems with a 300,000m^2-collecting area had been installed around the world.

10.1.2 Medium-Temperature Collector

These collectors produce at least half of the hot water needs of residential and commercial sectors in the U.S.A., where 30% of the system is bound to the federal tax credit, plus an additional credit in some states. In southern and northern climates, a noncomplex open-loop system installation might need 3–5 h and 4–6 h of labour, respectively. It needs further collector areas and more advanced plumbing to avoid the freezing of the collectors, thus the payback time per household in different states might be 4–9 years. European grants are similar. A solar plumber with 2 co-workers with basic knowledge can mount a system daily. Thermosyphon maintenance costs are negligible, although the use of antifreeze and main power might add to the costs. The operating costs of a typical American household is also cut by $6 per person monthly. Solar water heating might cut a four-member family's CO_2 emission by 1 tonne per year (natural gas replacement) or by 3 tonnes/year (electricity replacement).

Common designs of medium-temperature collectors are drain systems, pressurized glycol, batch and low-pressure freeze tolerant polymer water pipe having photovoltaic pumping (Gong Cheng *et al.* 2022). European and international operational innovations involve permanently wet manifolds that decrease or even omit stagnation (a non-flowing high-temperature stress), which would otherwise threaten a collector's life expectation.

10.1.2.1 Solar Drying

Solar thermal power is suitable for desiccating construction wood, wood fuel chips and other biomass forms, and food products such as fish, cereals and fruits. The process is high-quality, eco-friendly and inexpensive (Gardon, J.L 2022). Solar drying involved cost-efficient, black-fabrics-based breathable plate air sensors. By increasing heat, the solar thermal energy permits air to pass and discard moisture.

10.1.2.2 Cooking

In a solar cooker, the cooking, pasteurization and drying are done using sunlight while a cost-effective, eco-friendly and clean air process is accomplished. In 1767, H. Saussure invented a simple box cooker with an insulated container having a see-through lid, which is efficiently useable with partially overcast skies at 50 to 100°C. In a concentrating solar cooker, reflectors (with a flat-plate, disc or parabolic geometry) are used to concentrate light onto a vessel for cooking. The models provide faster cooking at temperatures up to 350°C, though their proper functioning needs direct light. The solar bowl is a distinctive technology used in solar kitchens in Auroville, India. In the solar bowl, unlike a usual tracking reflector/fixed receiver, a fixed spherical reflector is used along with a receiver tracking the light focus according to the sun movement in the sky. At 150°C, the receiver produces steam which allows it to cook 2,000 meals per day. Another concentrating technology in many Indian solar kitchens is the Scheffler reflector, which was introduced in 1986 by Wolfgang Scheffler. In this tracking dish, the daily movement of the sun is recorded based on the single axis tracking. Equipped with a flexible reflective surface, the device can alter its curvature in compliance with seasonal change in sunlight angle of incidence. By having a constant focal point, the reflector facilitates firing and can arrive at 450–650°C.

The greatest Scheffler reflector, which was created in 1999 in Rajasthan, India, can prepare over 35,000 meals daily. By 2008, more than 2,000 Scheffler designed large cookers had been made around the world.

10.1.2.3 Distillation

Solar stills may be applied to produce potable water in regions where drinking water is scarce. Solar distilling is required in such cases to supply individuals with clean water. The sun's energy warms the water inside the still. Water evaporation and condensation subsequently occurs on the base of protective glass.

10.1.3 HIGH-TEMPERATURE COLLECTORS

In cases where a temperature lower than approximately 95°C is good, for example for heating space, flat, non-centring collectors are usually employed. Since heat losses are quite high in the glazing, the flat plate manifolds cannot reach temperatures exceeding 200°C although the medium for heat transfer might be stagnant. Temperatures like this are too low to be efficiently converted into electricity. The performance of the heat generators improves as the heat source temperature increases. Accordingly, in solar thermal power systems, using the concentrated solar power (CSP) technique, sunlight concentration is achieved through lenses or mirrors to attain higher temperatures. The real impact of high efficiency is a reduction in the size of the plant manifold and the overall use of land per unit of energy produced, thereby reducing a power plant's environmental impacts and its expenses.

By rising temperatures, the conversion practically increases. To a maximum degree of 600°C, standard steam turbines are up to 41% efficient. At higher than 600°C, gas turbines show further efficiency. Further temperatures might cause problems as various materials and technologies are needed. To tackle these problems, liquid fluoride salts functioning in the 700°C to 800°C range can be exploited using multi-stage turbines to attain thermal efficacy of at least 50%. When operating temperatures are higher, the plant can utilize higher-temperature dry heat exchangers for the thermal exhaust gases, hence water use of the plant is reduced, which is of great importance in arid areas with larger solar power installations. At a high temperature, heat storage efficiency also increases with the increase of the storage of watt-hours per unit fluid. Since heat generation in the CSP occurs first, heat is stored earlier than electricity storage, an approach which is less expensive and more efficient. As such, electricity generation takes place day and night. When solar radiation in a CSP site is predictable, the CSP power plant reliability increases, particularly when a fossil fuel-based backup system is installed. As the CSP plant reusability under a standby system is practical, the system costs are reduced. Besides the problems of reliability, unexploited arid and semiarid areas, fuel cost and pollution, the large-scale deployment of CSP is hindered by factors like aesthetics, expenses, land usage and high-voltage connection lines. Despite the existence of more than enough desert areas to supply a global demand for electricity, power supply still requires the installation of many mirrors or lenses in such areas. If a simplified design is used, the costs can be cut. Sun positions change all day long. In structures with low temperature and low concentration, tracking may be prevented (or restrained to some positions each year) when using non-imaging optics. In a higher concentration, when lenses or mirrors do not have any movement, their focus varies; however, the optics still give the largest acceptance angles to a given concentration. Consequently, an urgent need is felt for a systematic tracking of the solar position (a sun tracker is possible for solar PV), though cost and complexity might soar. Different tracking systems are recognizable in terms of their method of focusing light and tracking the solar placement.

10.1.3.1 Parabolic Bowl Designs

In a parabolic trough power plant, a bended, mirrored trough is used to project direct sunlight onto a fluid-holding glass tube (called also a collector, receive, or absorber) along the trough length, located at the reflector's focal point. The trough is orthogonally linear while it is parabolic along the other axis. For changing the sun's daily position vertically to the absorber, the bowl tilts westward to concentrate irradiation directly on the absorber. Nonetheless, seasonal variations in solar angle relative to trough do not need mirror adjusting, as light is simply focused elsewhere on the absorber, hence tracking on a secondary axis is not needed (Figure 10.1).

A glass vacuum chamber can be designed to enclose the receiver, thus significantly cutting heat loss via convection. A heat transfer fluid (e.g., melted salt, artificial lubricant and pressurized steam) runs through the absorber and gets very hot, which is subsequently carried to a heat engine in which around 1/3 of the heat changes into electricity. In Andasol 1 in Guadix, Spain, a parabolic trough with long parallel rows of modular sun collectors is used (Solar Millennium AG. 2008). The high-precision reflection panels

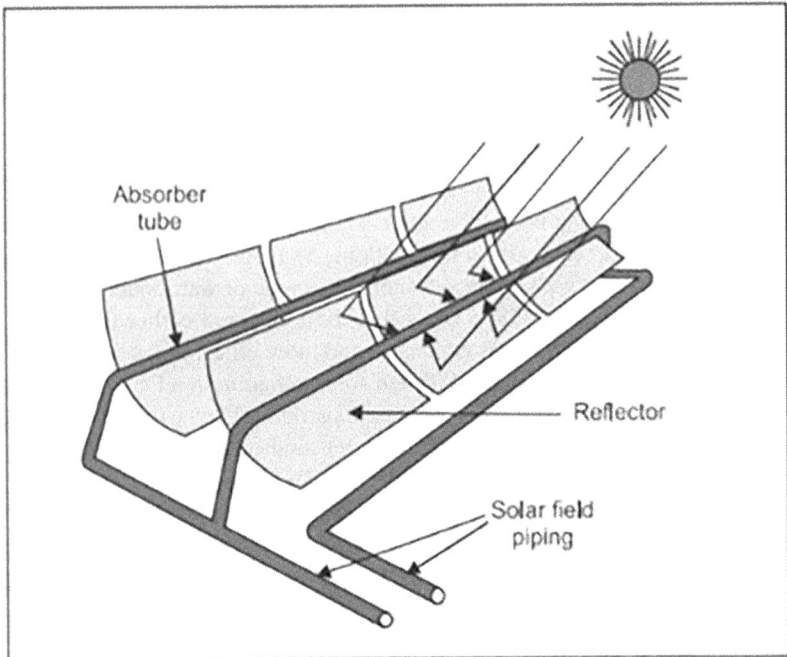

FIGURE 10.1 Design of a parabolic trough. Variations in solar position versus receiver do not entail adjusting mirrors.

track the east-to-west solar motion via rotation on one axis and thus concentrate sunlight on an absorption pipe along the collector's focus line. A heat transmission fluid, such as engine oil, circulates inside the absorption pipes (400°C maximally) and creates live steam to move the turbine generator. Large-scale parabolic systems have a great number of these troughs arranged in a parallel manner over a large region. A solar thermal system (called SEGS) based on this pattern has been functioning in California, U.S.A., since 1985. Parabolic troughs are currently the longest-established and experienced CSP approach. The solar power generation system (SEGS) involves 9 power plants with 350-MW capacity in total and is now the largest of all (thermal and non-thermal) operational solar systems. The more recent Nevada Solar 1 has a 64-MW capacity. Besides, Andasol sites 1 and 2 are under construction in Spain, each with a 50-MW capacity. These facilities, however, demand a larger array of sun collectors, compared to steam turbine-generator size, to keep heat and transfer it to steam turbine simultaneously [12], which leads to a more appropriate exploitation of the turbine. As the steam turbine operates day and night, Andasol 1 at its highest capacity outperforms the Nevada system at its peak capacity in terms of energy generation, which can be ascribed to the thermal energy storage of the old power plant and the larger solar field. A new capacity of 553 MW was also introduced in the solar park in Mojave, California. Additionally, a 59-MW hybrid power station with heat storage was designed in Barstow in California (Souliotis, M.K. *et al.* 2003). In Kuraymat in Egypt, a gas-fired power plant was designed which uses a 40-MW steam supply as input. Finally, a 25-MW supply of steam is utilized for a gas-fired power station in Hassi R'mel, Algeria.

10.1.3.2 Solar Towers

These towers (also called centre tower or heliostat power station) are equipped with thousands of the so-called heliostat mirrors to track and concentrate the thermal energy of the sun over a field of about 2 square miles, in the centre of which a tower is situated. Heliostats focus the solar light on a receiver on the tower top, where the concentrated light warms melted salt to > 538°C. The salt goes into a thermal reservoir, where 98% of thermal effectiveness is reserved. The product is subsequently pumped into a steam generator, and electricity is finally produced by a standard turbine that consumes steam. This allegedly called Rankine Cycle procedure resembles a standard coal-based power station, except for the fact that it is powered by the free and clean energy of the sun. In this design, compared to a parabolic bowl, a more effective conversion of thermal energy into electricity at higher temperatures and a cost-efficient energy storage are attained. In addition, the need for the ground surface flattening is minimized. A power tower can basically be constructed over hillsides. There are flat mirrors and concentrated plumbing in the tower. The system's shortcoming is that each of the mirrors requires a dual axis drive of its own, while in a parabolic design a shared axis is considered for many mirrors. In 2008, the Google-funded eSolar Corporation established by Bill Gross, CEO of Idea Lab, signed a power purchase agreement, PPA, with Edison to generate 245 MW of power annually. The next year, eSolar licensed its technology to NRG Energy Inc. (to cooperatively make 500 MW annually in concentrated solar thermal power stations) and to ACME Group (which began building the first eSolar station in 2012 to generate 1 gigawatt power over 10 years). The eSolar's patented solar tracking software incorporates the motion of 24,000 m^2 mirrors per 1 revolution via optic sensors for real-time adjustment and calibration of the mirrors. As a result, a high density of reflective materials is provided to develop modular CSP power stations in 46-MW units over 1 square mile land parcels, hence a land-power ratio of 16,000 m² (4 acres) to 1 MW. In March 2008, BrightSource Energy signed several PPAs with Pacific Gas and Electric Company to generate up to 900 MW power annually, the greatest solar energy promise committed by a utility. BrightSource has been constructing some solar energy stations in Southern California, the first of which began in 2009. In 2008, Solar Energy Development Center (SEDC) was opened in the Negev desert, Israel, by BrightSource Energy for 4 to 6 MW. Over 1600 heliostats are installed in the Rotem site to track the sun and reflect sunlight on a tower that is 60 m in height. The focalized energy is exerted to warm a tower-top boiler to 550°C, creating super-heated steam. The PS10 power tower in Spain has an 11-MW capacity. A 15-MW Tres system with heat keeping is going to be constructed in Spain. A 100-MW solar energy station has been designed in South Africa. In this design, between 4,000 and 5,000 heliostatic mirrors are used, each occupying an area of 140 m².

In 2012, 5 solar thermal energy stations were designed in Ouarzazate, Morocco, to generate about 2000 MW of energy annually, for which over 10,000 hectometres of land were required (Neslen *et al.* 2020). The 10-MW Solar One plants (later expanded and transformed into Solar Two) and the 2-MW Themis plants are out of service. NREL made a performance-cost comparison between power towers and parabolic troughs and proposed that by 2020, power tower and parabolic trough power generation would be 5.47 ë/kWh and 6.21 ë/kWh, respectively. The capacity factor of the electricity pylons was estimated at 72.9% and at 56.2% for the parabolic troughs. It is hoped that the costs can be cut by developing inexpensive, mass-producible, and durable solar plant components.

10.1.3.3 Parabolic Dish Collectors

A dish Stirling system has a big reflecting dish (which resembles a satellite TV dish antenna), in which all the coming sunlight is concentrated onto a single point above the dish and a receiver gathers the heat and changes it to a convenient form. The dish is usually coupled to a Stirling engine in a Plat-Stirling system though a steam engine is occasionally employed. Accordingly, a rotational kinetic energy is created, which is convertible into electricity through a power generator. A dishwashing system offers a better conversion into electricity because the dish can attain greater temperatures owing to the greater light concentrations (like tower protocols). However, some disadvantages also exist. Heat conversion into electricity needs a moving structure (including a heavy motor that demands a rigid framework and a robust tracking system) and involves maintenance. In the dish system, a centralized method of conversion is preferred over a decentralized one. Moreover, parabolic (rather than flat) mirrors are utilized and tracking is thus dual axis.

10.1.3.4 Fresnel

The linear Fresnel reflector power plant exploits a number of long, narrow, slightly curved (or flat) mirrors so that light can be focused onto 1 or multiple linear receivers above mirrors. A small-size parabolic mirror might be annexed above a receiver to increase light concentration. In order to reduce the overall costs, 1 receiver is shared between numerous mirrors in these systems (compared to cuvette and parabola designs), while a single-axis straightforward linear focus geometry is used for tracking—as is the case with troughs, as opposed to dual axis flats and centre towers. The need for fluid fittings is discarded since a stationary receiver is used (as in dishes and troughs). Mirrors are based on a simpler structure as they are not to support the receiver. Using appropriate aiming strategies (mirrors aiming at various receivers at dissimilar daily hours), a denser packing of the mirrors on the existing land surface is attained. Such systems have recently been constructed in Belgium (Solarmundo project) and Australia (CLFR). Solarmundo, piloted in Liège, was closed after effectively proving the linear Fresnel technique. The Solar Power Group GmbH (SPG) in Munich was subsequently established by several Solarmundo members. The SPG cooperated with the Aerospace Center (DLR) and built a Fresnel-based prototype for direct steam production in Germany.

10.1.3.5 Linear Fresnel Reflectors

One-axis tracking designs include linear Fresnel reflector (LFR) and compact LFR (CLFR). Unlike a parabolic trough, the LFR receiver is placed over a mirror field. Additionally, a reflector has numerous low-row parts, focusing collectively on a projected long-tower absorber parallel to the rotation axis of reflector. The design might cut the costs because the row of absorbers is common for several mirror rows. A basic problem with LFR technology is to avoid the shading and the blocking of the projected irradiation from the nearby reflectors. To decrease the shading and blocking problems, raised receiving towers might be used or absorber size should be enlarged; however, more land use is required, which increases the costs. Another solution has been offered by the CLFR. In the conventional LFR, just a linear receiver on a linear tower exists, which excludes any orientation direction of an individual reflector. As this system might be used in a wide area, it can be assumed that the system involves numerous linear absorbers. Consequently, when the absorbers are sufficiently close, the individual reflectors can direct the projected solar radiation to 2 or more absorbers. This offers potential for a denser array, as alternate reflector tilt patterns can be configured in a way that tightly packed reflectors are arranged without shadowing or blocking. The CLFR system cuts the costs in all the array elements, which is an encouragement for the advancing of this technique. The system's cost effectiveness compared to the parabolic trough, is in terms of construction costs, parasitic pumping losses and maintenance costs. The decreased structural cost is owing to the substitution of elastically curved or flat glass reflectors for the costly sagging reflectors near the surface level. Additionally, heat transfer loop is distinguished from the reflective field, which is different from trough systems incurring the high cost of high-pressure flexible lines. Decreased parasitic loss is thanks to using water for passive direct-boiling heat transfer fluid. The glass evacuated tubes certainly and inexpensively decrease radiative losses. Research on the CLFR systems has demonstrated that they provide a tracked beam at 19% electricity efficiency each year as a preheater.

10.1.3.6 Fresnel Lenses

Prototype Fresnel lens concentrators are designed to harvest thermal energy under international automated systems (Haberle, A. *et al.* 2002). In practice, there is not any large-scale thermal system based on Fresnel lenses, although products with Fresnel lenses accompanied by photovoltaic cells exist. In this protocol, lenses are cheaper than mirrors. Moreover, if a substance with some flexibility is selected, a less stiff frame is needed to resist the load of wind. A novel design of a lightweight, non-disturbing sun concentrator using asymmetrical Fresnel lenses that takes up minimal floor space and permits much concentrated solar energy is observed in the Desert Blooms project, although a prototype has not yet been attained (Clery, D. 2011).

10.1.3.7 Micro-CSP

Micro-CSPs are sunlight-based thermal systems with concentrating solar power (CSP) collectors on the basis of conventional concentrating sun energy systems in the Mojave Desert. These systems are smaller and lighter and work at temperatures ordinarily lower than 600°F (315°C), and are intended for modular installation in the field or on the roof to have easy protection from snow, wind and wet deployments. Sun panel maker Sopogy has made a 1-MW CSP design at Natural Energy Lab in Hawaii. Applications include community-size power stations (1–50 MW), industries, agriculture, and where much hot water is required, such as swimming pools, recreational resorts, water parks, laundries, distillation, sterilization, etc.

10.1.3.7.1 Thermal Energy Storage

Gain, transmission, storage, transit and insulation are the 5 principles that govern the heat in the system. As a result, heat is a measurement of an object's thermal energy, and it is determined by its mass, temperature and specific heat. Heat exchangers in solar thermal energy systems are designed to provide heat exchange in steady working conditions. The sun-accumulated heat in the design is referred to as heat gain. The greenhouse impact causes the entrapment of solar thermal heat. In this scenario, the greenhouse effect is caused by a reflective surface that can conduct short-wave light while also projecting long-wave radiation. If short-wave light strikes the receiver's plate and is trapped inside the collector, heat and infrared (IR) radiation are produced. The trapped heat is collected by the fluid, typically water, in the absorber tubes and is then transferred to a vault for storing heat. The transfer of heat is through convection or conduction. In the presence of hot water, kinetic energy is conducted to water molecules through the medium. The conduction process is responsible for the transfer of the thermal energy of hot water molecules, which occupy a greater space than the slowly moving cold molecules above them [9]. Convection is aided by energy transfer from rising heated water to flowing cold water. Conduction transports heat from the collector's absorber plates to the water. Collector fluid circulates from the carrier pipes to the vault, where heat is convectively transmitted throughout the medium. Because of heat storage, power plants can generate electricity for hours even when there is no sunlight. In sunny hours, heat is stored in an insulated tank and is released for making electricity in nonsunny times. Heat transmission ratio is associated with convective and conductive media and temperature differences. Heat is transferred faster when temperature differences are greater. The activity of heat transportation from a collector to a storage vault is referred to as heat transport. Thermal insulation is essential both in the heat transport tubes and, in the vault, as it precludes the loss of heat in relation to the decreased system efficiency or energy loss.

By storing heat, a thermal plant can generate electricity nightly and in cloudy weather. This enables solar power application for generating peak power and base load, while it can substitute coal and natural gas-based sites (Kalogirou, *et al.* 2004). In addition, generator utilization is higher, hence cutting costs. Heat is kept in an insulated tank in the daytime and is released for making electricity in non-sunny times. Among the thermal storage media are various phase change materials, concrete, pressurized steam and molten salts like potassium nitrate and sodium.

10.1.3.7.2 Steam Accumulator
The PS10 tower stores heat (in tanks) as pressurized steam at 285°C and 50 bars [9]. The vapour condenses and becomes vapour again when the pressure is reduced. Storage time is 1 h. A longer storage time is conceivable, though not realized in any power plant yet.

10.1.3.7.3 Storage of Molten Salt
Many fluids (e.g., air, water, oil and sodium) have been examined to carry solar heat, but molten salt seems to be the best choice in solar power towers as this non-flammable and non-toxic material is liquid under atmospheric pressure, provides an effective and low-cost means for thermal energy storage and its working temperatures are in harmony with steam turbines at today's high temperature and pressure. The melted salt system has already been employed in metal and chemical and non-solar environments as a heat transfer medium. Molten salt contains 60% sodium nitrate and 40% potassium nitrate. Technically viewed, the addition of calcium nitrate to the salt mix is advantageous and might decrease costs. The salt melts at 220°C and remains liquid at 290°C if stored in an insulated container. This system is unique due to the decoupling of solar power collection from energy production, the possibility of power generation under undesired weather conditions or in the night-time through thermal power stored in the well-insulated liquid salt tanks that store energy for 7 days. As a size reference, reservoirs with sufficient thermal storage to run a 100-MW turbine for 4 h are approximately 9 m in height and 24 m in diameter [18]. The Spanish Andasol station was the first commercial consumer of melted salt for heat reservation and nocturnal production. Two years later, on July the 4th, 2011, Torresol's 19.9-MW concentrated solar energy site in Spain incessantly generated electricity for 24 h by a molten salt-based heat reservoir.

10.1.3.7.4 Graphite Heat Storage
Direct: The power tower at Cloncurry in Australia stores energy in purified graphite which is placed on the tower top. The heliostats' heat is directly stored. Therefore, heat for electricity creation is simply obtained from graphite (Crawford, R.H *et al.* 2004).

Indirect: Heat transfer from reflectors to storage vaults is facilitated by molten salt coolants. The heat is transferred from salts to an auxiliary heat transmission fluid through a heat exchanger and subsequently to the storing space, or rather the salt is directly consumed for heating the graphite. Graphite is relatively cost-efficient and compatible with liquefied fluoride salt. Graphite offers a good storage medium considering its great mass and volumetric capacity.

10.1.3.7.5 PCM

Phase change materials (PCM) are another strategy for power storage. Under a similar heat transmission framework, PCMs can offer storage more effectively. An organic PCM does not involve corrosives, and needs slight, if any, sub-cooling and is chemically and thermally stable. However, it is flammable and suffers from a limited heat conductivity and a low enthalpy. Inorganic PCM, on the other hand, shows a greater phase-change enthalpy, but it also lacks thermal stability and is problematic in terms of corrosion, sub-cooling and phase separation (Kinga Pielichowska, *et al.* 2014). The higher enthalpy in inorganic substances causes the hydrate salt to be a viable choice in the solar power storage field.

10.1.3.7.6 Water Use

A protocol that needs water for cooling or condensation purposes is perhaps in opposition to solar stations siting in arid lands with scarce water supply. An example is the Solar Millennium Company's plan to construct a station in the Amargosa Valley, Nevada, consuming 1/5 of the existing water in the region. This is also the case with California's Mojave Desert where acquiring sufficient and suitable water rights is difficult or impossible (as regards the rules banning drinkable water consumption for cooling). Some models entail less use of water, for example the Ivanpah project in southeastern California is based on the application of air cooling to perform steam-to-water conversion, yielding a 90% decrease in water consumption though with a lowered power yield. The water circulates back again to a boiler under a closed, eco-friendly procedure.

10.1.3.7.7 Solar Energy Conversion Rate

In total, the solar dish/Stirling motor shows the greatest energy performance. A single flat Stirling sited at Sandia National Laboratories National Solar Thermal Test Facility maximally produces 25 kW of power, and a 31.25%-conversion efficiency. Parabolic trough solar power stations are constructed with an efficiency of around 20% (Y. Bravo, *et al.* 2013). Fresnel reflector efficiency is a little lower—which is balanced by a denser packing. Gross conversion effectiveness (considering the fact that only a slight part of the entire site area is occupied by solar troughs or pans) is determined by the net production capacity on solar power divided by the total plant area [4]. The 500-MW-SCE/SES would extract ~2.75% of radiation (1 kW/m^2) distributed over its 4,500 acres (18.2 km^2) (Diakoulaki, D. *et al.* 2001). Regarding the 50-MW plant under construction in Spain (total area 1300×1500 m = 1.95 km^2), the gross efficiency of conversion is 2.6% (Berenson, C. 2018). Also, efficiency is not related to the cost directly: when computing the total cost, efficiency and maintenance and construction costs must be regarded (Hampel, B.H. 2006).

10.1.3.7.8 Types of Solar Power-Based Water Heater

Water heaters based on solar energy are either active (with an electric pump for fluid circulation) or passive (without pump). The hot water quantity produced depends upon the system's size and type, the amount of the accessible sunlight, proper deployment, inclination angle and orientation of the sensors (Bhatia, S.C 2018). The systems are regarded as either open/direct or closed/indirect loop. An open-loop design flows drinkable water via the manifold. A closed-loop one employs a transport fluid (e.g., water) to gather warmth and a heat exchanger to change warmth to drinkable water.

10.1.3.7.9 Direct and Indirect Designs

A direct system circulates drinking water via collectors. Such systems are more cost-efficient and better transfer heat from collectors to the tank, and yet disadvantageous in that:

1. They are vulnerable to overheating.
2. They are vulnerable to frost.
3. Collectors accumulate scales in hard water regions.

Typically, they are unfit for cold weather because, upon freeze damage to the collector, the pressurized water pipes force water to spurt out of the frost-hit collector until the issue is detected and corrected (Figure 10.2). Indirect ones apply a heat exchanger to distinguish drinking water from a heat transfer medium (HTF) circulating via the collector. FHTs are commonly water or a water-antifreeze mixture based on a non-toxic propylene glycol. After warming HTF in panel, it goes to the heat exchanger for heat transfer to the drinking water. The indirect systems, although less cost-effective, might offer protection against both freezing and overheating.

10.1.3.7.10 Passive and Active Designs

A passive design depends upon thermal convection or heat pipes for circulating water or warming the fluid inside the system. Passive designs are less costly and require little, if any, maintenance service; however, they are comparatively less efficient and mainly suffer from freezing and overheating. Active heating designs apply pump(s) for circulating heating fluid and/or water through the system (Figure 10.3).

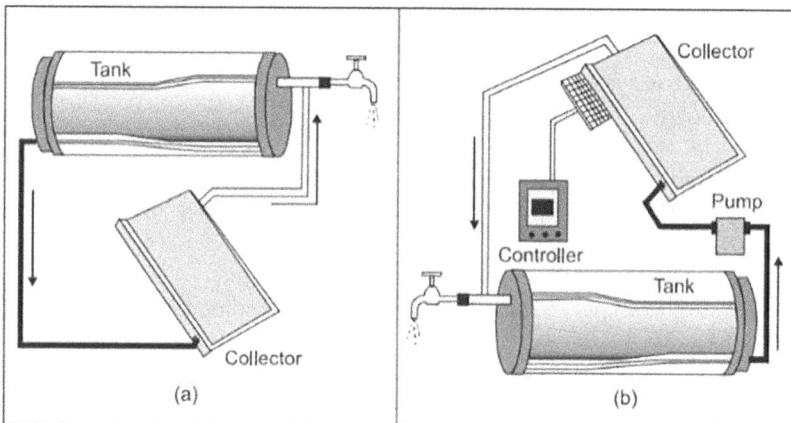

FIGURE 10.2 Direct design: (a) passive CHS having a tank over the collector; (b) active design with photovoltaic panel-driven controller and pump.

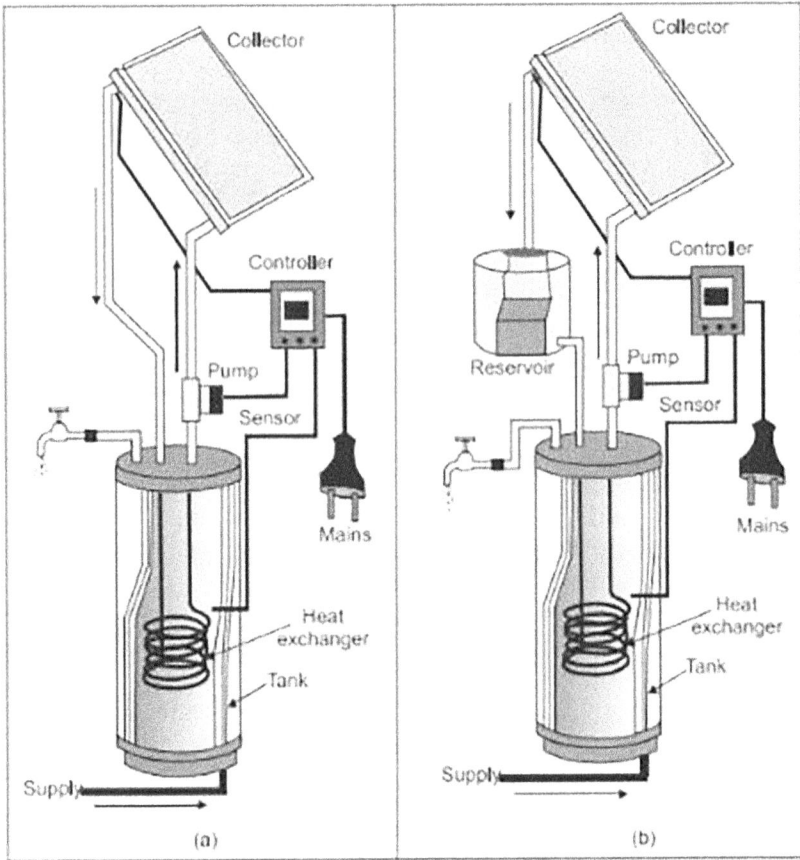

FIGURE 10.3   Indirect active designs: (a) using heat exchanger in tank, (b) drain design having drain tank. In the diagrams, both pump and controller are mains-powered.

While the active systems are likely to somewhat increase the costs, they have some
merits as follows:

1. The reserve tank might be located beneath the headers, and thus the design
   convenience increases and the already-available storage tanks are easily
   applicable.
2. If need be, the tank can also be concealed from sight.
3. It is possible to put the storage tank in a (semi)conditioned space, decreasing heat loss.
4. It is possible to employ dump tanks.
5. Greater effectiveness.
6. More system control.

Recent active water designs are equipped with electronic controllers that yield numerous functionalities like changing the system-controlling parameters, interacting with a

backup gas or electric water heater, calculating and logging of energy reserved by an SWH system, far access, security and numerous informative representations like reading temperature (Baum, B. *et al.* 2013). The highly popular differential pump controller detects temperature differences of water output of the collector and water inside the storing tank around heat exchanger. In a usual active design, the pump is turned on by a controller when the water inside the heater is ~8–10°C hotter than the tank water, and the pump is off as the difference in water temperatures reaches 3–5°C. Therefore, with the pump running, the water receives warmth from the collector and hence the pump is prevented from unnecessarily turning off and on. (In direct designs, such an on differential is reducible to ~4°C since the heat exchanger is not obstructed at all.) Some active SWHs employ energy from a small-sized photovoltaic panel (PV) to provide power for one or more speed-variable DC pumps. To ensure pump(s) longevity and proper performance, the DC pump and the photovoltaic panel need to be in harmony. These are generally of an antifreeze variety without controllers, because the manifolds are dominantly hot in the sunshine (while the pump(s) are running). Nonetheless, a differential regulator (driven by a photovoltaic panel's DC output) is applied to prevent the pumps from running when sunlight exists to drive the pump but collectors are even colder than stored water (U. Desideri *et al.* 2013). In a photovoltaic system, as an advantage, the solar hot water is collectable in electricity outages if the sun is still in the sky (Clauser, H.R 2009). An active design might also have a bubble or a geyser pump rather than an electrical pump. The geyser causes the HTF to flow between the collector and the storage tank using the sun's energy while no external source of energy is applied; this pump is appropriate for flat panel and vacuum tube systems. The closed HTF circuit in a geyser pump is under decreased pressure, and hence the fluid boils at lower temperatures when heated under the sunlight. The vapour bubbles create a bubble pump, hence the flow becomes upward. Accordingly, the bubbles leave the hot flow and condense at the peak circuit point, followed by the downward flow of the fluid to the heat exchanger considering the fluid level differences. FHT usually reaches the heat exchanger at 70°C and comes back to pump at 50°C. In frosty weathers, FHT contains water and propylene glycol antifreeze, typically in a 60:40 ratio. Pumping typically begins at ~50°C and rises at sunrise to a point of equilibrium, subject to the heat exchanger efficiency, hot water temperature and solar radiation strength.

10.1.3.7.11 Passive Direct Design

Integrated collector storage, ICS, or batch heating is based on a tank with both storage and collecting applications (Souliotis, M. *et al.* 2004). A batch heater is essentially a thin straight tank with a southward glass side. While having a straightforward and rather inexpensive design, compared to plate and tube collectors, some additional bracing might be needed in case of rooftop installation of the ICS, considering its high weight when filled with water [400–700 lbs]. Furthermore, the night-time losses of heat are considerable as the sunward side is mostly uninsulated, making the design only appropriate in moderate weather conditions. A convection heat storage unit (CHS) resembles the ICS, except for a convection-driven transfer between the collector and the tank as these two are physically separated (Nandi, B.R. *et al.* 2012). In a CHS design, standard flat-plate or evacuated tube headers are typically used, and the storing tank is located above the headers for a proper convection. The CHS is preferable to the ICS as it minimizes heat losses. In CHS, it is possible to insulate the tank more properly. Besides, as the panels are situated underneath, their heat loss

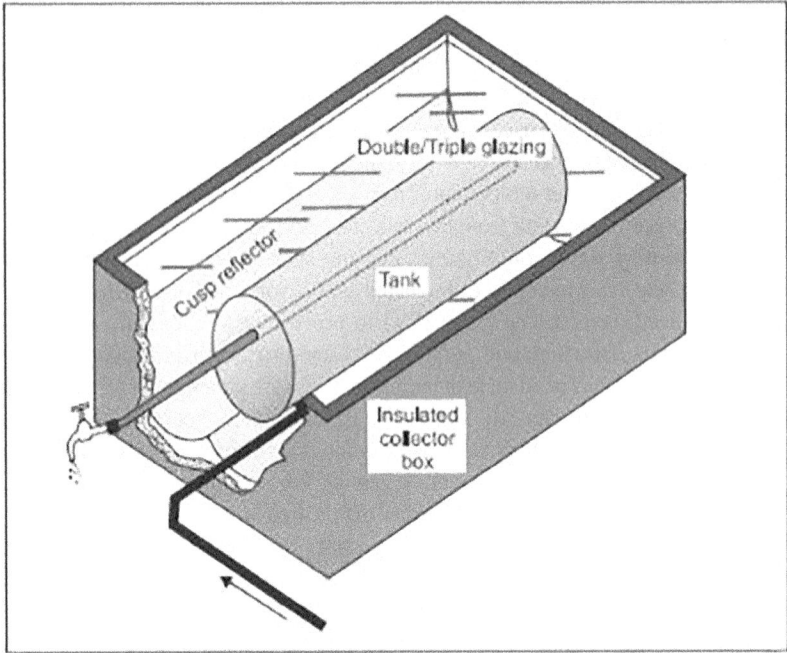

FIGURE 10.4 Integrated collector storage (ICS), system.

might not lead to convection as cold water is to remain at the lowermost section of the structure (Figure 10.4).

10.1.3.7.12 LCA of the Solar Thermal Energy

Many authors have extensively studied the benefits of using solar systems (solar collectors' LCA) and have comparatively analyzed various typologies of collectors. Nevertheless, they typically lack clarity of assumptions and data references and the results are commonly shown as aggregate indexes, making it difficult to compare different works or to analyze the dominance of each stage of the life cycle. In addition, some life cycle stages (e.g., installation and maintenance) are usually not elaborated and are even disregarded (Kuenlin, Aurélie *et al.* 2013). Research has, indeed, considered the complete LCA of a solar collector very expensive and also assumed that only the impacts relevant to material processing and the assembly of the collector are important. The principles of eco-design suggest using disaggregated knowledge to define the stages with topmost impacts and the greatest potential for improvement. The objectives of this section are to:

- Draw an ecological balance of a given equipment (here, a passive collector in Italy).
- Ensure the clarity of assumptions, design limitations and databases for comparison with other works.

- Disaggregate the results to the largest possible extent, to demonstrate the impact of each element and each of the life cycle stages and to circumvent the uncertainties associated with the weighting practices and impact evaluations.

The results presented are from the CS2 case study carried out in the framework of Task 27-Subtask C of the IEA on The performance, durability and sustainability of advanced windows and solar components for buildings. Data relating to generation, establishment and maintenance were directly gleaned; through the cooperation of the Italian company. Data gathering involved the Environmental Management System active at the site of production. The data on energy sources and raw material have been in agreement with Italian norms. Databases of other European states were consulted when Italian data were not accessible.

10.1.3.7.13 Functional Unit (FU)
The initial LCA stage is defining objectives and scope, particularly a description of a functional unit (FU). FU is the reference unit expressed in quantified performance of the product system. The analysis is the baseline for gathering data and comparing various research relating to a given product category. The FU choice might not be immediate. We have verified three alternatives:

1. FU is tantamount to the full equipment. Concerning the total collector, the results are given in the form of universal quantities. This design might be the most intuitive, and yet a misleading, choice causing misunderstandings. Generally, the collector typologies fall into two major classes: forced circulation collectors and natural circulation collectors.
2. An impacts per unit collecting area, which can also cause misunderstanding. By broadening the collecting area, specific environmental impacts like CO_2/S decrease. Thus, the specific impacts of two collectors with similar total impacts might be different. Accordingly, a collector with the largest surface area is presumably, but not necessarily, more ecological. Moreover, extension is not emblematic of a proportional growth in the yield of energy, since collector area and collected energy are nonlinearly related.
3. Impact per unit of energy production is a choice for power systems, as it shows environmental corollaries of a system's power performance. Nonetheless, application of this approach to a solar collector's LCA is difficult. The system's output is highly variable, contingent upon the solar energy input. The impact on the energy output might be misleading, as the ecological profile of a collector might vary based on the position.

Our LCA examination indicates the first FU alternative (Ardente, F. *et al.* 2004), and the environmental impact is associated with the manifold assembly.

The system under study is a thermal solar collector with the dimensions $2005 \times 1165 \times 0.91$ m and a total net surface of 2.13 m². FU consists of:

- Absorbing collector with main frame, absorbing plate and thermal fluid flow pipes.

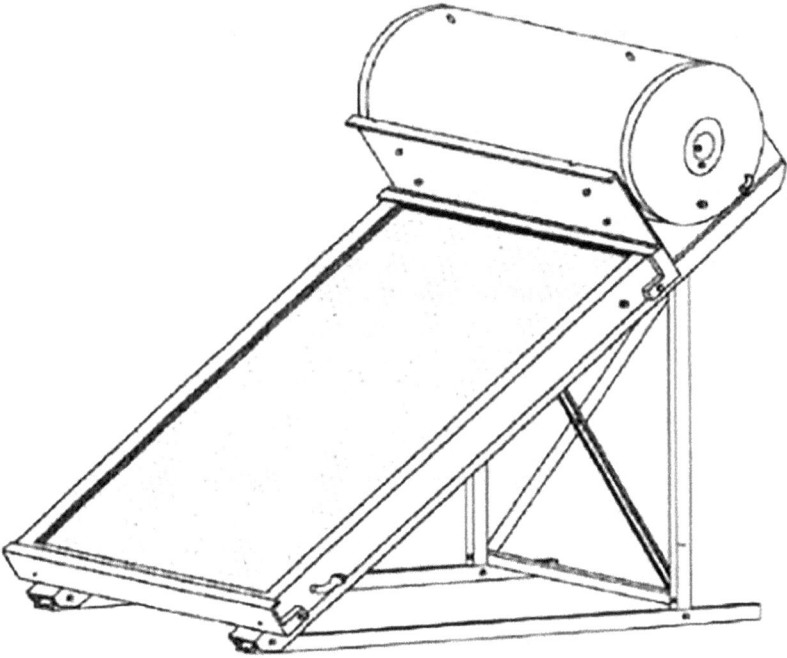

FIGURE 10.5   Solar thermal collector with water tank and support.

- Water tank with heat exchanger, cover, electrical resistance and interior pipes for the flow of sanitary water.
- External support for fixing the rooftop-mounted system.

The collector falls into the passive solar class. The tight connection between the water tank and absorbing surface constitutes a single unit, and thermal fluid is circulated through natural convection. No pump is needed to circulate the internal liquid and hence no energy is used up. The collector is highly recommended for small-sized local plants where the household demand for hot water might be low or average. A direct installation of the collector and the water tank on sloping roofs is also possible. A flat-roof installation is also possible since an optional steel bracket is provided by the production facility. The FU also features an external support (Figure 10.5).

10.1.3.7.14 Technicalities and Collective Details

Our collector is composed of painted galvanized steel that is 0.003 m wide. The absorption surface involves a copper plate painted black, soldered with thermal flow pipes. The plate is covered by an aluminium frame with a high reflectance coefficient to increase collector efficiency. Thermal insulation is ensured by high density polyurethane—PUR foam (width 0.03 m). The collector is protected by a single, highly transparent tempered glass with a low percentage of iron oxidation. The 0.004-m wide glass is impact-resistant and is fixed to the frame hermetically. To decrease heat loss in the collector, an internal vacuum is made. The 0.16-m^3 capacity water tank is mainly composed of a galvanized steel frame. It has a stainless-steel cover located on manifold top. The interior space and exterior cover of the water tank are occupied with high-density PUR foam. Fluid flow has two cycles: (i) heat transport and (ii) sanitary water. In thermal fluid, water (50–80%) is mixed with propylene glycol (20–50%) to escape freezing in cold weather. The use of a 50% mixture is recommended. A cylindrical gap functioning as a heat exchanger is a means of the mixture flow. In the water tank, magnesium anode (for corrosion minimization) and an electrical heater are utilized (Murty, P.S.R. 2017). We have not taken this assisting resistance as part of the water tank, but it is discussed in the other parts section. It is therefore more convenient to indicate this component's effect on global eco-profile. The other parts section also represents package substances (cardboard, low-density polyethylene, LDPE) and external pipes made of high density polyethylene, HDPE to link a collector to the water tank. The support is made of galvanized bars of steel, assembled and fixed to the collector via screws and bolt, as represented in Table 10.1.

10.1.3.7.15 Life-Cycle Phase Analysis
In what follows, research assumptions as well as environmental and energy issues regarding collector life cycles are discussed. The phases are energy and raw material generation and submission, generation procedure, establishment, maintenance, waste discarding, and transport.

TABLE 10.1
Materials and Masses

Absorbing Collector		Water Tank		Support		Other Parts	
Material	Mass (kg)	Material	Mass (kg)	Material	Mass (kg)	Material	Mass (kg)
Galvanized steel	33.9	Galvanized steel	49.9	Galvanized steel	27	Card-board	3.0
Glass	10.5	Stainless steel	21.5	Stainless Steel	0.5	LDPE	0.8
Copper	8.2	Rigid PUR	4.8			HDPE	0.87
Stainless steel	6.1	Thermal fluid	5.4			Copper	0.46
Rigid PUR	4.2	Copper	3.8				
Aluminium	4	Epoxy dust	0.7				
Thermal fluid	0.9	Steel	0.4				
Epoxy dust	0.3	Welding rod	0.2				
Welding rod	0.1	Brass	0.1				
Brass	0.04	Magnesium	0.2				
Flexible PUR	0.01						
PVC	0.01						
Total	68.2		86.2		27.5		5.1

10.1.3.7.16 Transport

FU is mostly made of metal and plastics.

Due to the impossibility of defining the exact quantity of trips for the collector production, TKM is considered a functional unit for transport trucks. It shows the energy and environmental effects concerning the transport of 1000 kg of product over 1 km of road. Impact calculation is performed using distances and masses. Transport is supposed to be performed by 28,000-kg capacity trucks. A foreign-purchased glass is also assumed to be conveyed by trucks with medium/high capacity. Specific effects concerning trucks have been referred to research by the Italian Environmental Protection Agency. Regarding all the inputs used during the cycle stages of life and taking into account the average values of distance, the overall transport load was computed to be 154 TKM (Table 10.2).

10.1.3.7.17 Process of Generation

The production process data of the manifold has been collected using field analysis. The production process principally involves the processing of metals and their assembly with other externally worked things like plastic or auxiliary metallic parts. An absorbent collector, a water tank and support have been generated at various times, and are finally packed and stored in the warehouse—Figure 10.6. Thus, external retail firms take care of transporting collectors and install them for end users.

10.1.3.7.18 Absorbing Collector Realization

The collector includes a frame, an absorber plate having pipe for thermal fluid flow and glass. A zinc steel plate has been used to provide the frame.

Following cut and bend, it is enamelled with epoxy powders. The plate and pipes are copper-made. Pipes are worked discretely and subsequently welded to the plate

TABLE 10.2

Transport-Caused Air Pollutions Estimated

Air Emissions Caused by Transport	Amount
$C_6H_6(mg)$	5.9
$C_{20}H_{12}(mg)$	0.03
CO_2	20
$Cd(mg)$	0.3
$NMVOC(g)$	27.4
$CH_4(mg)$	890.4
$CO(g)$	56.4
$NO_x(g)$	259.7
SO_2	16.9
$Pb(mg)$	2.1
$Particalate(g)$	14.2
$N_2O(g)$	2.8
$Zn(g)$	0.9

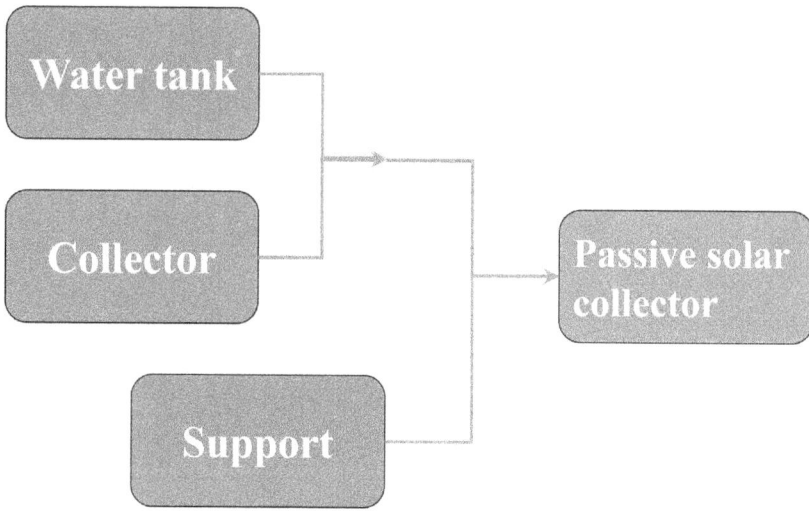

FIGURE 10.6 Collector production flow.

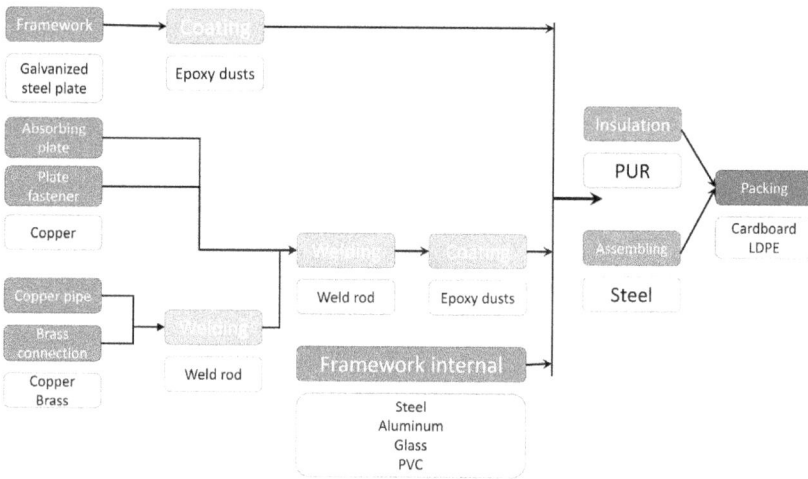

FIGURE 10.7 Collector manufacturing, process flow diagram.

via acetylene welding. In the absence of data on acetylene welding, atmospheric emissions were not calculated. However, small amounts of acetylene were employed, and therefore air emissions were ignored. For a better absorbency, pipes and absorber were then coloured in black. The absorbent plate, the frame and the glass were assembled successively. Finally, the PUR insulation is blown, and external frame is coloured with powdered epoxy. Figure 10.7 displays the flowchart of the production process with material succession and numerates the related sub-processes. Sub-process analyses indicate energy and mass flowing. The framework is expounded in Figure 10.8.

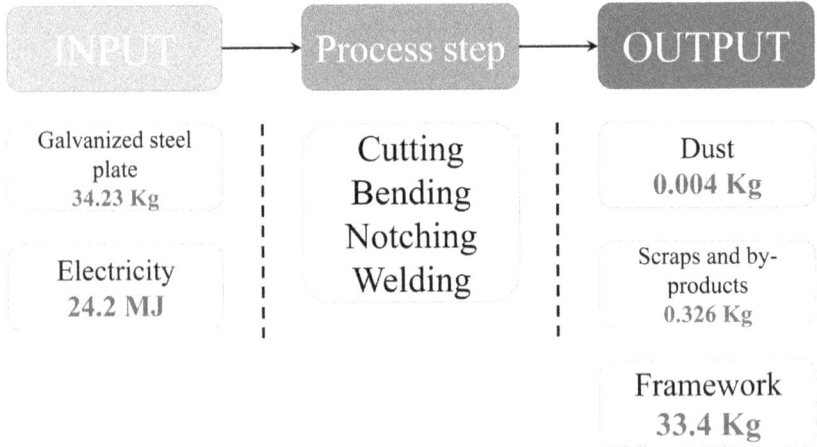

FIGURE 10.8 Collector's framework creation details.

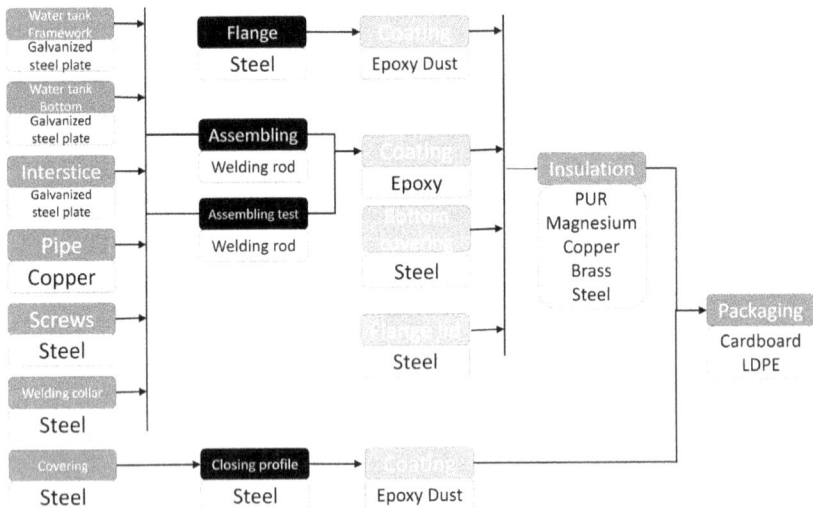

FIGURE 10.9 Process diagram.

10.1.3.7.19 Production of Water Tank

It comprises the frame, interstice and outer coating. The tank is 0.444 m in diameter and is formed by galvanized sheet steel in a cylindrical shape, to which the sides are welded. A flange is attached to a side to support electrical resistance, copper pipe and magnesium anode of the coil. Another steeled cylinder is also on the outside welded to the tank. The fluid circulates in the interstice and has thermal exchange with the tank. The outer coating is created and painted separately. Water tank components are subsequently assembled, and PUR insertion in vacancies is finally accomplished. In Figure 10.9, the process along with mass details and sub-processes are shown.

10.1.3.7.20 Installation
It includes the following steps:

1. FU Transport from factory to warehouse for retail.
2. Transport from warehouse to the user.
3. Component Installation.

The first step involves the use of many trucks, and destinations depend upon the locations of the sales companies. Accordingly, mediocre conditions are considered as follows. Functional unit: 1 TKM of truck with a capacity of 28,000 kg; a 100 km-distance travelled (two-way). FU is moved from the warehouse to the user per van with a 3500-kg capacity. The firm makes a trip (30 km; two-way) for any of the collectors. The installation involves fixing the support on the rooftop with collector and water tank being fixed to support.

Specific transport-related impacts refer to Italian works. Table 10.3 shows emission quantities from transport. Concerning drilling operations throughout fastening, 0.5-MJ low-voltage power was consumed.

10.1.3.7.21 Maintenance
FU is estimated to have a 15-year useful life on average. If unpredictable damage like glass breakage would not occur, frequent maintenance is unneeded. Cycles typically have an operation each of 4 to 5 years (within 2 to 3 operations in the FU useful life). Maintenance assumptions are:

- 2 maintenance operations (5 and 10 years after purchase).
- Technicians travelling (80 km by diesel vehicles in total).

TABLE 10.3
Installation-Transport Contaminants

Transport-Induced Air Contaminants	Amount
$C_6H_6(mg)$	0.72
Benzopirene(g)	0.004
$CO_2(g)$	10.17
Cd(mg)	0.06
NMCOC(g)	12.79
$CH_4(mg)$	318
CO(g)	46.24
$NO_x(g)$	61.23
$SO_2(g)$	4.44
Pb(mg)	0.25
Particulate(g)	12.43
$N_2O(g)$	0.87
Zn(g)	0.11

TABLE 10.4

Maintenance Phases—Atmospheric Emissions from Transport

Transport's Air Emission	Amount
Global distance (Km)	80
$NH_3(mg)$	80
$CO_2(Kg)$	9.2
Cd(mg)	0.03
Cr(mg)	0.14
$CH_4(mg)$	480
CO(g)	28.0
$NO_x(g)$	22.0
$SO_2(g)$	2.88
Zn(g)	0.003
Particulate(g)	3.2
$N_3O(g)$	0.8
Ni(mg)	0.2
COV(mg)	4.40
Cu(mg)	0.005
Se(mg)	0.03

- The following components are substituted in each maintenance practice:
- PVC seals.
- Magnesium anode.
- Waterproofing.
- Electrical resistance.
- Thermal fluid with 1/2 water and 1/2 propylene glycol.

Transport-caused emissions are summarized in Table 10.4.

10.1.3.7.22 Arrangement

The manufacture of FU leads to the formation of scrap and waste (4.4 kg, excluding raw material wrappings). From time to time, the waste is delivered to a disposal utility.

Regarding the elimination of FU, data is unavailable. The company launched the solar collector production some years before, and therefore the sold collectors are still in their useful life condition. Moreover, the manufacturing company has not launched any plan concerning the collectors' recycling while components such as metal with over 80% of the total mass are recyclable (Blengini, G. 2009). It is therefore assumed that collectors would be gathered and discarded at the closest landfills via trucks (50 km in both directions). Assuming that the transport is done by some 28,000-kg trucks, it is estimated that 1.4 kg CO_2 et of other contaminants is released. Effects of landfill management are not considered.

10.1.3.7.23 Environment
The environmental impacts include the following:

- Resource usage.
- Atmospheric contamination.
- Aquatic emission.
- Waste and solid contaminants.

Environmental impacts are either direct (pertaining to production and transport) or indirect (relevant to process input creation of raw substance and energy resource). The FU manufacture directly caused air contamination and slight waste. Diesel-based emissions are slight and thus ignorable, compared to fuel combustion emissions (Heckert, W.W. 2005).

10.1.3.7.24 Resource Uses
Based on the analysis of life cycles, around 415 kg of resources have been consumed. The material used (ferrous minerals mainly) is summarized in Table 10.5. FU is mainly composed of steel.

10.1.3.7.25 Air Contamination
Direct and indirect contaminations of air are shown in Table 10.6. Approximately 650 kg of CO_2 is emitted (Mirasgedis, S. *et al.* 1996). Emissions are usually indirect and are largely linked to raw material production (incidence: 80 to 90%). Production and transport emissions have a total impact of 10 to 20%. Direct emissions of metallic contaminants (e.g., Fe, Cr, Mo and Mn) linked to the process of production is prevalent, which is related to the cutting and welding phases.

TABLE 10.5

Resource Usage

Main Resources Consumption	Amount
Ferrous minerals (kg)	293.5
Water (m³)	31.6
Iron scraps (kg)	47.3
Bauxite (kg)	14.9
$CaCO_3$ (kg)	14.3
Copper minerals (kg)	8.2
NaCl (kg)	7.7
Zinc (kg)	6.8
Sand (kg)	6.4
Copper scraps (kg)	5.4
Lime (kg)	2.0
Clay (kg)	1.3
Nitrogen (kg)	1.2

TABLE 10.6
Direct/Indirect Atmospheric Contaminants

Impacts Parameters	Indirect		Direct		Total
	Raw Materials	Electricity	Transport	Production	
CO_2 (kg)	580.4	35.8	40.8		657.0
CO (kg)	4.4	0.01	0.1		4.5
SO_2 (kg)	3.3	0.2	0.03		3.6
CH_4 (kg)	2.1	0.05	0.002		2.2
NOx (kg)	1.3	0.1	0.4	0.1	1.8
Dust (kg)	0.5'	0.02	0.03	0.07	0.6
NMCOV (kg)	0.2	0.05	0.03	0.01	0.3
Mn (kg)	0.0001			0.3	0.3
Fe (kg)	0.0004	5.6E-7		0.1	0.1
N2O (g)	20.9	1.5	1.9		24.3
HCl (g)	35.0	1.1			36.1
Cr (g)	0.01	0.004	0.0003	10.7	10.7
Ni (g)	0.3		0.0004	4.7	5.1
Cu (g)	0.01		0.005	3.4	3.4
Zn (g)	2.3		1.0		3.3
HF (g)	2.5	0.2			2.6
NH_3 (g)	2.5	0.03			2.5
Mo (g)	0.003			0.6	0.6
Pb (g)	0.5	0.012	0.003		0.5
PAH (g)	0.2	0.001			0.2
Benzene (mg)	529.0	87.3	7.1		623.4
Cd (mg)	125.4	1.3	0.1		126.9

Water contaminants are exclusively indirect (neither production process nor transport has any contact with water). Table 10.7 presents the main pollutants. Organic discharges represent 18 kg of chemical oxygen demand (COD); other emissions are usually small amounts of metal ion.

10.1.3.7.26 Waste
The direct wastage of the company is 4.4 kg. All the waste created is summarized in Table 10.8.

10.1.3.7.27 Potential Impact on Environment
Ecological profile of collectors is delineated based upon the potential environmental consequence:

- Global Warming Potential GWP.
- Acidification Potential AP.

TABLE 10.7
Water Contamination

Water Contaminants	Amount
COD (kg)	18.1
Fe (g)	49.8
Mg (g)	16.4
K (g)	7.8
NH3 (g)	4.8
Phosphorus (g)	1.4
Cr (g)	1.1
Pb (g)	0.5
Na (g)	0.4
Ni (g)	0.4
Mn (g)	0.3
Cd (mg)	5.4
Hg (mg)	4.0

TABLE 10.8
Waste Generation

Wastes Generation	Amount
Normal waste (kg)	59.5
Special waste (kg)	5.2
Ash (kg)	6.8

TABLE 10.9
Potential Effects on the Environment

Potential Effects on the Environment	Amount
GWP ($kgCO_2$ eq.)	721
AP ($kgSO_2$ eq.)	5
ODP (kgCFC-11 eq.)	Negligible
NP ($kgPO_4$ 3K eq.)	0.7
POPC (kgC_2H_4 eq.)	0.4

- Ozone Depletion Potential ODP.
- nitrification potential NP.
- Photochemical ozone creation potential POPC.

The indices were computed using the characterization factors for compiling the Italian Environmental Product Declaration, EPD (Table 10.9).

10.1.3.7.28 LCA of Concentrate Sun Power Plant

Here, energy analysis and LCA for CSP plants are considered. Integrated analysis is utilized to assess a specific CSP consisting of a series of mirrors for tracking solar energy and reflecting it back to the tubes of receivers for steam generation for running a turbine joined to a power generator. LCA is to measure environmental effects of such a particular CSP throughout its life cycle, considering each life cycle stage's contribution to the environmental load. LCA helps to identify environment hot spots and gives opportunities for process improvement and optimization of the product or a specific stage of its life cycle.

10.1.3.7.29 Methodology

The methodology is as follows: developing LCA via Eco-indicator 99 and ELCA via Cumulative Exergy Demand Approach. The commercial SimaPro8 professional software, single-user version [38] was used to corroborate results. Thermoeconomic analysis via SPECO, specific exergy costing, was performed to assess the cost level incurred by exergy devastation in any component and in the whole design.

10.1.3.7.30 Eco-Indicator 99

For the environmental effects quantification, various indicators are practical, classified into two classes which relate to: problem (midpoints) and damage (endpoints). The first class categorizes effects into environmental subjects (e.g., potentials for global warming, acidification, and ozone depletion), whereby a better image of the ecological impact is provided though entailing LCA knowledge for the results comprehension. The second class renders health, natural resources and environment. Here results are more comprehensible but with a limited transparency. The damage-focused Eco-indicator 99, as used here, is a common approach in the LCA environment, which is reinforced by a very strong dataset. Damage has 3 items: (i) HUMAN HEALTH (HH) which fixes, for example, disabled and lost years of life, DALY containing carcinogens, respiratory organic substances, respiratory inorganic substances, climate change, radiation and ozone layer; (ii) ECOSYSTEM QUALITY (EQ) that represents species loss in an area in a given period (species/m^2 year) comprising eco toxicity, acidification/eutrophication and land use; and (iii) RESOURCE CATEGORY (R), the resource impact category damage coefficient is presented by excess energy for extraction in MJ, including mineral and fossil fuel. Impact categories are addable to the aforementioned categories, weighted and unified into a single score, representing overall environmental load in points. As recorded in the literature, one point (Pt) can be interpreted as one thousandth of the annual environmental load of an average European inhabitant [41]. Weights assignment to impact categories is defined by three methods, which are known to be Equality perspective (HH 30%, EQ 50%, and R20%); Individual perspective (HH 55%, EQ 25%, and R20%) and Hierarchical perspective (HH 40%, EQ 40%, and R20%). The hierarchical method is used here, and the results are determined by a singular combined score, based on environmental effects of different elements (Frischknecht, R. *et al.* 2007). The rating for each category of impact is:

$$IMP_J = \sum_k d_{kj} \cdot LCI_K$$

10.1.3.7.31 Cumulative Exergy Demand

Exergy budgets are defined for different unit procedures as exergy contents are assigned to inward/outward mass/energy flows. Exergy terms cover chemical-physical components. Combining two diverse viewpoints in a composite technique has much capability to exploit the merits and omit the demerits of separate devices; in SimaPro setting, the Cumulative Exergy Demand Method, CExD is attained straight from Ecoinvent. Exergy is assumed to be a metric of the probable losses of useful energy resource. CExD is more wide-ranging than Cumulative Energy Demand, CED by subsuming energy and addition of non-energy resources (Ehtiwesh *et al.* 2016). All proposed categories contribute significantly to CExD in at least one product. In product/service evaluations and comparison affirmations, vigilant selection of the suitable CExD is needed considering energy and resource quality demand to be stated via CExD. Bösch et al.'s work is the foundation for executing CExD. The CExD indicator was developed to illustrate the overall omission of exergy from nature to deliver a product, summarizing the exergy of all required resources (Szargut, J. *et al.* 1997). Exergy is used for resources in the Ecoinvent database, taking into account kinetic, chemical, hydropotential, solar-radiative, nuclear, and thermal exergies. The impact category indicator has 8 classes of resources: fossil, nuclear, hydro, biomass, other renewables, water, minerals and metals. The CExD in SimaPro has ten impact categories (Table 10.9). Consequently, the CExD indicator determines the exergy of nature-discarded resources, hence no longer accessible for a future exploitation.

10.1.3.7.32 Stock Analysis

In LCA and ELCA research, the plant boundary needs to be utterly clarified. For this purpose, we must analytically specify the production procedures and the input and output flows within the system limits. Within the limits of the system, the stages of construction and dismantling are similarly taken into account. The analysis covers the whole plant lifetime of 25 years. The construction involves the assemblage of the constituents, taking into account raw material and manufacture processes. To carry out an

TABLE 10.10
CExD Impact Categories in Ecoinvent Data

Category	Subcategory	Name
CExD	Fossil	Non-Renewable Energy Resources, Fossil
	Nuclear	Non-Renewable Energy Resources, Nuclear
	Kinetic	Renewable Energy Resources, Kinetic (in the Wind)
	Potential	Renewable Energy Resources, Solar, Converted
	Water	Renewable Energy Resources, Potential (in Barrage Water), Converted
	Primary Forest	Non-Renewable Energy Resources, Primary Forest
	Biomass	Renewable Energy Resources, Biomass
	Water Resources	Renewable Material Resources, Water
	Metals'	Non-Renewable Resources, Metals Material

TABLE 10.11
Andosol Power Plant Life Cycle Inventory

Material	Quantity			
	Solar Field	Storage System	Power Block	Total
Chromium Steel [Kg]	361,889	112,276	44,050	518,215
Concrete [M3]	19337.5	1628	83.6	21049.1
Synthetic Oil [Kg]	1,995,000'	X	X	1,995,000
Flat Glass Coated [Kg]	6,148,846		X	6,148,846
Molten Salt [Kg]	X	25,600,000	X	25,600,000
Reinforcing Steel [Kg]	15,168,192	386,578	593,258	16,148,028
Carbon Steel [Kg]	1,916,292	X	X	1,916,292

approximate sizing of the plant studied, data must be gathered on different equipment components, weights, materials, waste and processes involved in the manufacture of each device in the plant (Ardente, F *et al.* 2003). The selected inventory is principally based upon plant data Andasol 1 (Viebahn, P. *et al.* 2008). Using the suggested

TABLE 10.12
Andasol Components versus Ecoinvent—V. 3 Database

Component	Ecoinvent V. 3 Equivalence	Note
Chromium steel [kg]	Steel, chromium steel 18/8 RER, steel production, converter, chromium steel 18/8, Alloc Def, U	
Concrete [m3]	Concrete, sole plate and foundation CH, production, Alloc Def, U	
Synthetic oil [kg]	Diphenylether-compound, RER, production, Alloc Def, U	
Flat glass coated [kg]	Flat glass, coated, RER, production, Alloc Def, U Molten salt Potassium nitrate, RER, production, Alloc Def, U	
Molten salt [kg]	Potassium nitrate, RER, production, Alloc Def, U	
Reinforcing steel [kg]	Reinforcing steel RER, production, Alloc Def, U	
Carbon steel [kg]	Sheet rolling, steel GLO, market for, Alloc Def, U	
Chromium steel manufacturing	Metal working, average for chromium steel product manufacturing, RER, processing, Alloc Def, U	
Concrete excavation	Excavation, hydraulic digger RER, processing, Alloc Def, U	
Reinforcing steel manufacturing	Metal working, average for steel product manufacturing, RER, processing, Alloc Def, U	
Diesel burned in construction	Energy, from diesel burned in machinery/RER Energy	19.99E+6 MJ
Diesel burned in dismantling	Energy, from diesel burned in machinery/RER Energy	8.8E+5 MJ

baseline, the materials with key impacts have been selected, as given in Table 10.11. Data components are all returned to functional unit. The unit used as a reference for analysis is the Andasol plant, which is the first parabolic trough power plant in Europe (South Spain) and the first in the world with storage. The plant components are: parabolic collectors, a traditional power generation cycle and a molten salt storage system. The energy cycle is arranged at a nominal generation capacity of 50 MW, and a 2-tank indirect thermal storage system with a thermal capacity of 1010 MWH, corresponding to ~28,500 t of liquid salt 60% sodium nitrate, 40% potassium nitrate adequate for 7.5 h of power production. In the lack of data on the manufacture and removal of $NaNO_3$ in the Ecoinvent database, data on KNO_3 were considered as a good alternative. The agreement between experiment and Ecoinvent database v. 3 is presented in Table 10.12.

Each row has two collector units, each with 12 collectors. Each manifold has 28 mirrors and 3 absorption pipes. The heat transfer fluid (HTF) is a synthetic oil (Terminol VP-1) of the Dowtherm A type, as a mix of 2 highly stable organic compounds biphenyl (C12H10) and diphenyl oxide (C12H10O). The disposal of the main materials is illuminated in Table 10.13 (Erik Pihl, *et al.* 2012).

This study is about a specific geographical spot in Libya. However, shipping is supposed for material made in America and Europe; internal transport is by truck. Table 10.14 presents transport stages, where TKM points to the mass of materials in tonnes multiplied by the distance in kilometres (Ortiz, O. *et al.* 2009).

TABLE 10.13
Recycle Fractions to Estimate End-of-Life Impacts

Materials	Ecoinvent V. 3 Equivalence	Fraction %
Reinforced steel	Waste reinforcement steel (waste treatment) CH, treatment of waste reinforcement steel, recycling, Alloc Def, U	90
	Waste reinforcement steel (waste treatment) CH, treatment of waste reinforcement steel, collection for final disposal, Alloc Def, U 10	10
Concrete	Waste concrete, not reinforced (waste treatment) CH, treatment of, recycling, Alloc Def, U	95
	Waste concrete, not reinforced (waste treatment) CH, treatment of, collection for final disposal, Alloc Def, U	5
Glass	Waste glass (waste treatment) CH, treatment of waste glass, municipal incineration, Alloc Def, U	100
Chromium steel	Waste reinforcement steel (waste treatment) CH, treatment of waste reinforcement steel, recycling, Alloc Def, U	90
	Waste reinforcement steel (waste treatment) CH, treatment of waste reinforcement steel, collection for final disposal, Alloc Def, U	10
Molten salt	Salt tailing from potash mine (waste treatment) CH, treatment of, residual material landfill, Alloc Def, U	100
Synthetic oil	Waste mineral oil (waste treatment) GLO, market for, Alloc Def, U	100

TABLE 10.14

Materials Transport

Name	Ecoinvent Corresponding to Each Process	Amount (TKM)
U.S.A.	Transport, Transoceanic Freight Ship, OCE	2.21E+08
Europe	Transport, Transoceanic Freight Ship, OCE	9,223,269
Local	Transport, Lorry >16 T, Feet Average, RER	107852.8

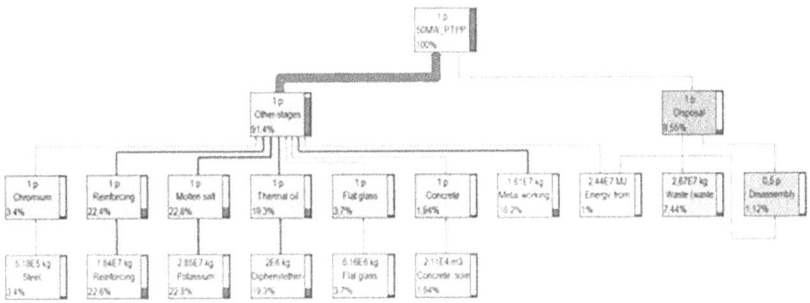

FIGURE 10.10   Top process network diagram.

10.1.3.7.33 Results and Discussion

A typical energy and environment life cycle analysis of CSP was done via the material inventory of the Andasol (F. Dinter *et al.* 2014). The analysis has stages of material production, construction and discarding. In Figure 10.10, a network diagram is presented for the top process of the analysis procedure.

10.1.3.7.34 Assessing Life cycle

Categories of environment damage for all the components (%) are shown in Figure 10.10; this allows the assessment of the contributions of all unit processes to damage categories. Table 10.15 shows the damage category for all the various unit components (Amersfoort 2001).

Figure 10.12 presents impacts on a single scoring scale (MPt, mega points) for all classes. Accordingly, the greatest impact -14.4MPt (69%) belongs to human health, followed by resource -5 MPt (24%) and ecosystem quality -1.4MPt (7%). Based on the macro-impact categories analysis, respiratory inorganic substance category has the highest impact -45.43%, followed by fossil fuels -20.4% and carcinogens category -14.3%, while other categories have relatively small impacts. Also, for materials, the largest impact belongs to steel -9.77 MPt (46.9%) followed by molten salt 5.19 MPt (24.9%) (Nils Breidenbach, *et al.* 2016) and then synthetic oil 4.27 MPt (21%).

TABLE 10.15
Damage Impacts Concerning the Damage Categories

Damage	Human Health (DALY)	Ecosystem Quality (Pdf*M² Y)	Resources (Mj- Surplus)
Total	315.2391629	19768284.09	191329191.7
Steel	154.0400461	11022881.53	74394439.69
Molten Salt	70.04195319	4474356.085	63304671.99
Synthetic Oil	70.54339569	3465114.623	35309810.2
Glass	11.06639576	479431.1153	11490645.69
Concrete	7.359603958	241546.4502	3450221.048
Diesel	2.187768226	84954.28983	3379403.033

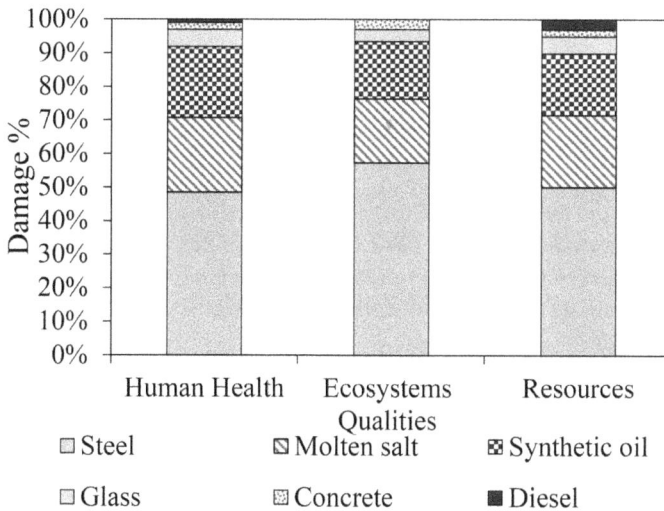

FIGURE 10.11   Damage assessment phase.

TABLE 10.16
LCA-Derived Environmental Impact

Impact Category	Life cycle 1 mwh (pt)	
	Desideri et al.	This study
'Human health (hh)	1.11	1.3
'Ecosystem quality (eq)	0.805	0.463
Resource (r)'	0.405	0.13
'Total	2.32	1.9

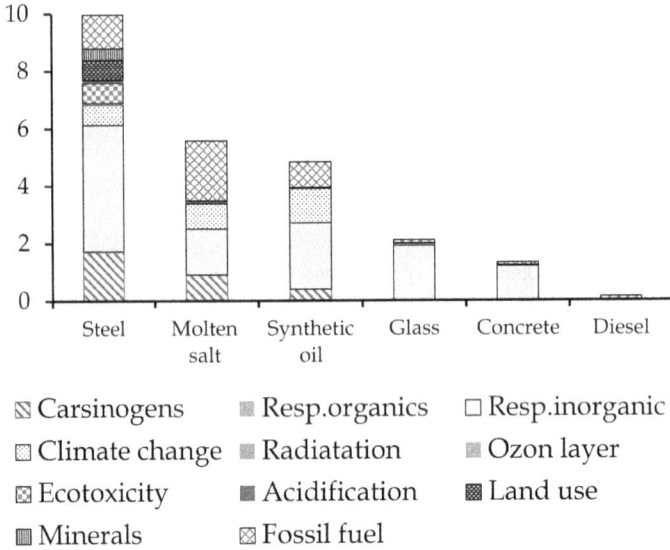

FIGURE 10.12   Single scores of individual effect categories.

Concrete has the least impact. The single environmental impact life cycle score is compared to Desideri's team in Pt per 1 MW h. According to our comparative analysis, we have observed lower total impacts than those in. The slight differences might be ascribed to our application of an enhanced material inventory version as well as the Ecoinvent database.

REFERENCES

Allen, J.A., Rogers, D.F., *Principles of Energy Conversion*, McGraw-Hill, 2014.

Alliger, G., *Renewable Energy Resources*, Interscience, 2011.

Ardente, F., Beccali, G., Cellura, M., Eco-sustainable energy and environmental strategies in design for recycling: the software ENDLESS. Ecological Modelling, 2003.

Ardente, F., Beccali, G., Cellura, M., Lo, Brano V., Life cycle analysis of solar thermal collector (first part: life cycle inventory). *First report of the International Energy Agency (IEA)-Task 27 Performance of Solar Facade*, Subtask C Project C1 Environmental Performance; March 2003.

Ardente, F., Beccali, G., Cellura, M., Lo, Brano, V., The environmental product declaration EPD with a particular application to a solar thermal collector. *Proceedings of conference ECOSUD* 2003, Siena, 4–6, 2003.

Ardente, F., Beccali, G., Cellura, M., Lo Brano, V., Life cycle assessment of a solar thermal collector. Renewable Energy, 30, 1031–1054, 2005. 10.1016/j.renene.2004.

Ardente, F., Beccali, M., Cellura, M., FALCADE: plowboy interview with Mother Earth News, Harold R. Hay 1976.

Aurelie, K., Germain, A., Leda, G., Francois, M., Life Cycle Assessment and Environomic Optimization of Concentrating Solar Thermal Power Plants. *In: Proceedings of the 26th international conference on efficiency, cost, optimization, simulation and environmental impact of energy systems*, Guilin, China, EPFL—CONF—186393, 2013.

Baum, B., Parker, C.H., *Energy Security Issues*, Reinhold, 2013.

Bauer, E., *Energy Conservation Equipment*, Van Nostrand Reinhold, 2012.

Berenson, C., *Solar Energy*, Wiley-Interscience, 2018.

Bhatia, S.C., *Advance Renewable Energy Systems*, 2014.

Bravo Y., Carvalho M., Serra, L.M. Monné C., Alonso, S. Moreno, F. Muñoz M., *Environmental evaluation of dish-Stirling technology for power generation*, Solar Energy, Volume 86, Issue 9, 2012.

Brydson, J.A., *Waste Heat Recovery*, Elsevier, 2020.

Blengini, G., Life cycle of buildings, demolition and recycling potential: a case study in Turin, Italy. Department of Architectural Engineering, Faculty of Urban Architecture No. 2501162, 2009.

Clauser, H.R., *Photovoltaic Solar Devices*, Reinhold, 2009.

Clery D. Environmental technology. Greenhouse-power plant hybrid set to make Jordan's desert bloom. *Science*. 2011.

Collins, P., The Beautiful Possibility by Paul Collins Cabinet 6, 2002 Spring.

Crawford, R.H., Treloar, G.J., Net energy analysis of solar and conventional domestic hot water systems in Melbourne, Australia. Solar Energy, 2004.

Desideri, U. Zepparelli, F. Morettini, V. Garroni E., Comparative analysis of concentrating solar power and photovoltaic technologies: Technical and environmental evaluations, Applied Energy, Volume 102, 2013, Pages 765–784.

Diakoulaki, D., Zervos, A., Sarafidis, J., Mirasgedis, S., Cost benefit analysis for solar water heating systems. Energy Conversion Management, 2001.

Duffie, J.A., Beckman, W.A., *L'energia solare idle applicazioni termiche (Italian language)*, Liguori ed., 1978.

Efthymiou, E., Cöcen, Ö. N., & Ermolli, S. R. (2010). Sustainable Aluminium Systems. Sustainability, 2(9), 3100–3109. https://doi.org/10.3390/SU2093100

Ehtiwesh, I.A.S., et al. Renewable and Sustainable Energy Reviews, 56, 2016.

Erik Pihl, Duncan Kushnir, Björn Sandén, Filip Johnsson, Material constraints for concentrating solar thermal power, Energy, Volume 44, Issue 1, 2012, Pages 944–954.

F. Dinter, D. Mayorga Gonzalez, Operability, Reliability and Economic Benefits of CSP with Thermal Energy Storage: First Year of Operation of ANDASOL 3, Energy Procedia, Volume 49, 2014, Pages 2472–2481.

Frischknecht, R., Editors, N., Althaus, H., Bauer, C., Doka, G., Dones, R., et al. Implementation *of life cycle impact assessment methods, ecoinvent report no. 3*. Swiss Centre for Life Cycle Inventories, 2007.

Gong Cheng, Xinzhi Wang, Zhangzhou Wang, Yurong He, *Heat transfer and storage characteristics of composite phase change materials with high oriented thermal conductivity based on polymer/graphite nanosheets networks*. International Journal of Heat and Mass Transfer, 183, Part B, 2022.

Gardon, J.L., *Energy, the Biomass Option*, Interscience, 2020.

Häberle, A., Zahler, C., Lerchenmüller, H., Mertins, M., Wittwer, C., Trieb, F., Dersch, J., *The Solarmundo line focussing Fresnel collector. Optical and thermal performance and cost calculations*, 2002.

Hampel, B.H., *Biological Energy Conversion*, Reinhold, 2006.

Heckert, W.W., *Combustion Engineering*, Wiley-Interscience, 2005.

ISO14040. Environmental management—Life cycle assessment—Principles and framework; 1998.

Mirasgedis, S., Diakoulaki, D., Assimacopoulos, D., Solar energy and the abatement of atmospheric emissions. Renewable Energy, 1996.

Morley, David., AICP, Editor, *Planning for Solar Energy*, 2014.

Murty, P.S.R., Chapter 24 — Renewable energy sources, Editor(s): P.S.R. Murty, *Electrical Power Systems*, Butterworth-Heinemann, 2017.

Nandi, B.R., Bandyopadhyay, S., Banerjee, R., Analysis of high temperature thermal energy storage for solar power plant. IEE ICSET, Nepal, 2012.

Neslen, Arthur (2015–10–26). *Morocco Poised to become a Solar Superpower with Launch of Desert Mega-Project.* The Guardian. ISSN 0261–3077. Retrieved 2020.

Nils Breidenbach, Claudia Martin, Henning Jockenhöfer, Thomas Bauer, Thermal energy storage in molten salts: overview of novel concepts and the DLR test facility TESIS. *Energy Procedia,* 99, 2016.

Norton, Brian. Harnessing Solar Heat. Springer. ISBN 978-94-007-7275-5, 2013.

Ortiz, O., Castells, F., Sonnemann, G., Sustainability in the construction industry: a review of recent developments based on LCA. Construction and Building Materials, 2009.

PRé. The Eco-indicator 99—a damage oriented method for life cycle impact assessment, Methodology report. PréConsultants B.V., Amersfoort, The Netherlands; 2001.

Kalogirou, S.A., *Solar thermal collector and applications.* Progress in Energy and Combustion Science, 2004.

Kinga Pielichowska, Krzysztof Pielichowski, *Phase change materials for thermal energy storage.* Progress in Materials Science, 65, 2014.

SimaPro8 World's Leading LCA Software Package, PRé Consultants. PRé Consultants, (www. pre-sustainability.com/simapro) 2019.

Souliotis, M., Tripanagnostopoulos, Y., Experimental study of CPC type ICS solar systems. Solar Energy, 2004.

Solar Millennium AG. The parabolic trough power plants Andasol 1 to Germany, 2008.

Szargut, J., Morris, D., Steward, F., *Exergy Analysis of Thermal, Chemical and Metallurgical Processes,* Hemisphere Publishing Corporation, 1997.

Tripanagnostopoulos, Y.G.T., Souliotis, M.K., Battisi, R., Corrado, A., Application aspects of hybrid PV/T solar systems. In: ISES Solar World Congress 2003, Go Teborg, Sweden; June 14 19, 2003.

Tsilingiridis, G., Martinopoulos, G., Kyriakis, N., Life cycle environmental impact of a thermosyphonic domestic solar hot water system in comparison with electrical and gas water heating. Renewable Energy, 2004.

Veenstra, A., Oversloot, H.P., Spoorenberg, H.H.R., The environmental performance of solar energy systems and related energy saving installations. In: Dissemination workshop of the International Energy Agency (IA) Task 27 Performance of Solar Facade, Copenhagen, April 2002.

Wagner, H., Ermittlung des primaerenergieau/wanfes und Abschaetzung der emissionen zur Herstellung und zum Betrieb von augewaehlten alsorberanlangen zur Schimmtadwasserwaermung und von Solarkollector-anlagen zur Brauchwassenerwaermung (German language). VDI Berichte, Reihe 6, no. 325: 1995.

Viebahn, P., Kronshage, S., Trieb, F., Lechon, Y., *Final report on technical data, costs, and life cycle inventories of solar thermal power plants.* NEEDS, Project no: 502687, 2008.

Part IV

---

*Modern Control Systems
and Government Polices
for Solar Industrial Heat*

11 Role of Modern Tools in Solar Thermal System Design

C. Shanmugam and T. Meenakshi

CONTENTS

11.1 INTRODUCTION

Solar energy is the preferred alternate energy source for its cleanliness and sustainability. The solar resource available from the sun designated as global solar radiation (GSR) is captured and used for various system operations with its state of the

DOI: 10.1201/9781003263326-15

art technology. The GSR is critically valued for climate monitoring, prediction of weather, selection of site for solar thermal plants or for design of solar heating systems [1, 2]. The potential of the solar energy at a particular place is evaluated with the solar radiation data measured in the specific region. The geographical location, weather conditions and the season are the dominant parameters that decide the volume of solar radiation reaching the earth's surface. Conventionally, solar radiation components are measured with the use of a pyranometer, solarimeter, pyrheliometer, etc. The conventional method is expensive and the measuring techniques are difficult as more measurement devices are required to be placed at different locations for more accuracy [3, 4]. However, the practical viability and efficient functionality of these technologies at a selected location depends on the available solar radiation. This results in lack of adequate data for solar radiation for instituting solar energy generation systems.

11.2 DISCRETE GROUND STATIONS FOR GSR ESTIMATION

Global solar radiation is estimated with discrete ground stations that collect the meteorological variables related to sunshine like temperature, sunshine duration, cloud cover, humidity, etc. The meteorological stations available at the ground observe the GSR of the location but the challenging component of accuracy and estimates of resolution are scant and inadequate to represent the indigenous features of GSR. The reliability of the results depends on the interpolation of data for continuous estimates of GSR. The role of geostationary satellites is to provide continuous data in time series at the rate of pixel level and from the continuously varying signals, details about the atmospheric condition and the earth's surface are reflected. The satellite image collected undergoes several interactions like diffusion and absorption and it is contained in the image; this approach is considered to be more accurate than the classical method [5].

In continuation with the geosatellite observatory, some studies have tried to directly link satellite-captured signals sent to ground and the measured readings through empirically calculated expressions or machine learning approaches to obtain spatial estimations of GSR [6–8]. The artificial neural networks (ANN) models have been widely used for the collection of GSR for variable time periods and this has been widely used for renewable energy systems [9].

11.3 ARTIFICIAL NEURAL NETWORK

An artificial neuron network is a computational system model that imitates nerve cells in the human brain. The ANN has three or more added layers that are connected with each other. Figure 11.1 depicts the structure of a typical neural network. The first layer comprising of input neurons pass the input data to the deeper layers called hidden layers, which adaptively change the information received and pass the resulting output data to the last output layer. The ANN uses learning algorithms that can independently train and estimate. The weight and bias function for different input used in back propagation is given in Figure 11.2. The learning rules make use of back propagation algorithm, a method through which the ANN can regulate its output results by considering the errors. By the process of back propagation, every

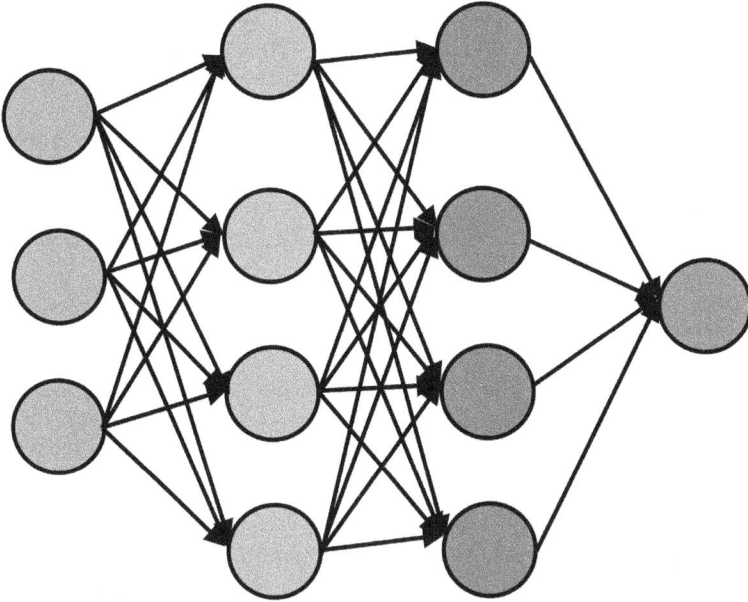

FIGURE 11.1 Layout of artificial neural network.

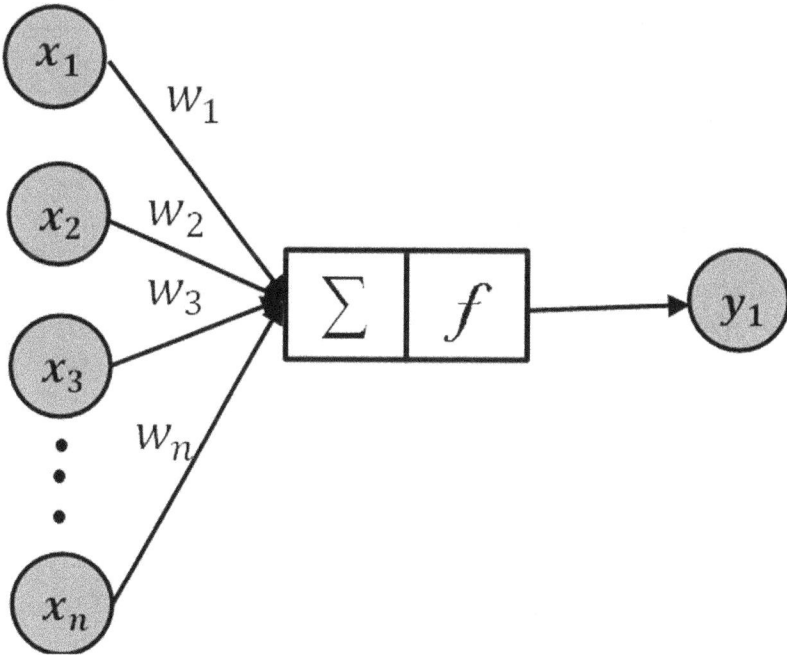

FIGURE 11.2 Back propagation methodology: a neuron is represented with the given equation (1).

time the weights and bias are automatically updated to minimize the error. The most predictable advantage of ANN is it can very well learn from perceiving data sets. The ANN recognizes data samples for learning rather than whole data sets to arrive at its decisions, which is considered as one of the simplest mathematical models to enrich prevailing data analysis technologies and it also saves time and money. Of the different types of ANN, deep learning ANNs play a significant role in machine learning (ML) and is one of the promising artificial intelligence (AI) technology that yields accurate results. The ANN finds its recent application in the GSR data estimation by which it estimates the data in the region of interest [10].

$$o = f\left(wx + b\right) \tag{1}$$

Where w denotes the weight, x denotes the input and b denotes the bias. The bias is to enable better learning. For GSR estimation, the prominent inputs considered are speed of wind, normal atmospheric temperature and sunshine period.

11.4 PREDICTION OF GLOBAL SOLAR RADIATION AND AVAILABLE THERMAL ENERGY USING MACHINE LEARNING ALGORITHMS

Different researchers have developed their ANN model of which the deep learning algorithm with multi-layer perceptrons has been providing promising data of GSR for spatial patterns, and the process includes training of the ANN and data estimation. The training section is the unit where the network learns and stores obtained spatial models based on the association between input and desired target data. In the section of estimation, the network recalls the stored information to produce output for given input data sets. The ground measurements of hourly GSR are collected from different stations with the pyranometer and are used for model training. But, the ground measurements shall contain errors due to operation and equipment, which would considerably affect the final accuracy of GSR. Research is carried out with different types of ANN to optimize the GSR data accuracy. Convolution neural networks for data estimation were preferred for its advantages over conventional systems [10–14].

11.4.1 CONVOLUTION NEURAL NETWORKS

The convolutional neural network (CNN) is a topology used in ANN and these CNN are biologically stimulated variants of multi-layer perceptrons through which low-level and high-level featured data can be extracted easily. Figure 11.3 shows the flow diagram of the typical CNN with machine learning program for data extraction of GSR. The structure of a CNN is made of a convolutional layer, non-linear layer and a merging layer [15, 16].

The feature maps extracted from the GSR data are fed as mathematical array structures for the input layer. The convolution layer manipulates a sequence of convolution operations on the input feature maps with small kernel weights. The dot product of the kernel weights and the input from the local region is performed by

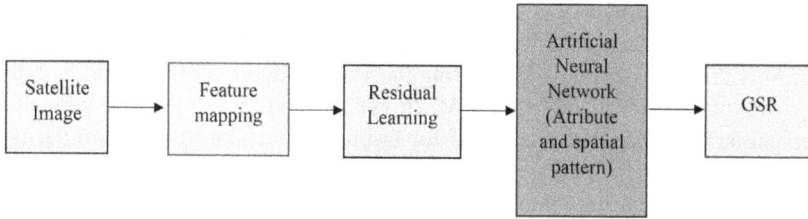

FIGURE 11.3 Flow diagram of GSR estimation using Deep Neural Network.

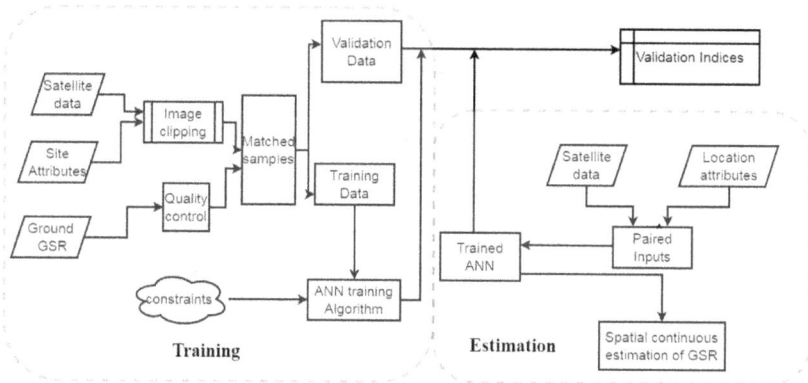

FIGURE 11.4 Flowchart of training ANN and estimation.

the convolution layer. To increase the learning efficiency convolution kernels are repeated across the entire layer, which uses the same weights and biases. The next layer is the non-linear layer which uses non-linear activation function to process each data value in the characteristic map. Of the different non-linear activation functions, the rectified linear unit (ReLU) is preferred as it overcomes the problems due to gradient and it speeds up the learning process. The ReLU has zero as the non-positive value and the positive value is unchanged. The action performed by the merging layer is the convolution of features along the spatial dimensions. The input to the CNN is the satellite image and it performs primitive mapping and learning, and finally produces conceptual illustrations of spatial pattern. The spatial pattern with additional attributes is combined by the MLP and is related to the target GSR values for exploring different sets of non-linear relationships. Figure 11.4 is the flowchart used to train the CNN and estimate the GSR. The training is done with the satellite data, site attributes and the ground data of the GSR. With this training data set, the CNN is trained along with constraints. The trained data is used for estimation of the spatial GSR with the satellite data and time location attributes.

11.4.2 Extreme Learning Machines

An extreme learning machine (ELM) is a new machine learning based algorithm for a feed forward neural network and is a very fast training technique used to train multi-layered neural networks [M16]. ELM is gaining popularity for its high performance, high selection sensitivity compared to conventional ANN. The data modeling in ELM is done by randomly structuring the weights and bias of the middle hidden layer and the output layer of the ELM. In general, the Moore-Penrose generalized inversion data matrix is used in ELM for estimating weights. Two dissimilar types of statistics are used for the estimation methods in ELM which includes the data recorded from a local station and the second includes data received from a different meteorological station farther away.

11.4.3 Modern Control of Solar Thermal Heating Plants

Solar thermal energy is the heat generated from the solar energy and is used in industrial applications, built-up areas and commercial regions. Solar thermal collectors are used to accumulate the heat energy and the classification is made as low-, medium- and high-temperature collectors based on the perceived temperature [17, 18]. The role of low-temperature collectors is to heat the ventilating air. The flat plate collectors fall under the type of collectors based on medium temperature and are used for heating water or air in residential and commercial applications. The collectors based on high-temperature use focusing mirrors to concentrate sunlight and are used for meeting the heat supplies in industries and electric power production. The closed loop structure of a solar thermal heating unit is given in Figure 11.5.

A modern solar thermal system integrates a solar based thermal collector, accumulation tanks and a gas heater. The closed loop control of the solar thermal heating unit is achieved by collecting the measured output from the solar thermal heating unit and feeding it to the accurate validated model which predicts the input to the controller. The ANN is used to estimate the GSR from the solar irradiance and is one of the prime inputs for the controller. The third input for the controller is the set

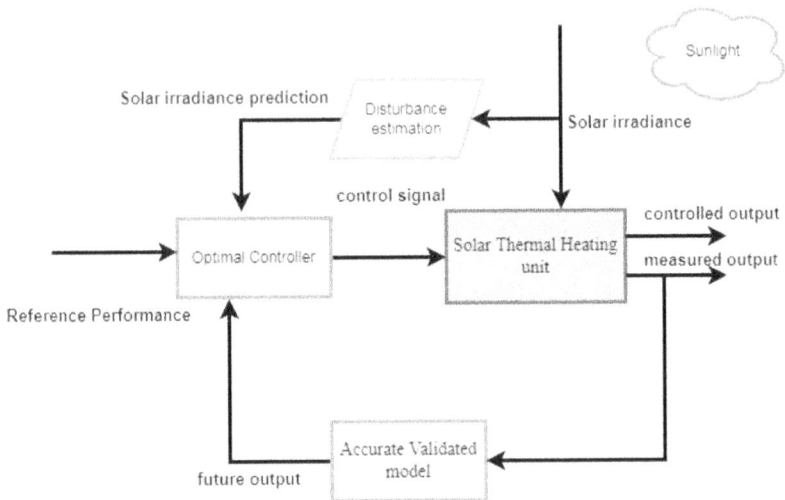

FIGURE 11.5 Block diagram of solar thermal heating unit.

reference value. Based on the inputs, the controller decides the control value for the solar thermal heating plant.

The role of AI in the ST system is to

1. Predict the ST performance.
2. Optimize the design of ST and have control over the operation.

The targets were achieved using AI-based prediction and optimization algorithms, along with fuzzy logic. Research was carried out to study the role of AI methods in the solar thermal systems that include solar water/air heaters, heating ventilation of buildings and air conditioning (HVAC) systems and also in power generating plants. The most frequently used ANN model for the ST system is the multi-layered feed forward neural network (FFNN). For the system modeling, the hybrid methods were employed to enhance the prediction accuracy and minimize the training period.

11.4.4 FUZZY LOGIC

Fuzzy logic is also comprehensively used in the decision making process of the solar thermal storage system and is mainly used for unpredictable parameters like atmospheric conditions or vague decision criteria. The fuzzy logic system (FL) is multi-valued logic and is expressed by the number value between the interval [0, 1] with a characterized membership function. The implementation of FL is done with the following steps: fuzzification in which all of the inputs are converted into linguistic variables using natural language like "low", "medium" and "high". The second step is the one in which the desired output is obtained according to the IF-THEN rules to derive the relation between the input combinations and possible outputs. Finally the defuzzification process is performed, in which the fuzzy logic output is converted to crisp value based on the membership function.

11.4.5 GENETIC ALGORITHM

Genetic algorithm (GA) is an evolutionary algorithm for solving optimization problems by genetic mechanism. The initial population for the optimization is randomly generated and each binary string stands for a possible solution. The operators work by imitating the genetic recombination and mutation phenomenon and results in new generation. GA may be less effective when compared with few conventional optimization algorithms and is less likely to fall into the local optimal peak.

11.4.6 IMPROVED EFFICACY OF SOLAR FLAT PLATE COLLECTORS

Figure 11.6 shows the model of a thermal system consisting of a flat plate solar based collector along with a Rankine heat cycle which was linked through a heat storage tank. The system was to be studied for its efficiency and state of charge and the prediction is done using learning algorithms. The estimation of the state of charge (SOC) and efficiency (η) was performed for several days with the help of ANN. The efficiency and the state of charge of the system were obtained by considering solar

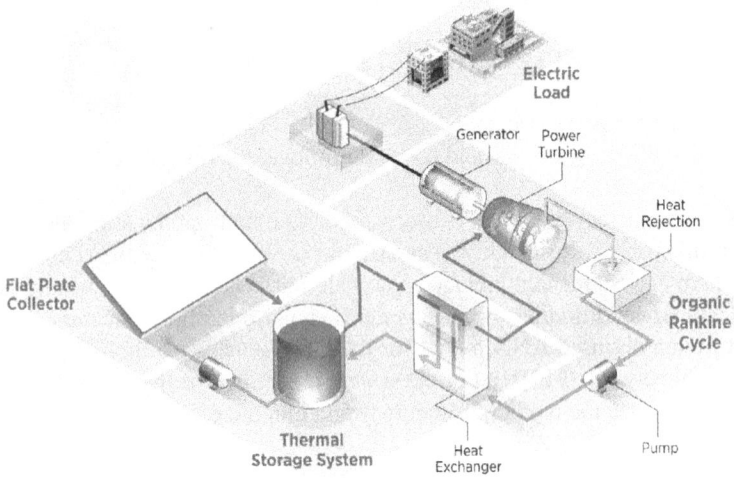

FIGURE 11.6   Schematic of Solar thermal system [19].

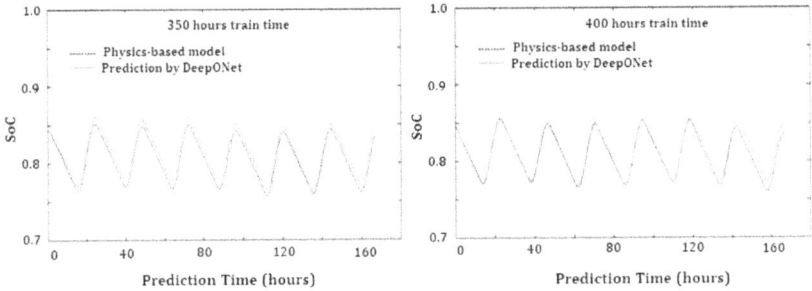

FIGURE 11.7   DeepONet convergence study for different time periods [19].

irradiance, air temperature and different load profiles as the input parameters for prediction and estimation. The results of the prediction were then employed to train the DeepONet neural network for about 20 days. Once trained, the developed DeepONet data-driven network was able to predict the η and SOC of the thermal system for several days [19].

As mentioned earlier, the training with the solar irradiance and different load profiles was executed and Figure 11.7 shows the results of the convergence for the physical based model and the DeepONet network for the training data. As the train time is increased, the convergence is higher.

From Figure 11.8 it was also observed that the mean square error of 0.076 after a training period of 150 hrs was reduced to 1.85e-5 after training for more than 300 hours.

11.4.7 MULTIVARIABLE OPTIMIZATION OF PARABOLIC TROUGH COLLECTORS

The parabolic trough collector (PTC) is a solar thermal collector with a parabolic structure lined with a polished mirror and a focal line. The sunlight hitting the mirror

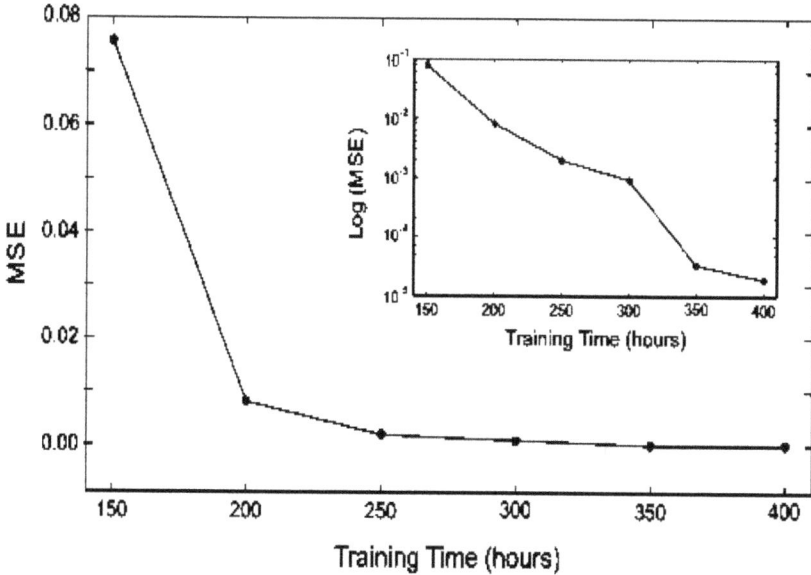

FIGURE 11.8 Mean squared error produced by the DeepONet as a function of the training time [19].

parallel is focused on the focal line which heats the object inside the focal line. The thermal efficiency of the PTC is considerably improved by using ANN in its control circuit. In the case study of a PTC with hydraulic cycle shown in Figure 11.9, a composition of 5 PTCs were used to preheat water for a heat storage system. The system described consists of a thermal storage tank with electrical water heating system and a hydraulic circuit. The working temperature of the tank is set at $90 \pm 1.0°C$. During the start of the operating cycle, water was heated by an electrical system and subsequently, the preheated working fluid is fed to the PTC receiver tube with constant monitoring of temperature and pressure of the storage tank [20].

Recirculation of the water is done to maintain uniform temperature; when achieved the valves V9 and V10 are opened and V7 is closed. Continuous data acquisition is done for temperature, wind speed and solar beam radiation and is recorded periodically.

The desired performance is achieved by using an ANN work to optimize on multiple input variables. The following six input parameters—rim angle (φ_r), input temperature (T_{in}), ambient temperature (T_{amb}), water volumetric flow rate (F_w), direct solar radiation (G_b) and wind speed (V_v)—were used in the input layer. TANSIG and LOGSIG functions were used in the hidden layer, PURELIN linear function was used in the output layer, and the Levenberg-Marquardt training algorithm were used for estimation.

Genetic-algorithms (GA) along with particle-swarm-optimization (PSO) were used for manipulation of multivariable objective function. The results portrayed that by using the multivariate inverse ANN methodology, the performance of the PTC was improved by optimizing to a maximum of three variables at the same time. In optimizing three variables simultaneously, it was noted that a thermal efficiency of 67.12% was obtained which was higher than optimizing one variable or two variables and this method was found to be a promising one to improve the thermal efficiency of PTC [20].

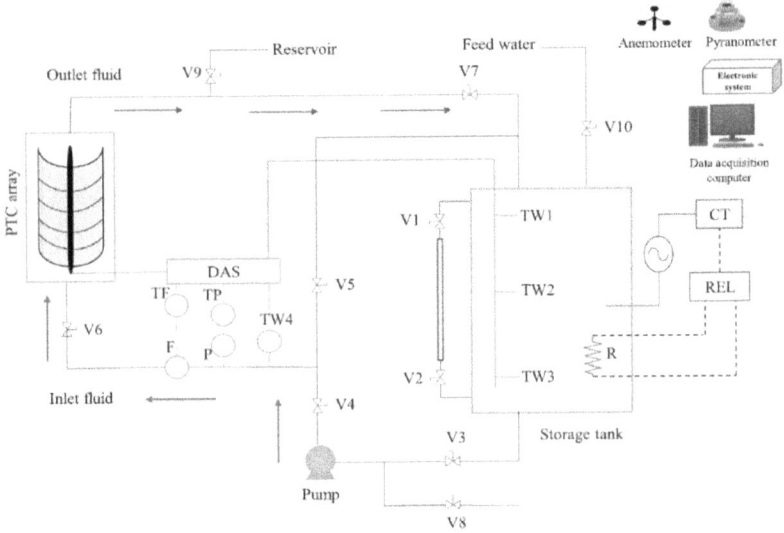

FIGURE 11.9 Schematic of the hydraulic cycle of a PTC [20].

11.4.8 Design of High-Performance Water-in-Glass Evacuated Tube Solar Water Heaters by a High-Throughput Screening Based on Machine Learning

Solar energy is prevalently used in water-in-glass evacuated tube solar water heaters (WGET-SWH). The process of the proposed solar water heater incorporating extrinsic properties to yield high collection rates using ANN is shown in Figure 11.10. The steps include generating and preprocessing data sets, training and validating ANN, providing new independent variables into ANN and finally recording the best data sets of heat collection rates.

The forecasted variables of two designed WGET-SWHs for various numbers of tubes and tank volumes along with the heat collection rate is given in Table 11.1. The experimentation was performed with the experimental setup in Figure 11.11. The test was carried out for four days outdoors and records were maintained for each day for a duration of seven hours during the day time. Water mixing was done to maintain uniform temperature and 11.32 and 11.44 MJ/m² are the average heat collection rates of two designs, respectively. 1.35% and 1.90% were the rate of error between the experimental and predicted values, respectively. The comparison findings demonstrate that the two new designs' experimental heat collection rates and their ANN-predicted values correspond quite well [21].

In order to measure the data groups, for the training and testing sets of the modeling technique, the normative measuring techniques of GB19141–2011 (GB/T 19141, 2011) were used [21]. As a result, it may be argued that the ambient conditions were the same for the database that was measured experimentally. The two newly constructed heaters were measured and operated at similar ambient temperatures in the

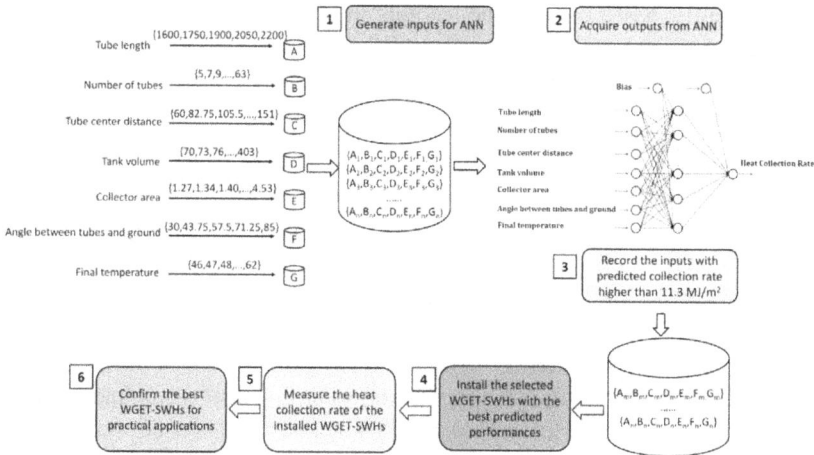

FIGURE 11.10 Flow diagram of SWH with new combinations of inputs for high heat collection rates.

TABLE 11.1

Forecasted Variables of Two Designed WGET-SWHs [21]

	Length of Tube (mm)	Number of Tubes	TCD (mm)	Volume of Tank (kg)	Area of Collector (m²)	Angle of Inclination (°)	Final Temperature (°C)	Heat Collection Rate
Design A	1800	18	105.5	163	1.27	30	52–62	11.6–11.8
Design B	1800	20	105.5	307	1.27	30	52–62	11.7–11.8

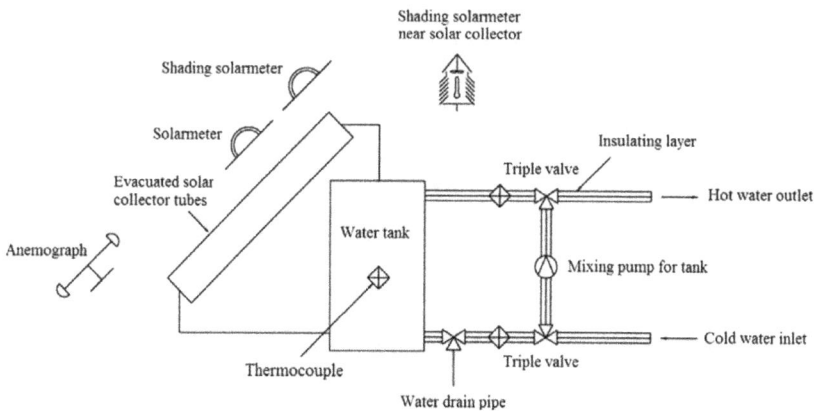

FIGURE 11.11 Experimental setup of SWH [21].

spring in Beijing based on machine learning models established, and the measure-ment and the testing results are equivalent to the predictive model and database. This is the key justification for why the majority of ambient conditions can be ignored during modeling.

11.4.9 SOLAR THERMAL SYSTEMS PROGNOSIS USING DEEP LEARNING

Machine learning and deep learning techniques have been used to define and forecast the performance of thermal systems. The ANNs have been used to study thermal sys-tems, like solar thermal systems, heat pumps and refrigeration systems. These tech-nologies have produced outstanding performance predictions and anomaly detection during thermal system operation.

Considering Figure 11.12, the system's hot water is shown in red lines, while the cool water is shown in blue lines. The sanitary water is intended to be preheated, from 10°C to 40°C (in summer) and to 30–35°C (in winter), by SHW. When the available radiation is inadequate to reach these temperatures, the heat-recovery chiller acts as a separate heat load, which regulates the temperature of the indoor swimming pool. The additional heat is transferred from both sources to the 4 m³ tanks, which are acting as intermediate heat storage with temperature value of 35–40°C. Four heat pumps in the heating portion provide the energy needed to heat the water to 60°C, after which it is kept in four tanks. The output temperature is controlled using mains water. In the preheating section, main water is utilized as make-up water also. At the two first water tanks, the remaining hot water finally enters the system again (Preheat Tank 1–2 in Figure 11.12). The temperature and operational state of cen-trifugal pumps, heat exchangers with flat plate type, and the aforementioned parts are all monitored. The gathering of data is a crucial phase in creating deep learning models, and synthetic data can be used to save time. Deep learning models use data generated by the Transient System Simulation program (TRNSYS) to estimate outlet temperature future values of the solar collector. This methodology is broken down into three primary steps: data collecting, temperature prediction model construction, and evaluation of the performance prediction ability of various model configurations.

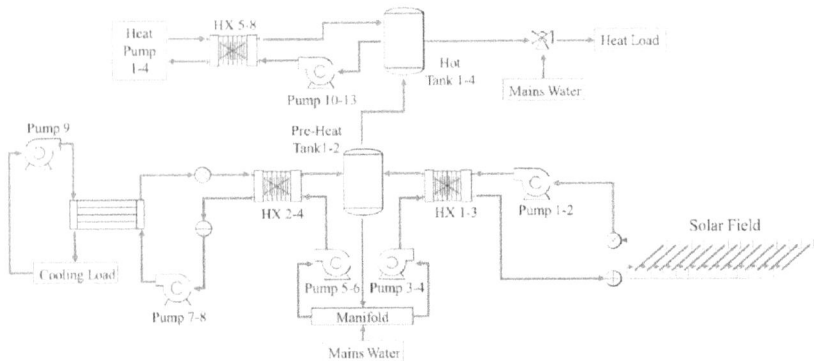

FIGURE 11.12   Preheating Section of SHW System [22, 23].

Synthetic data was produced under nominal design parameters using the TRNSYS simulator program. In order to construct and choose the network's hyperparameters, various configurations based on Recurrent Neural Network (RNN), Deep Neural Network (DNN), Long Short-Term Memory Recurrent Neural Network (LSTM) designs were trained and tested. Each model's predictive performance is rated based on two different prediction goals. First, for a single future temperature value, each model's performance is assessed using the root mean-squared error measure. The second objective is to make multiple time step predictions based on the previous prediction values with a two-hour time frame, recreating temporal behavior of the sequences over brief future time periods, and reporting both root mean squared error and mean absolute error scores. The LSTM model produced more precise results than RNN and more accurate predictions for temperature sequence than any other architecture among the many ones examined [22, 23].

11.4.10 ML in Solar Energy Systems Utilizing Nanofluids

By adding nanoparticles to the working fluid, the performance of solar collectors/ photovoltaic/thermal (PV/T) systems based on thermal-optical properties were enhanced; however concern about particle agglomeration, sedimentation and suspension stability still exist. Utilizing nanofluids typically results in a penalty for increased viscosity and pressure decrease. Cost, time and operational constraints are factors that affect experimental experiments. Furthermore, difficult mathematical and physical models are avoided by machine learning (ML). Nanofluid based solar collectors have been modeled and optimized using it. The two methods that are most frequently used are ANN and GA. In this study, gas-phase nanofluid based solar transparent parabolic through collector was optimized using a GA approach [24].

The results showed that a nanoparticle with 0.3% volume fraction and 650°C outlet temperature produced the best heat transfer efficiency of 62.5%. Cu_2O-water nanofluid based evacuated tube solar collector's exergy and energy efficiency were also examined. Comparing the results, the MLP model performed better than the RBF model. Optimization of the construction of a PV/T collector using nanofluid and determination of the link between input/output parameters is performed using a hybrid ANN and particle swarm optimization (PSO). A block diagram of this technique is shown in Figure 11.13.

The ideal reference value for the prediction procedure was discovered using the PSO method. The PSO's schematic was used as the basis for the MLP, ANFIS and RBF ANN to search for optimal mapping and predict the targets. For each approach, the optimum parameters were acquired individually. The trained model was evaluated using variance, root mean square error, and mean absolute percentage error after the neural network had been tuned. One hundred thirty experiment-based training points were used to train the model. Distinct experimental conditions with 13 points were used for validation. The fluid inlet temperature, ambient temperature and incident radiation were the input variables. Temperature at the fluid outflow and system electrical efficiency were the output variables [24].

A number of studies using the MLP ANN to examine the effectiveness of PV/T systems were performed. The ANN output and experimentally measured values

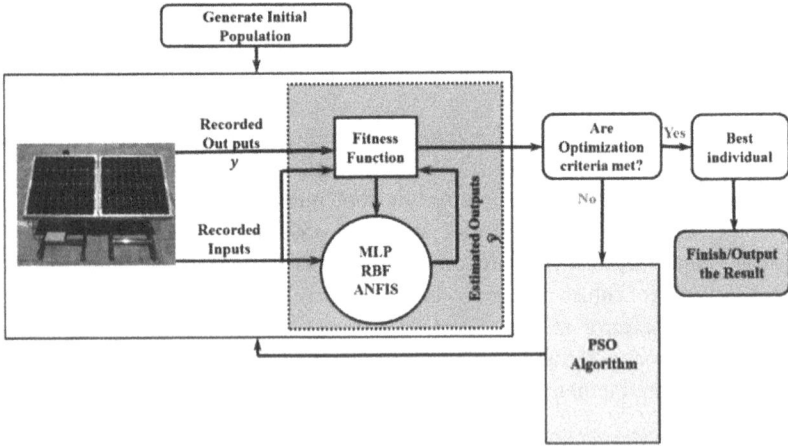

FIGURE 11.13   PV thermal nanofluid based collector based on ANN and PSO [24].

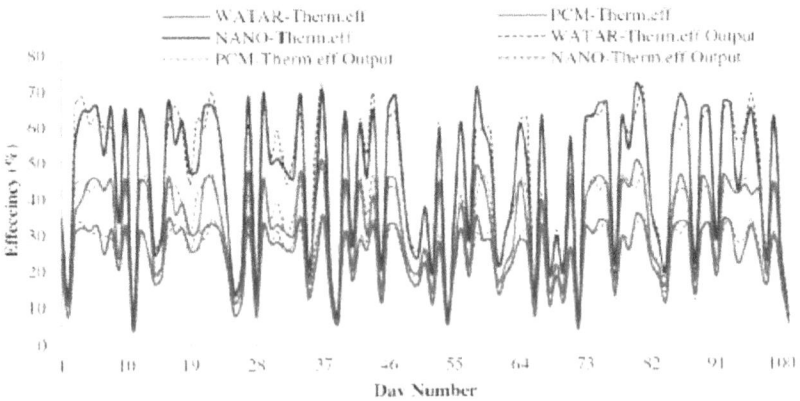

FIGURE 11.14   The ANN's prediction of nano based PV/T systems on thermal efficiency.

of thermal heat efficiency and thermal efficiency output of different PVT models operating under similar conditions were recorded and displayed, as shown in Figure 11.14. The different models are based on a nanofluid system referred to as NANO, water-based system referred to as WATER and a paraffin wax system referred to as PCM.

On comparing the experimental and ANN outputs, the NANO based system was found to be most effective. The electrical efficiency was increased by using nanofluid and nano-PCM from 8.07% to 13.32%, while the thermal efficiency was increased to 72%. The ANN's predictions and the experimental results were in agreement [24].

The significant problems that require attention during the energy system's assessment and optimization are

1. The ML models' predicted findings are only applicable within the boundaries of their own experimental range.
2. In order to enable larger applications of ML models, the thermal performance of different heat exchangers, large databases for the thermophysical characteristics of different nanofluids, and the optical and radiative performance of nanofluids in the solar spectrum must be established.
3. Determination of the relative errors of the ML algorithms is challenging.

11.4.11 ML IN CONCENTRATING SOLAR THERMAL

Machine learning is applied in concentrating solar thermal (CST) systems in forecasting of solar radiation, components modeling, performance prediction, consumption forecast, fault detection, fault diagnosis, prediction of mirror cleaning, study of material durability, etc. [25].

11.4.12 SMART HELIOSTAT TRACKING SYSTEM

Using computer vision and machine learning methods, a smart heliostat tracking system is created, which is a closed loop tracking system, shown in Figure 11.15. A closed loop tracking system was developed using deep learning. The key parts of the device are an embedded microprocessor, a wide-angle camera for tracking the sun at sunrise and sunset, and a portable solar battery to supply power for the device's minimal energy needs. To find potential objects of interest, the camera's live video stream can be examined (sun, receiver, heliostats and clouds). Heliostats can be automatically controlled using this information [25].

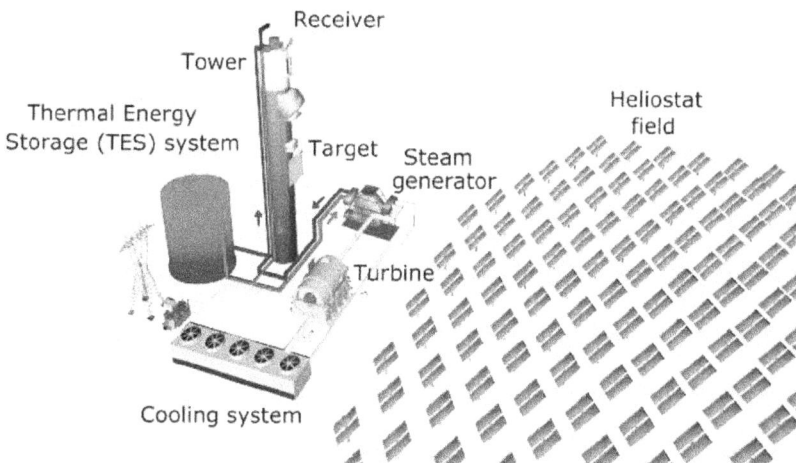

FIGURE 11.15   Smart heliostat tracking system.

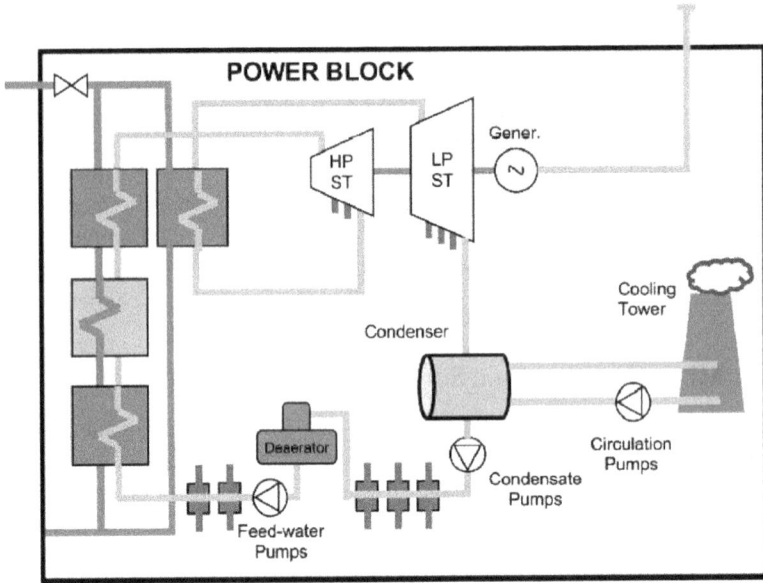

FIGURE 11.16   Power Block diagram [26].

11.4.13 CONCENTRATED SOLAR POWER (CSP) SOLAR STEAM GENERATOR SIMULATION USING MACHINE LEARNING

Concentrated solar power facilities (shown in Figure 11.16) must be operated and designed more efficiently if they are to compete more favorably with other renewable energy sources. It is essential to have an accurate performance model for both the initial stages of bid presentation and the subsequent phases. This model has numerous blocks, but the Power Block is by far the most important one (PB). Accuracy, generalization assurance and transient reproduction capability are required for PB modeling. Training and testing are done using the Support Vector Regression (SVR) approach. One year's worth of data at one-minute intervals is being used for model training. SVR performs well with a polynomial kernel [26].

11.4.14 ECONOMIC MODEL PREDICTIVE CONTROL (EMPC)

A system of solar thermal cooling is considered with storage to minimize backup energy consumption, which is depicted in Figure 11.17. Genetic algorithm is used to solve the optimization problem, based on perfect weather forecasts [27].

11.5 CONCLUSION

This chapter has explained the different modern soft computing tools available for solar insolation prediction, estimation and optimization of solar thermal storage

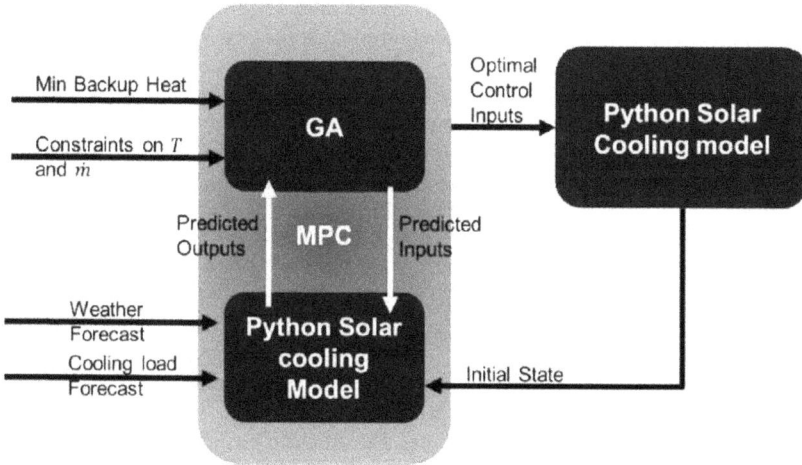

FIGURE 11.17   EMPC system [27].

systems. The control of the solar heating process with machine learning algorithms for industrial applications was illustrated. The role of the modern tools in the solar heat assessment of parabolic trough collectors, solar thermal heaters, solar flat plate collectors with the parametric control were presented. It was observed that the deep artificial neural networks show an enhanced performance with an increase in effi- cacy of about 15%. The multivariate estimation and control using GA and PSO reduces the mean square error to 0.0001 and yields in better convergence for train- ing for a longer period. Hence the use of modern tools for training and estimation in the solar based systems is found to be more realistic and improves the system performance.

REFERENCES

[1] Oliver M, Jackson T. Energy and economic evaluation of building-integrated photovolta-ics. Energy 2001;26:431–9.

[2] Kumar R, Umanand L. Estimation of global radiation using clearness index model for sizing photovoltaic system. Renew Energy 2005;30:2221–33.

[3] Chen L, Yan G, Wang T, Ren H, Calbo J, Zhao J, et al. Estimation of surface short-wave radiation components under all sky conditions: modeling and sensitivity analysis. Remote Sens Environ 2012;123:457–69.

[4] Mellit A, Kalogirou SA. Artificial intelligence techniques for photovoltaic applications: a review. Prog Energy Combust Sci 2008;34:574–632.

[5] Hay JE, Hanson KJ. Evaluating the solar resource: a review of problems resulting from temporal, spatial and angular variations. Sol Energy 1985;34:151–61.

[6] Perez R, Seals R, Zelenka A. Comparing satellite remote sensing and ground network measurements for the production of site/time specific irradiance data. Sol Energy 1997;60:89–96.

[7] Hassan GE, Youssef ME, Mohamed ZE, Ali MA, Hanafy AA. New temperature-based models for predicting global solar radiation. Appl Energy 2016;179:437–50.

[8] Janjai S, Laksanaboonsong J, Nunez M, Thongsathitya A. Development of a method for generating operational solar radiation maps from satellite data for a tropical environment. Sol Energy 2005;78:739.

[9] Zarzalejo LF, Ramirez L, Polo J. Artificial intelligence techniques applied to hourly global irradiance estimation from satellite-derived cloud index. Energy 2005;30:1685–97.

[10] Lu N, Qin J, Yang K, Sun J. A simple and efficient algorithm to estimate daily global solar radiation from geostationary satellite data. Energy 2011;36:3179–88.

[11] Ömer Ali Karaman, Tuba Tanyıldızı Ağır, İsmail Arsel. Estimation of solar radiation using modern methods. Alex Eng J 2021;60(2):2447–55, ISSN 1110–0168

[12] Meenal R, Selvakumar AI. Review on artificial neural network based solar radiation prediction, In 2017 2nd International Conference on Communication and Electronics Systems (ICCES), Coimbatore, 2017.

[13] Senkal O, Sahin M, Pestemalcı V. The Estimation of solar radiation for different time periods. Energy Sources 2010;32:1176–1184.

[14] He, Zhaoyu, Guo, Weimin, Zhang, Peng. Performance prediction, optimal design and operational control of thermal energy storage using artificial intelligence methods. Renew Sustain Energy Rev 2022;156:111977, ISSN 1364–0321.

[15] Jiang, Hou, Lu, Ning, Qin, Jun, Tang, Wenjun, Yao, Ling. A deep learning algorithm to estimate hourly global solar radiation from geostationary satellite data. Renew Sustain Energy Rev 2019;114:109327, ISSN 1364–0321.

[16] Anicic, Obrad, Jović, Srđan, Skrijelj, Hivzo, Nedić, Bogdan. Prediction of laser cutting heat affected zone by extreme learning machine. Opt Lasers Eng 2017;88:1–4.

[17] González-Roubaud, Edouard, Pérez-Osorio, David, Prieto, Cristina. Review of commercial thermal energy storage in concentrated solar power plants: steam vs. molten salts. Renew Sustain Energy Rev 2017;80:133–48.

[18] Kalogirou S. The potential of solar industrial process heat applications. Appl Energy 2003;76:337–61.

[19] Osorio, Julian D., Wang, Zhicheng, Karniadakis, George, Cai, Shengze, Chryssostomidis, Chrys, Panwar, Mayank, Hovsapian, Rob. Forecasting solar-thermal systems performance under transient operation using a data-driven machine learning approach based on the deep operator network architecture. Energy Convers Manag 2022;252:1–14.

[20] Ajbar W, Parrales A, Cruz-Jacobo U, Conde-Gutiérrez RA, Bassam A, Jaramillo OA, Hernández JA. The multivariable inverse artificial neural network combined with GA and PSO to improve the performance of solar parabolic trough collector. Appl Therm Eng 2021;189:116651, ISSN 1359–4311.

[21] Liu Zhijian, Li Hao, Liu Kejun, Yu Hancheng, Cheng Kewei. Design of high-performance water-in-glass evacuated tube solar water heaters by a high-throughput screening based on machine learning: A combined modeling and experimental study, Solar Energy 142 (2017) 61–67.

[22] Correa-Jullian C, Miguel Cardemil J, Droguett E L, Behzad M. Assessment of Deep Learning techniques for Prognosis of solar thermal systems, Renewable Energy 145 (2020) 2178–2191.

[23] Correa-Jullian C, López Droguett E, Miguel Cardemil J. Operation scheduling in a solar thermal system: A reinforcement learning based framework, Applied Energy 268 (2020),114943.

[24] Ma T, Guo Z, Lin M, Wang Q. Recent trends on nanofluid heat transfer machine learning research applied to renewable energy, Renewable and Sustainable Energy Reviews.

[25] Bonilla1 J, Carballo J A, Berenguel M, Fernández-Reche J, Valenzuela L. Machine Learning Perspectives in Concentrating Solar Thermal Technology, July 2019,

Conference: EUROSIM 2019.10th Congress of the Federation of European Simulation Societies.

[26] Gonzalez Gonzalez A, Alvarez Cabal JV, Vigil Berrocal MA, Peón Menéndez R, Riesgo Fernández A. Simulation of a CSP solar steam generator, using machine learning. Energies 2021;14:3613.

[27] Untrau A, Sochard S, Marias F, Reneaume J-M, Le Roux Galo AC, Serra S. Analysis and future perspectives for the application of dynamic real-time optimization to solar thermal plants: A review. Solar Energy, 2022:275–91.

12 Global Energy Model and International Solar Energy Policies

Mirhamed Hakemzadeh, Kamaruzzaman Sopian, Bhuvana Venkatraman, and T.V. Arjunan

CONTENTS

12.1 WORLD ENERGY MODEL

Since 1993, the International Energy Agency has used the World Energy Model (WEM) to anticipate energy for various time periods. The World Energy Model (WEM) is a large-scale simulation that looks at how energy markets work. The WEM is critical for producing accurate sector-by-sector and region-by-region forecasts for the World Energy Outlook (WEO) scenarios [1]. The World Energy Outlook (WEO) examines future energy trends based on WEM by looking at modelled scenarios based on a new set of basic assumptions regarding energy system dynamics [2].

DOI: 10.1201/9781003263326-16

12.1.1 THE SUSTAINABLE DEVELOPMENT SCENARIO (SDS)

The Sustainable Development Scenario (SDS) depicts a path to achieving the Paris Agreement's goals by bolstering clean energy policies and increasing investment in order to achieve the key goals of sustainable development, such as successfully limiting climate change, improving air quality, and ensuring universal access to modern energy.

According to SDC, all present net-zero commitments are fully satisfied, and advanced economies will achieve net-zero emissions by 2050, China by 2060, and the rest of the world by 2070. Furthermore, significant efforts are undertaken to control climate change, cut greenhouse gas emissions in the short term, and limit global temperature rise to 1.65°C by 2100 [3].

Industrial greenhouse gas emissions should be decreased by 1.2% per year to 7.4 $GtCO_2$ by 2030, according to SDS projected in Figure 12.1. Governments may meet the SDS targets through boosting energy efficiency, attracting renewables, and providing innovative financing, as well as enacting mandated CO_2 emissions reduction and energy efficiency legislation [5]. Renewable heat consumption has increased by an average of 2% per year from 2010 to 2018, which has been able to provide up to 6% of industrial energy demand, which should be increased to 9% by 2030 with the implementation of the SDS plan, as shown in Figure 12.2.

Figure 12.3 shows renewable heat for industries mainly obtained from bioenergy, which is only consumed in limited sectors of industries. The second-largest share of renewable heat consumption in the industry comes from renewable electricity, which is not economically justifiable due to the low efficiency and current price of solar photovoltaic panels. In addition, solar electricity systems require a large area to achieve heat energy demand.

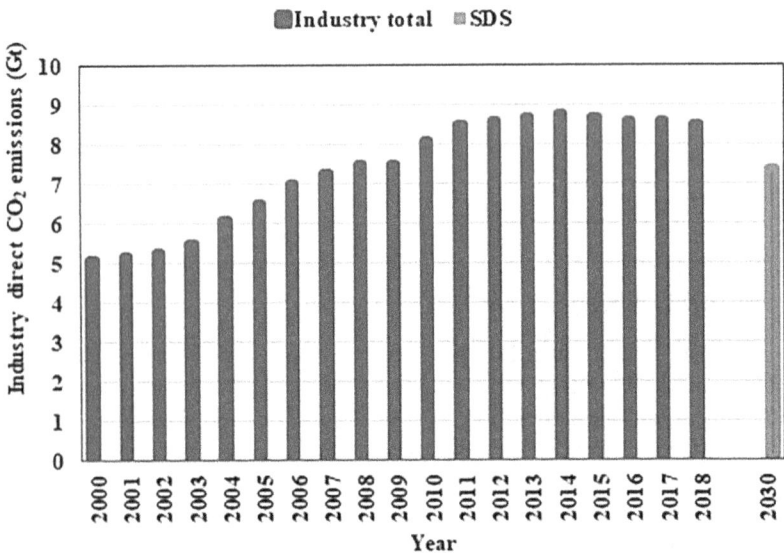

FIGURE 12.1 Industry direct CO_2 emissions in the SDC, 2000–2030 [4].

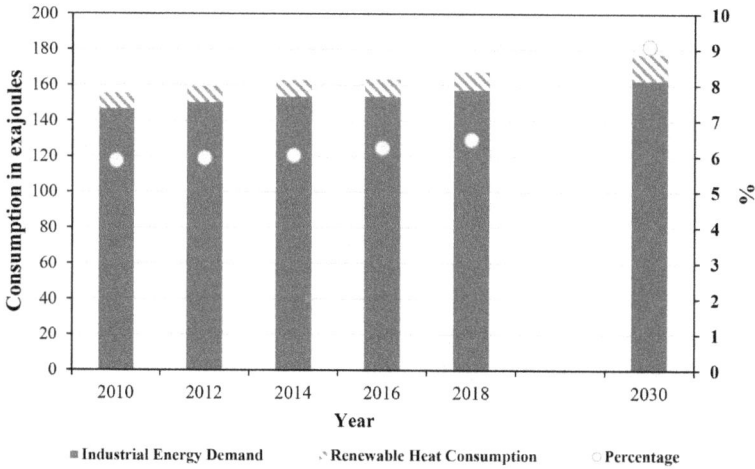

FIGURE 12.2　Final industrial energy consumption and fuel shares in the SDC, 2010–2030 [6].

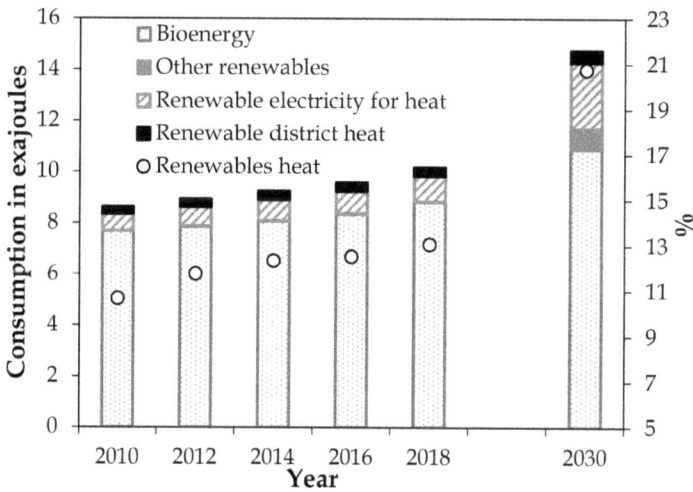

FIGURE 12.3　Renewable heat in the industry in the SDC, 2010–2030 [7].

The contribution of solar heating systems to meet the heat energy required by industrial processes has been only up to 3.5% between 2010 and 2018. Although heat energy produced using solar heating systems will be increased up to 43% and 23% compared to 2010 and 2018 by 2030, the share of renewable district heat in the energy supply will be decreased to 3.1% with regards to SDS, which is very small compared to other renewables due to the lack of government support and regional climates as the main obstacles to the development of these systems.

12.1.2 THE GLOBAL ROADMAP SCENARIO (REMAP)

Renewable energy can provide the best path to reach the targets outlined in the Paris Agreement by lowering greenhouse gas emissions and thereby mitigating climate change, according to a global roadmap (REmap) published by the International Renewable Energy Agency (IRENA). As a result, all nations must considerably raise the proportion of renewable energy in their total energy consumption, with the total share of renewable energy rising from roughly 15% in 2015 to around two-thirds by 2050. Furthermore, REmap demonstrates that the cost savings from reduced air pollution, improved health, and less environmental damage will far outweigh the additional costs of long-term and comprehensive energy transmission by 2050, resulting in a significant improvement in the global socio-economic status of the energy system, GDP, welfare, and employment. Furthermore, environmental advantages are comparable in all nations, and regional net profit on employment varies over time, but it has a beneficial influence in virtually all areas and countries [8].

To date, despite all the benefits of renewable energy, REmap figured out that industrial sectors have made the least use of renewable energy, while due to the high energy requirements of some industries and the high carbon content of some products, having the socio-economic benefits of renewable energy is inevitable in industry. Hence, according to the REmap scenario, by 2050, the use of renewable energy in the industry should more than quadruple from 15% to 63%, while the use of non-renewable energy in the industry should decrease from 86% to 37%, which will lead to halving greenhouse gas emissions by 2050. As shown in Table 12.1, industrial use of solar heating systems for low-temperature processes will increase dramatically from 1 million to 3.4 billion square metres of solar thermal collectors, which will meet 6% of the industry's heat demand. In addition,

TABLE 12.1
Final Industrial Energy Consumption and Fuel Shares in the REmap, 2015–2050 [8]

Energy Source	Energy Consumption	2015	2050
Renewables	District heat	<1%	2%
Renewables	Electricity:	7%	36%
Renewables	Geothermal heat	n/a	2%
Renewables	Solar thermal	n/a	4%
Renewables	Biomass	7%	19%
Non-Renewables	District heat	4%	5%
Non-Renewables	Electricity:	20%	6%
Non-Renewables	Gas	20%	15%
Non-Renewables	Oil	11%	5%
Non-Renewables	Coal	31%	6%
Total Renewables		**14%**	**63%**
Total Non-Renewables		**86%**	**37%**

80 million units of heat pumps will be installed to meet low-temperature heat needs [8].

The research method in REmap is based on the analysis of the global energy system according to the current and planned policies of countries to achieve optimal ways to make changes in the global energy system, which is mainly based on renewable energy. Hence, achieving the REmap goals requires political action to steer the global energy system towards a sustainable pathway, and the focus areas that make achieving these goals technically feasible and economically viable are as follows [8]:

1. Strong cooperation between energy efficiency and renewable energy must be established to meet more energy-related decarbonization needs by 2050 in a cost-effective manner.
2. Renewable energy should make a significant contribution to the power sector, which necessitates the timely deployment of infrastructure and the redesign of sector regulations to allow for the cost-effective integration of large-scale solar and wind generation as the backbone of power systems until 2050.
3. The use of electricity in transportation, construction, and industry should be increased to deep and cost-effective decarbonization of the transport and heat sectors through electrification. Also, modern bioenergy, solar thermal, and geothermal should be used in places like transportation, industry, and buildings where energy services can't be powered by electricity.
4. Innovations in technology and the development of new technologies must happen all the time across the system for renewable energy to move forward.
5. Social and economic structures and investments need to be in sync so that the energy transition can happen faster, adapting to climate change costs less, and there is less social and economic disruption.
6. Energy transmission, the costs of energy transmission, and the benefits of energy transmission must be fairly distributed at both the micro and macro levels in order for energy services to work well in all regions and converge.

12.1.3 The Regional Climate Scenario (RCS)

The Regional Climate Scenario (RCS) investigates ways to improve the share of solar heating systems to supply the thermal energy needed by industry. RCS emphasizes that latitude, clearness index, solar irradiance, ambient temperature, humidity, wind velocity, precipitable water, and unit operation are the significant parameters influencing the commissioning of thermal power plants. This scenario can be combined with the SDS and REmap scenarios as a complement and catalyst.

The Regional Climate Scenario (RCS) has been presented by National University of Malaysia researchers on a small scale, which can be generalized to solar thermal plants. The main objectives of this scenario are as follows:

1. Improve the technical-economic potential of the systems.
2. increasing the amount of solar radiation received by the absorber.
3. Achieve uniform solar energy distribution.
4. Solar heating system optimization.

5. Increase system efficiency and reliability.
6. Improving the capacity and storage systems for heat energy.
7. Ensuring the supply of water resources.
8. Accelerate progress and development.

This scenario considers three indicators of regional policies, climate regions, and techno-economic potential for the development of the integration of solar heating-systems in industrial sectors.

12.1.3.1 Climate Regions

A detailed and comprehensive assessment of the performance and lifespan of solar heating systems in various regional climates and the impact of microclimates and changes in global or regional climate patterns on solar energy distribution is critical. In RCS, the climate zones are specified according to the Köppen classification system. There are five major climate regions and 13climate types with regard to the Köppen classification system, as shown in Figure 12.4. The climate regions consist of tropical, dry, mild, continental, and polar, which are determined according to climate features, such as the average temperature and precipitation, windiness, humidity, cloud cover, atmospheric pressure, and fogginess, which depend on the latitude, elevation, topography, and distance from the ocean.

According to the RCS, four climatic zones consisting of tropical, Mediterranean, arid, and semiarid regions are the most suitable areas for the operation of solar thermal power plants to provide the thermal energy required by industrial processes in the low-temperature range. Figure12.5 shows the comparison of the conditions of

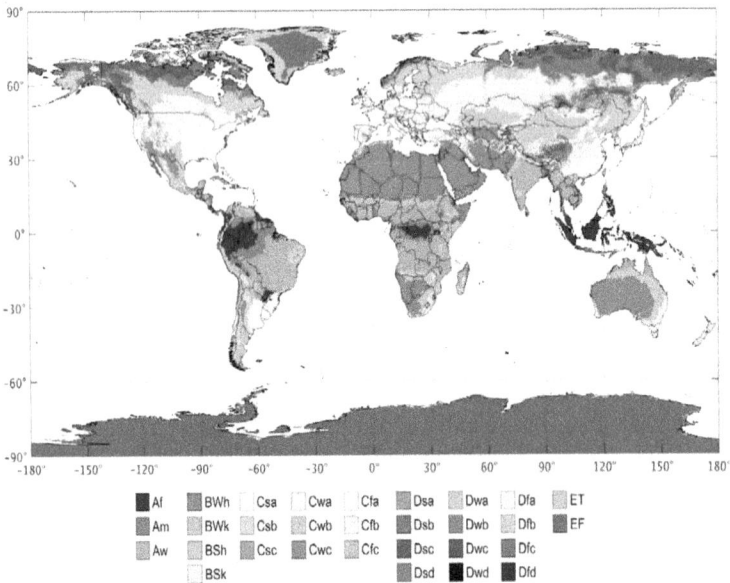

FIGURE 12.4   Köppen classification system [9].

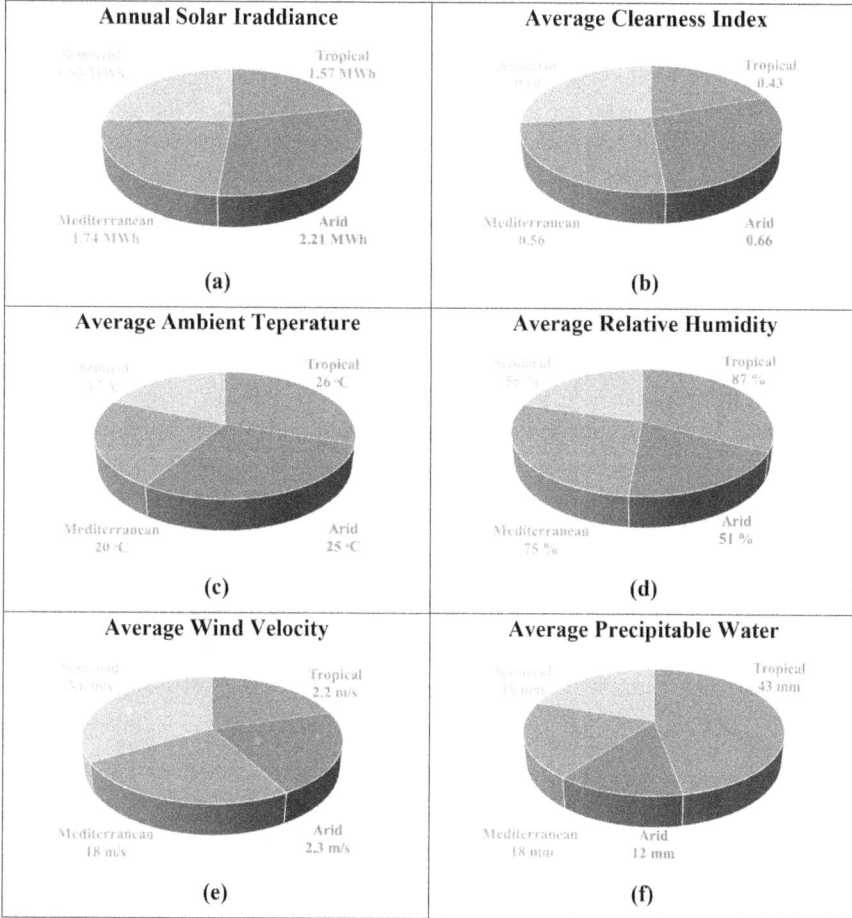

FIGURE 12.5 Comparison of the conditions of four climatic zones consisting of tropical, Mediterranean, arid, and semiarid regions for the integration of the solar heating systems in industrial sectors.

these four climatic zones for the integration of solar heating systems into industrial sectors.

Figure 12.5 (a) shows the highest and lowest distribution of solar energy is in arid and tropical climate zones, respectively, while the highest extraterrestrial solar radiation belongs to the tropical climate region with the lowest clearness index due to the attendance of atmospheric gases, aerosols, and clouds that lead to the processes of absorption, scattering and reflection of sunlight, as shown in Figure 12.5 (b) [10]. Figure 12.5 (c) shows the highest average ambient temperature belongs to the tropical and arid climate zones, while there is relatively little seasonal variability in the average monthly temperature in the tropics [11].

Furthermore, the intensity of solar radiation is affected by relative humidity, and it decreases as relative humidity rises and vice versa [12]. Arid and tropical regions have the highest and lowest average relative humidity, as seen in Figure 12.5 (d). According to Figure 12.5 (e), the Mediterranean and semiarid climate zones have the highest average wind speed, and convective heat transfer energy increases with wind speed due to an increase in the heat transfer coefficient, which increases heat loss from the glazing cover and bottom of the collector to the surroundings [13, 14]. The tropics have the highest average rainfall, as illustrated in Figure 12.5 (f), and higher rainfall lowers solar radiation intensity and consequently hot water outlet temperature [15].

12.1.3.2 Techno-Economic Potential of Integrated Solar Heating System

The RCS attributes the development of solar heating systems to supply the energy required by industry to the techno-economic potential of these systems, which is very favourable in industrial processes that require heat in the low-temperature range. RCS evaluates the performance of solar heating systems integrated into the industry through three techno-economic indicators, converted solar energy to heat, wasted heat energy, and profitability. Figure 12.6 indicatesthe comparison of the techno-economic potential of solar thermal plants integrated into industrial sectors under four climatic zones consisting of tropical, Mediterranean, arid, and semiarid regions.

Figure 12.6 (a,b) shows that the highest percentage of heat energy supply through solar heating system occurred in the arid region, while Figure 12.6 (c,d) discovers that the least heat energy wasted occurs in the tropics due to lower solar radiation intensity and higher ambient temperatures. Moreover, the highest system profitability is related to solar heating systems in arid climates, as shown in Figure 12.6 (e,f). In addition, the performance of the solar air heating system is dramatically poorer than that of the solar water system.

Despite the advantages of the dry climate, including arid and semiarid climate types for solar heating systems due to the high distribution of solar energy, low humidity, and low precipitation and coverage of about 26% of the world's area, only 14% of solar thermal plants exist in the arid and semiarid areas. Hence, the RCS scenario emphasizes the expansion of solar thermal plants in arid and semiarid areas to achieve the goals of the SDS and REmap scenarios.

12.2 SOLAR ENERGY POLICIES FOR DIFFERENT COUNTRIES

As everyone knows, every country has its own energy policy where the government's emphasisis on the production, transmission, distribution, and consumption of energy so as to address any problems relating to energy development. While framing the energy policy, every country and government keeps in mind the impact of global warming and environmental protection [16]. To avoid the adverse impact of climate change, the global warming has to be restricted to not more than 1.5°C and has to reach to net zero by 2050 as per the Paris Agreement [17]. In this regard, the significance of solar energy is based on abundant, non-toxic natural resources and solar energy being the cleanest, most cost-effective, efficient, and globally accessible energy source, which has substituted the conventional type of energy production through oil and coal for more than a decade now. Globally, the need for solar energy was felt because of the rapid demand for power supply and the growing shortage of

Converted Solar Energy in Water
Heating System

Tropical
53.3%

Semiarid
53.9%

Mediterranean
52.6%

Arid
63.5%

(a)

Converted Solar Energy in Air Heating
System

Tropical
10.4%

Semiarid
10.8%

Mediterranean
10.8%

Arid
27.0%

(b)

Heat Energy Wasted in Water Heating
System

Tropical
8.1%

Semiarid
14.9%

Mediterranean
13.4%

Arid
29.8%

(c)

Heat Energy Wasted in Air Heating
System

Tropical
33.6%

Semiarid
37.4%

Mediterranean
34.9%

Arid
45.1%

(d)

Profitability of Water Heating System

Tropical
0.860

Semiarid
0.861

Mediterranean
0.858

Arid
0.875

(e)

Profitability of Air Heating System

Tropical
0.711

Semiarid
0.712

Mediterranean
0.712

Arid
0.768

(f)

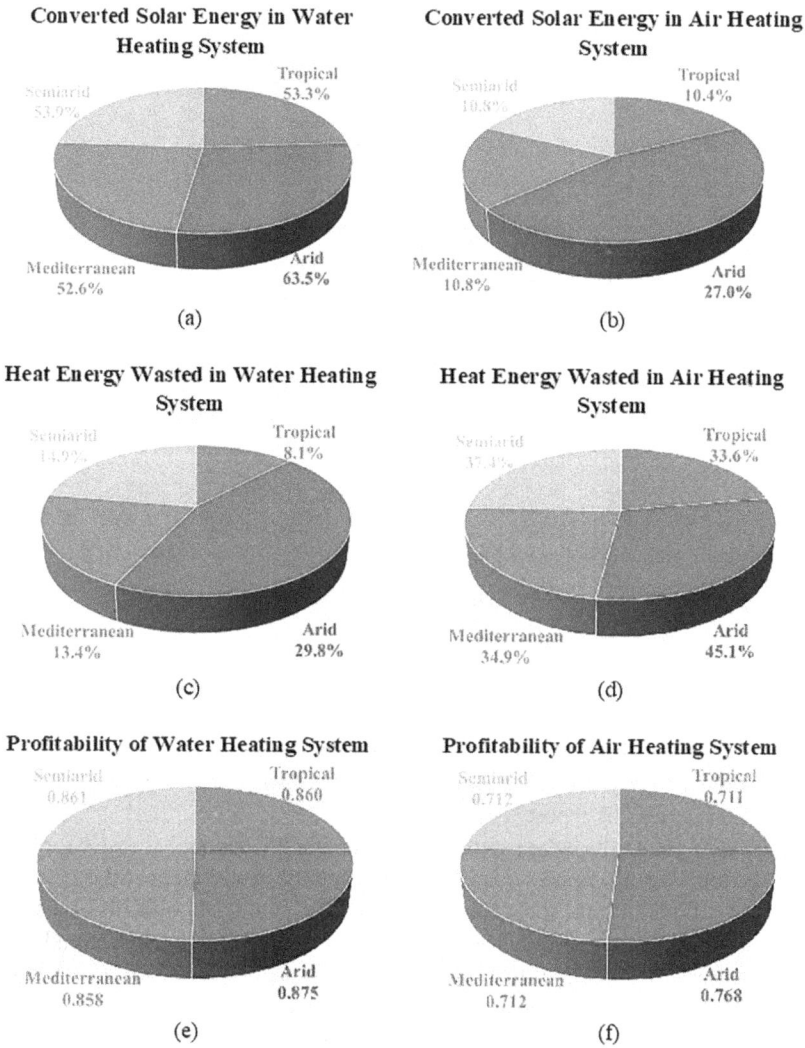

FIGURE 12.6   Comparison of the techno-economic potential of solar thermal plants integrated into industrial sectors under tropical, Mediterranean, arid, and semi-arid regions.

natural resources to generate the same. The covid pandemic did leave its impact in the energy markets andboth the primary energy and carbon emissions fell very rapidly. But the solar power grew largely despite the fall in the overall energy demand. The global energy demand fell by 4.5% in the year 2020 during thecovidpandemic [18]. The reason for the same was due to the lack of demand for power by the industries because of the lockdowns in almost all the countries. Studies say that although there has been a drift in the fall of energy demand globally during the covidpandemic

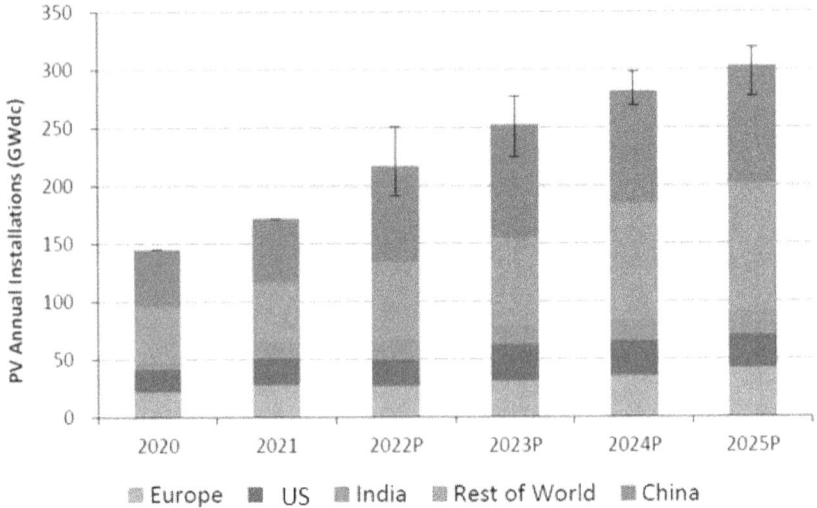

FIGURE 12.7   Annual Global PV Demand.

Source: U.S. Solar Market Insight: 2021 YIR, Spring 2022 Solar Industry Update.

period, there had been increasing growth of wind energy and solar energy recorded at 173 TWh and 148 TWh respectively during this period. In a span of five years, between 2015 and 2020, it is said that the wind and solar capacity has doubled. The average annual growth rate of the capacity of solar energy is 18% as per BP's Rapid and Net Zero scenarios report [19].

Figure 12.7 shows the actual and projected annual PV demand globally for 6 years where the top 4 countries are separately shown and the rest of the world is considered as a whole. China, Europe, the USA, and India contributed to two-thirds of the global PV installations in this period also. The projections are showing an uptrend but the growth rate was slowing over this period, especially in the USA. Further studies will attribute reasons and solutions for the downfall trend in the USA.

From Figure 12.8, it is clear that the energy use of Americans in 2011 in tonnes of oil equivalent per capita was quite high with 50 GDP per capita followed by countries like Australia, Germany, and Canada with nearly 42, 41, and 40 GDP per capita respectively. While a cluster of countries was somewhere closer to or below the world average energy use per capita. When it comes to India, the country finds its place below the world average energy use per capita which is a matter of concern. Whereas, big economies like Russia, China, Japan find themselves on the better side when compared to India.

Despite the global pandemic situation, the capacity of both solar and wind energy increased up to 50% more than at any time in history. China is the major contributor to this. The reason may be the major changes in the Chinese subsidies and accounting practices. Also, this has directly impacted reducing the cost of solar and wind power across the world. The renewable energy solar capacity for the selected countries is given in Table 12.2.

Energy use in tonnes of oil equivalent per capita

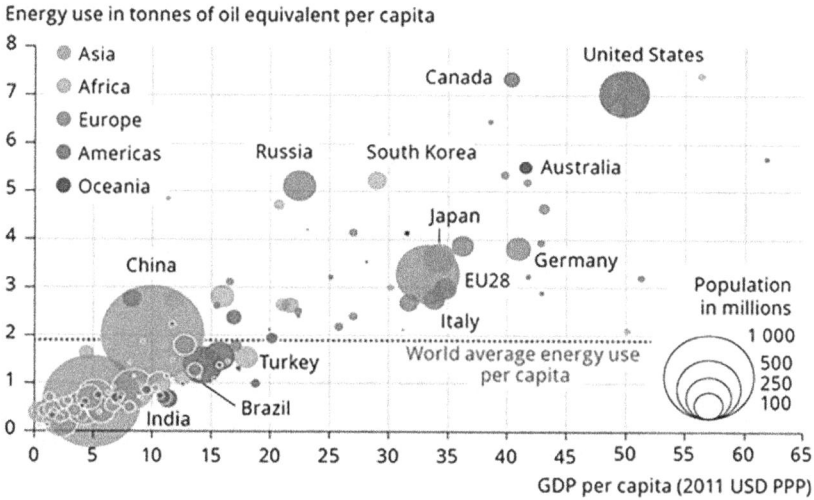

FIGURE 12.8 Energy consumption vs per capita GDP, PPP [20].

Australia's installed photovoltaic (PV) power gigawatts (GW) was recorded highest in the year 2020 with 17.6 GW, while the least was recorded in the year 2010 with 1.1 GW of solar capacity. China recorded its highest solar energy production in the year 2020 with 253.8 GW, whereas, the least of its production of solar energy capacity was recorded in the year 2010 with 1 GW. When it comes to India, the highest production in solar energy was recorded in the year 2020 with 39 GW, while the least for it was recorded in the year 2010 with 0.1 GW. Japan recorded its highest solar energy capacity of 67 GW in the year 2020, whereas, the least of it was recorded at 3.6 GW in the year 2010. When it comes to Malaysia and Pakistan, the solar energy capacity was nil in the years 2010, 2011, and 2012. Whereas, the highest capacity of solar energy was recorded in the year 2015, for both Malaysia and Pakistan with 1.5 GW and 0.7 GW respectively while the least capacity was recorded in the year 2013 with 0.2 GW for both these countries. Amongst these countries, China's capacity for solar energy was the highest with 253.8 GW, while the least was that of India, Malaysia, and Pakistan with 0.1 GW each. For the purpose of study, countries that are leaders in solar power generation like India, the USA, China, Japan, Germany, Australia, and Pakistan are considered. The objective is to understand the global scenario, progress, and policies related to solar energy in leading countries.

12.2.1 UNITED STATES OF AMERICA (USA)

The United States of America (USA) solar energy capacity generation has grown from 0.34 GW in 2008 to 97.2 GW in 2021. More than 3% of US electricity comes from solar energy by way of solar photovoltaics (PV) and concentrating solar thermal power (CSP). The main reason for this tremendous growth is the decrease in the price of PV panels. The prices of which were dropped by 70% since 2014. This has made solar energy prices competitive to conventional energy sources like coal, hydro, and others.

TABLE 12.2
Renewable Energy Solar Capacity for the Selected Countries

Installed Photovoltaic (PV) Power* Gigawatts Countries/Year	2010	2011	2012	2013	2014	2015	2016	2017	2018	2019	2020	Growth Rate per Annum 2020	2009–19	Share 2020
Australia	1.1	2.5	3.8	4.6	5.4	6.1	6.9	7.6	8.9	13.6	17.6	29.6%	44.7%	2.5%
China	1.0	3.1	6.7	17.7	28.4	43.5	77.8	130.8	175.0	204.6	253.8	23.7%	85.9%	35.9%
India	0.1	0.6	1.0	1.4	3.4	5.4	9.7	17.9	27.1	34.9	39.0	11.5%	97.1%	5.5%
Japan	3.6	4.9	6.4	12.1	19.3	28.6	38.4	44.2	55.5	61.5	67.0	8.6%	37.2%	9.5%
Malaysia	+	+	+	0.1	0.2	0.2	0.3	0.4	0.5	0.9	1.5	68.8%	109.6%	0.2%
Pakistan	+	+	+	0.1	0.2	0.3	0.6	0.7	0.7	0.7	0.7	3.1%	69.2%	0.1%

Source: Full report—Statistical Review of World Energy 2021 (bp.com).

By 2030, it is planned to have solar rooftops for 1 out of 7 homes. With CSP installations, the cost of solar power is estimated to be lower by 50% in another 2–3 years and by the trend, the CSP installations will generate 158GW of power by the year 2050 [21]. Around 70 GW of the installed solar capacities is contributed by California, Texas, Florida, Nevada, New Jersey, Massachusetts, in the USA. In the USA, the solar panels cost on average between US$3500 to US $35000 depending on the capacity, model and requirements [22]. Despite this, the challenges are manifold. Being a vast landscape country, the USA is hasdifferent efficiency criteria across various geographic areas. In this country, the amount of energy the sun provides varies from region to region and bytime of year. For example, Southern California, Arizona, and New Mexico receive 5.75kw hours per day of solar energy on average while other states get around 4kwh per day. Also, the hottest month, July, receives more energy as compared to other months. The government policies are designed to accommodate the requirements of this industry. The US Congress offered a significant Solar Tax Credit of 30% for panels installed before December 2019, 26% for panels installed in 2020, and 22% for panels installed in 2021. The tax credits are to expire by the end of 2022. The expiration of the tax credit could hamper the demand in the households unless the government acts on it. However, the decreased energy bills may sustain the solar panel installations across the country. The President of the USA has pledged to allot more money in its budget for clean energy research and implementation and hasfurther plans to improve the clean energy sources. The government's intentions will translate and bring growth to this sector as planned [23]. Table 12.3 shows the various policies supporting solar energy in United States.

TABLE 12.3
United States Policies to Support Solar Energy [24]

	Policies	Key Characteristics
1	Renewable Portfolio Standards (RPS)	Supply from the electric utilities by the electric provider to the customers as per the specified minimum demand.
2	Funding and Financial Incentive Policies	More attractive policy due to reduction in cost barriers and cost of regulatory compliance, lowering risk, etc.
3	Policy Considerations for Combined Heat and Power (CHP)	With a single fuel source it produces both electrical and heat services.
4	Electric Utility Policies	Encourage more investments in solar power and combination of heat and electrical energy.
5	Net Metering	It motivates the customers to use solar photovoltaic for generating electricity to get subsidiesfor the generated electrical energy.
6	Feed-In Tariffs	Motivates solar energy utilization through the payment of pre-established market rates for solar power generation.
7	Property Assessed Clean Energy (PACE)	Promotes solar energy installations for residential property rather than individual borrower.
8	Financial Incentives	The state governments are motivating solar energy utilization through loans, grants, tax credits, and rebates to the customers.

12.2.2 CHINA

China is the leader in the deployment of solar power having one-thirds of the global solar capacity. Also, China has contributed to around two-thirds of global solar module production. The solar power generation accounted for 3% of total China's total energy generation in the year 2018. China became the first country to sell its electricity generated through solar power for a rate lesser than the price for electricity generated from coal in 2018. PV technologies dominate China's solar industry as compared to other solar power projects which are in developing stages in the country [25].

Figure 12.9 shows that in the year 2010 the solar capacity of rest of the world was nearly 19 GW, which started to increase from the year 2012. China's contribution was also depicted with the rest of the contribution in the annual change, nearly 2 GW in the year 2011. The solar capacity contribution of China started to increase in the total contribution for the same, where the contribution reached its zenith in the year 2017, with almost more than 50% contribution in solar capacity for that particular year. While the rest of the world's contribution to the total solar capacity contribution was recorded highest in the year 2020. The contribution of China in terms of solar

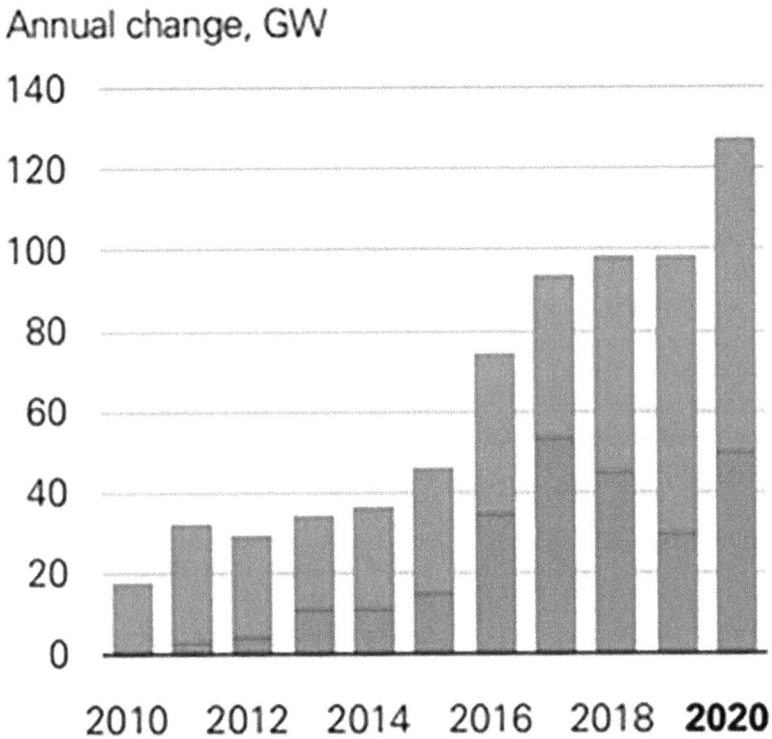

FIGURE 12.9   Solar capacity in China.

Source: Full report—Statistical Review of World Energy 2021 (bp.com).

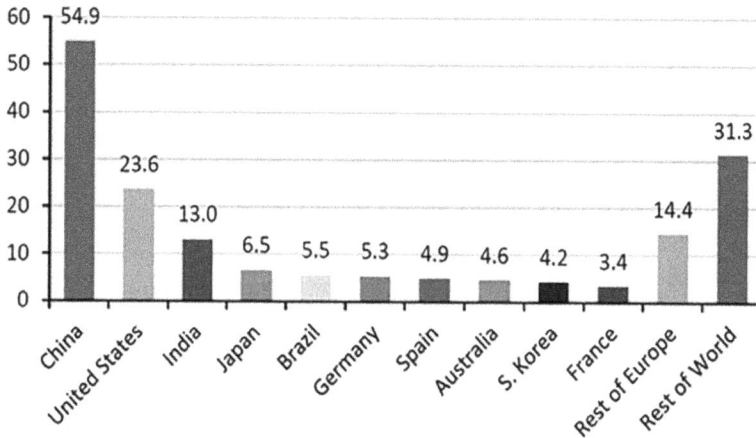

FIGURE 12.10   Annual PV deployment, 2021 [172 GWdc].

Source: IEA, Snapshot of Global PV Markets: 2022; Wood Mackenzie/SEIA: U.S. Solar Market Insight: 2021 YIR, Spring 2022 Solar Industry Update.

capacity to total contribution was least in the year 2011. Whereas, the least contribution in terms of solar capacity for the rest of the world when compared to the total contribution was recorded in the year 2017. As a whole, the highest solar capacity was recorded in the year 2020.

It is understood from Figure 12.10 that China's annual PV installations are growing at a speed of 14% every year in 2021 which is one-third of the total annual global deployment while the US stands in the second position in terms of both cumulative installations and annual installations in the PV Market.

It is seen from Figure 12.11 that the leading PV markets in the world are China, the United States, India, Japan, and Germany in 2021. The Chinese government made major amendments to its solar policies in 2018. The changes were actually said to be made to control the cost of subsidies and address the issues relating to overcapacity in power markets. Also, the Chinese government expends a lot of money on research and development activities of solar power. Despite the covid-19 situation, the subsidy was removed for the Chinese solar companies. In 2021, studies say that solar contributed 30% to new generation capacity in the country. China, being the largest market, installed a 55 GWdc record of solar capacity. The capacity of China has grown by approximately 120% in just a decade. The government of China has declared that no further subsidies will be provided for new rooftop PV systems. On the contrary, the Chinese central government has asked the state-owned Independent Power Producers to achieve a capacity share of 50% or even more by 2025. So, it is now a big challenge for the independent power producers of China to achieve this goal, keeping in mind the other attributes of profits with the huge pricing of equipment needed for the same. As a result of this, the RET may or may not be achieved within the said period. The various polices and theirsignificance for supporting solar energy in China are given in Table 12.4.

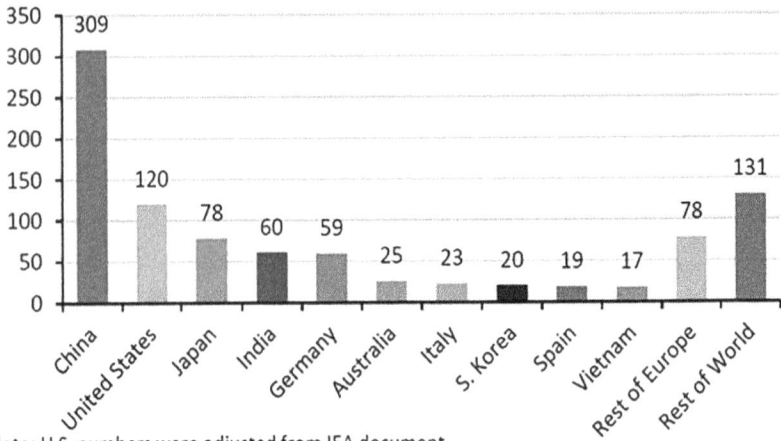

Note: U.S. numbers were adjusted from IEA document.

FIGURE 12.11   Cumulative PV deployment, 2021 [939 GWdc].

Sources: IEA, Snapshot of Global PV Markets: 2022; Wood Mackenzie/SEIA: U.S. Solar Market Insight: 2021 YIR, Spring 2022 Solar Industry Update.

TABLE 12.4
China Policies to Support Solar Energy/Renewable Energy [26]

Year	Policy	Details
2005	The Renewable Energy Law of the People's Republic of China	The policy outlines the mid-range of government policy goals.
2005	Renewable Energy Industry Development Guidance Catalogue	It covers the field of renewable energy such as wind, bio, solar, ocean, geo, and hydropower.
2006	Interim Measures for Renewable Energy Power Price and Cost-Sharing	Direction to fix the price of solar, geothermal, and ocean power.
2006	Management Rules of Renewable Energy Power Generation	Motivates the power generating and the grid companies to understand their responsibilities for the development of renewable energy power generation.
2006	Interim Measures for Management of Special Fund for Development of Renewable Energy	Supporting the key areas, like management, funding, and evaluation.
2007	Medium and Long-Term Development Plan for Renewable Energy in China	The renewable energy share must reach 10% by 2010 and 15% by 2020.
2008	The Renewable Energy Development Planning during 11th Five Year Planning Period	Targets and priorities for the development of renewable energy under 11thfive year plan.
2009	The Amendment to the 2006 Renewable Energy Law	It mandates that the power generated through renewable energy must buy from the power grid operators within the region.
2000	10th Five Year Plan for Energy Conservation and Resources Comprehensive Utilization	Focuses mainly on energy conservation, efficiency, and utilization of resources.

12.2.3 GERMANY

Germany is one of the most developed countries that pushes intensely itsrenewable energy sources, particularly solar energy. It is worthwhile to mention here that this country has the least sunshine hours. But then the country is showing tremendous growth in solar PV projects. In the year 2020, the country's installed solar power capacity was 60.07 GW. The country ranked fourth in the installed capacity in the year 2021 (International Renewable Energy Agency (IRENA)). The government of Germany has imposed a few laws to promote and develop the renewable energy in the country. Because of this, there had been a significant share of solar technology in the total power generation of the country. The share of power generated through solar was 11% in the total electricity production in 2020. The Federal Ministry of Economic Affairs and Climate Action also has new legislation to improve the solar share of power in the country. The Renewable Energy Act (EEG) brought new reforms in this sector in 2021. New legislation to reduce or cut the gas emissions was also introduced during this period by the European Union to reduce global warming. Also, they planned to tax foreign companies for the pollution that they cause. The government has set a target of 80% in the total electricity mix to be achieved through renewable sources by 2030. It was observed that even the pandemic situation did not impact much on this solar energy market even during the lockdown. The German solar market is well developed due to the availability of sophisticated infrastructure for solar technology to get connected with the national grids and also the existence of high level and high-quality renewable energy companies in the country. Despite these advantages, the country's climatic conditions are not amicable at all times. The country is aiming at removing the coal power and nuclear power segments from the entire electricity mix by 2030. The German government has made rooftop solar mandatory for new private buildings and new commercial buildings [27]. The changes in renewable capacity deployment with the support of policies is shown in Figure 12.12.

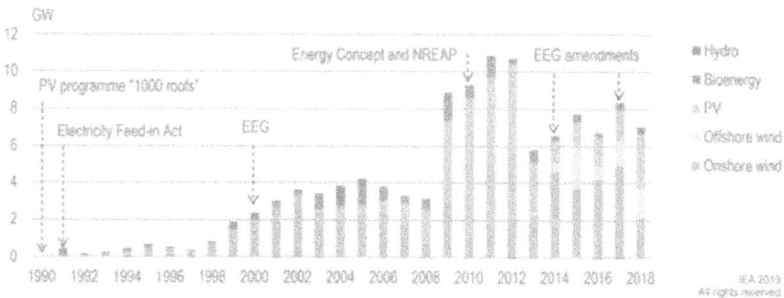

Renewable power capacity deployment in Germany can be classified into three periods: the early phase, the PV boom phase, and the more recent phase driven by competition.

FIGURE 12.12 Policy support changes and renewable capacity deployment 1990–2017.

Source: Germany policy 2020 review, International Energy Agency.

It is observed from Figure 12.12 that renewable power capacity deployment in Germany can be classified into three periods (the early phase followed by the PV boom phase, and finally reaching on to the recent phase which is driven by competition). We can say that in the initial years the renewable capacity deployment was low, while the maximum contribution is from hydro. The deployment started to increase from the year 2009, while in 2011 it reached its highest point with the maximum contribution from PV followed by onshore wind. In the recent phase, the deployment of renewable capacity saw some ups and downs with EEG amendments.

12.2.4 JAPAN

Japan is one of the most developed countries in the world. With the support of the Japanese public, Japan's Feed-In Tariff (FIT) policy for renewable energy came into force in July 2012 with an incentive structure to reduce costs and promote innovation. As per Figure 12.13, which shows the trends of renewable energy in the country since 2003, there has been a significant increase in solar energy since the introduction of FIT in 2012. The present level of solar energy capacity is 71.7 GW and annually it hopes to add 8 GW every year upto2030. It has already crossed the Cumulative Solar Capacity target of 64GW by 2030. Japan's Environment Ministry has proposed for the country to aim for 20GW Solar PV installations by the end of 2030 which will take its capacity to 108 GW. To give a boost to the sector, nearly 6GW capacity is to be installed on 50% of the government buildings and the balance from the private areas. Japan is giving thrust on rooftop solar installations as land is scarce in this country of tiny islands [28]. The government of Japan has initiated the following working policies:

1. Electricity generated by PV power for non-household customers was reduced from 27yen kWh to 24yen/kwh.
2. For household customers it was reduced from 33 yen/kWh to 31 yen/kwh [29].

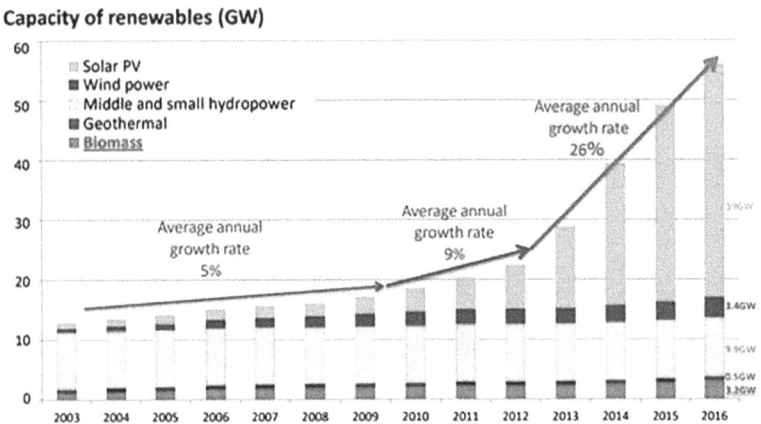

FIGURE 12.13 Trend of renewable energy since 2003 in Japan.

The government of Japan, despite the various anticipated problems and challenges which are incidental due to the seasonal changes, is still pushing aggressively the generation of solar energy capacity with the hope to join the 100GW club. For a solar power industry, the climate is the key. If that is not favourableat all times, the grid at the country level will be in trouble. Demand and supply issues will persist depending on the climatic conditions. It is expected that such problems can be handled through developing battery technologies and other storage mechanisms by the country.

From Figure 12.13, it is seen that the capacity of renewable energy was highest in the year 2016 where the contribution was highest from solar PV with about 39 GW followed by middle and small hydropower with 9.9 GW. In the year 2003, the renewable energy contribution was the lowest among these previously mentioned years with around 13 GW where the highest contribution is from middle and small hydropower. Over these 14 years, the average annual growth varies, while the maximum of it was observed between 2012 to 2016 (26% annual growth rate).

12.2.5 India

The government of India's Ministry of New and Renewable Energy (MNRE) is the nodal ministry for dealing with matters relating to new and renewable energy. The main objective is to develop renewable energy for providing the energy requirements of the country. The country provides tax benefits by offering tax exemptions in excise duty and customs duty. The solar mission 2010, being a part of India's national action plan, aims to develop the solar systems of the country. To achieve a target of 100 GW installed capacity through solar, the Indian government announced various policies to promote and develop solar energy in the year 2014 [30].

It is observed from Figure 12.14 that since 2003 with the enactment of the Electricity Act, 2003, there have been continuous rules, regulations, and policies made by the government, especially on the renewable energy side.

Out of 500 GW generation from renewable energy expected in 2030, 300GW is expected from solar power. India has achieved 50GW of cumulative installed solar capacity as of 28th February 2022. India stands fifth in solar power deployment in capacity additions with a contribution of around 6.5% to the global cumulative capacity. The MNRE, the Ministry of Power, and the government of India provide around 30% capital subsidy on solar projects. Both the central as well as state governments offer multiple subsidy schemes to the people for installing rooftop PV Systems in India. Also, there is a provision that the customer can sell back the excess power generated from thehouse or industry to the power grid. The subsidies, allowances, and schemes offered by the government encourage the public to venture into this form of energy in India. The initial cost of installation of solar is high but over a period of time, the power generation will become less expensive. The only difficulty that is felt by a common citizenis that the procedural delay it takes in getting subsidies from government agencies is tough in India. Not much financial support is available for installing rooftop solar for residential consumers, and small and medium enterprises. India does not have the potential to make solar wafers and polysilicon despite the fact that it had a 3 GW capacity and 8GW for solar cell production and solar panel production respectively [31]. India's solar power capacity increased more than 11 times between 2014–2019. It became the fourth most

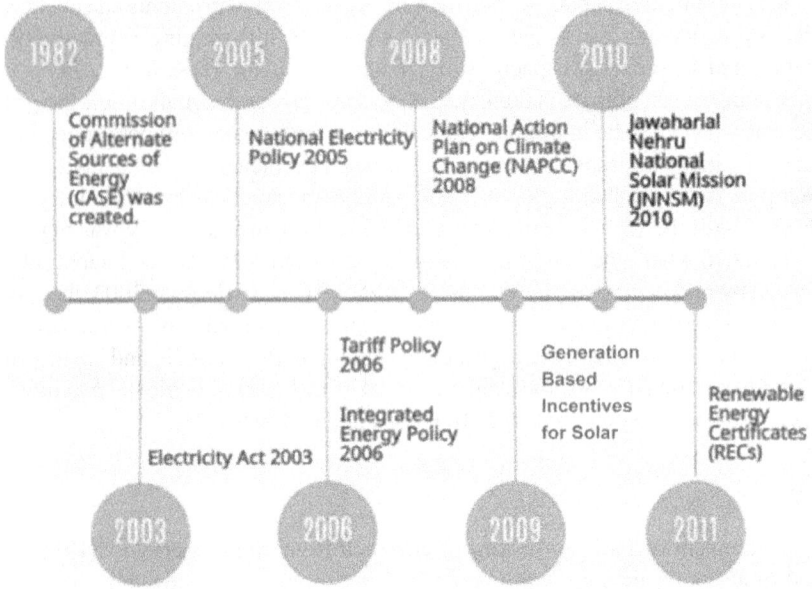

FIGURE 12.14 Government policies and regulations for solar energy in India [31].

Source: Government policies and regulations for solar energy in India—Solarify.

attractive renewable energy market in the world, with the fifth largest installed solar capacity at 29 GW towards the end of 2019 with a target of 100 GW of solar energy production by 2022 and 450 GW of renewable power by 2030. But after the outbreak of coronavirus in 2020, there had been serious disruptions in global supply chains of solar equipment and also due to heavy migration of labour to their hometowns due to the continued lockdown situation which caused delays in project completion. But as the situation and lives of the people have returned to normal, the country has added to its capacity. As per a study, India's total installed capacity in the solar sector inclusive of off-grid and rooftop has surpassed the milestone of 50 GW this February 2022. This increase in its installed capacity has beenseen over the last three years. Although there is a significant growth in the installed solar capacity, the contribution of this sector has not been reflected considerably in the entire country's power generation as expected. Also, till now India has not been instrumental in providing cheaper power for the end users despite the fact that there is a low tariff for solar power generation in the utility-scale segment. International Solar Alliance was established by India and France in 2015 with an objective to bring the countries closer to help collaboration to solve issues on analytics on solar energy, capacity building, and mobilizing investments. Presently there are 101 member countries in the ISA framework agreement with an objective to address collectively all the issues to scale up the solar energy applications as per need and to mobilize investments of more than USD 1000 billion by 2030 [32]. Definitely India is turning into being a major contributor for the renewable energy sector provided it gets investors to invest around 25 lakh crore to achieve the desired targets of 2030. The various enforcing policies and theirkey characteristics are tabulated in Table 12.5.

TABLE 12.5

Renewable Energy Policies in India [33]

Policy	Year	Key Characteristics
Renewable energy investment	2021	• Factor the socio-economic benefits of renewables into economic recovery stimulus funding. • Accelerate the coal phaseout in the power sector to unlock further investment in renewables. • Strengthen policy frameworks for storage, green hydrogen and system flexibility solutions. • Foster dialogue between investors and policy makers.
National Wind-Solar Hybrid Policy	2018	• The policy seeks to promote both hybrid projects and hybridization of existing renewable projects by the proper integration of both wind and solar energy. • There should be at least 25% of the rated power capacity of either wind or solar components in a hybrid project.
Ujwal Bharat—Power Sector Reform & Energy Access	2016	• Power sector reform and electrifying stranded and disconnected places. • Main mission is to provide uninterrupted affordable environmentally friendly power for all by 2019.
Uttar Pradesh Mini-Grid Policy 2016	2016	• The government supports by way of a 30% capital grant in return for its distribution company which will indicate the location of the project, fix tariffs, and involve in the technical specifications of the project to the mini-grid developing companies. • However, the sale of excess supply of electricity can be made to any customer of project owner's choice.
Andhra Pradesh Solar Power Policy 2015	2015	• Incentives for 10 years from the date of commissioning of the solar power projects and distribution losses will be exempted for such projects at 33kV or less than that. • Under this policy, the customers are given an option to choose either a gross or a net metre for the sale of power to distribution companies. • The charges for transmission and distribution will be exempted for SPP power. • Throughout the year, banking of 100% of energy shall be permitted for all Captive and Open Access/Scheduled Consumers.
Rajasthan net-metering policy 2015	2015	• This policy lets the eligible parties who install the rooftop PV take advantage of the net metering tariff of INR 7.5 per installed unit.
Uttar Pradesh net-metering regulations for rooftop solar PV	2015	• A support for around 25 years of in-tariff will be provided for rooftop PV installations.
Karnataka Solar Policy 2014–2021	2014	• To achieve a 3% contribution from solar of the total energy consumption. • To add 2000 MW of solar generation by 2021 in a phased manner
Uttar Pradesh renewable energy feed-in tariff 2014–2019 levels	2014	• The main focus was to increase renewable energy generation in the state.

Continued

TABLE 12.5 (Continued)

Policy	Year	Key Characteristics
Uttar Pradesh Solar Energy Policy 2013	2013	• Policy was made for 4 years with a target of 500 MW of solar project installation. Encourages private investment in the project.
Rajasthan Solar Policy	2011	• The policy aims at installing solar water heating systems on all industrial buildings, large residential buildings, government and private hospitals, hotels, and swimming pools and to develop solar parks with 1000 MW capacity and frame regulation for the same. • To promote more installation of solar PV and solar thermal plans for direct sale with a target for 2013 and 2017.
Jawaharlal Nehru National Solar Mission (Phase I, II, and III)	2010	• This mission has 3 different phases, each phase being supported by different targets and policies with an objective to make India a leader in solar energy globally.
Solar and Wind Power Generation-Based Incentives	2008	• The developers sell electricity to the state government-run electricity companies and will pay incentives based on the tariff amount set by the utilities. Subsidy—12 rupees/kilowatt hour for solar PV power and 10 rupees/kilowatt hour for solar thermal power.

12.2.6 PAKISTAN

The Pakistan Government initiated the Development of Alternative and Renewable Energy (ARE) under the policy for the Development of Renewable Energy for Power Generation, in 2006. The Ministry of Energy has developed strategies with an objective to address energy security, affordable electricity, and environmentally friendly and protective energy for its citizens. This policy covers all projects which are produced for sale to a public utility or for a private party if the producer wishes to avail of any incentives available in this policy. This policy covers solar, wind, geothermal, and biogas energy storage system technologies. The policy aims to provide low-cost power generation, ensure the utilization of indigenous resources, and encourage private sector investment [34]. The government had fixed a target of 20% renewable energy generation by capacity before 2025 and 30% of renewable energy generation by capacity before 2030 [35]. The power generation contribution from the solar sector was only 1% in the month of June 2022 with a generation of only 90kWh. The country is expected to announce its National Solar Energy Policy on 1st August 2022. Also, the Prime Minister has invited Chinese companies to invest in solar power recently [36]. The key feature of the energy policies available to reach the set target in Pakistan is given in Table 12.6.

12.2.7 AUSTRALIA

Australia is one among the countries that had signed the Paris Agreement regarding limiting global warming rise to 1.5 to 2°C which requires a sudden transformation from the conventional methods of power generation like oil and coal to renewable

TABLE 12.6

Key Features of Energy Policies in Pakistan [37]

Energy Policy	Key Characteristics
RE 2006 Policy	• There is no income tax charged on renewable energy projects and there is also relief on customs duty or sales tax for machinery imported on such projects. • Also raising of national and foreign finance is permitted and the non-Muslims and non-residents need not pay Zakat.
2012 NCCP of Pakistan	• This policy promotes not only the renewable energy resources but also emphasizes on expanding nuclear power and performance of coal based power with good efficiency levels. • Introduction of carbon tax and providing incentives for developing technology for carbon capture and storage systems.
National Power Policy of 2013	• This policy focuses on lowering the electricity cost by bidding and reducing subsidies. • It attempts to make the independent power producers responsible for the supply of fuel and to effectively manage theft of electricity.
ARE Policy of 2019	• Aims at a target of 20% power generation from ARE technologies by 2025 and30% by 2030 and also minimizes the total generation cost in the country. • The policy also focuses on creation of a trading scheme for national carbon credits and usage of open bidding process for tariffs.

energy by 2050 [38]. The federal, as well as state governments of Australia, provide many grants to appropriate small and medium-sized businesses investing in wind and solar systems. The energy policy of Australia has taken significant steps toward encouraging the deployment of lower-emission energy generation [39]. The country also is running an ongoing programme, "Advancing Renewables Program", wherein the eligible business houses can apply for a grant at any time throughout the year. Australia's Renewable Energy Target was designed to achieve 33000 GWh from renewable sources by 2020 comprising of 2 components of RET. (i) LRET—Large Scale Renewable Energy Target and (ii) SRES—Small Scale Renewable Energy Scheme. As the RET was achieved in September 2019 itself, the government decided to extend the same till 2030 [40]. The renewable energy targets for different parts of the country like Tasmania, Queensland, Victoria was set at levels varying from 25% to 100% before 2025. The energy targets for South Australia is set at 50% by 2025 and New South Wales and Western Australia do not have any such targets and would reach the goal by 2050 [41]. The Small Scale Renewable Energy Scheme of the country provides a financial incentive for individuals and business houses to install solar rooftops, water heaters, heat pumps, etc. The country has already met its RET which is more than a year ahead of its schedule. The prices are expected to drop down by 2030 due to increased generation and oversupply position in the market. The clean energy council has opposed the abolition of orto make any changes in the SRES (Small Scale Renewable Energy Scheme). SRES is a must for the Australian solar industry for the reason that households and the business community can substantially reduce their power bills through the use of rooftop solar. Also it provides huge employment opportunities and generates investment in the home industries [42]. Through the Power2U programme launched by Ausgrid, government of New

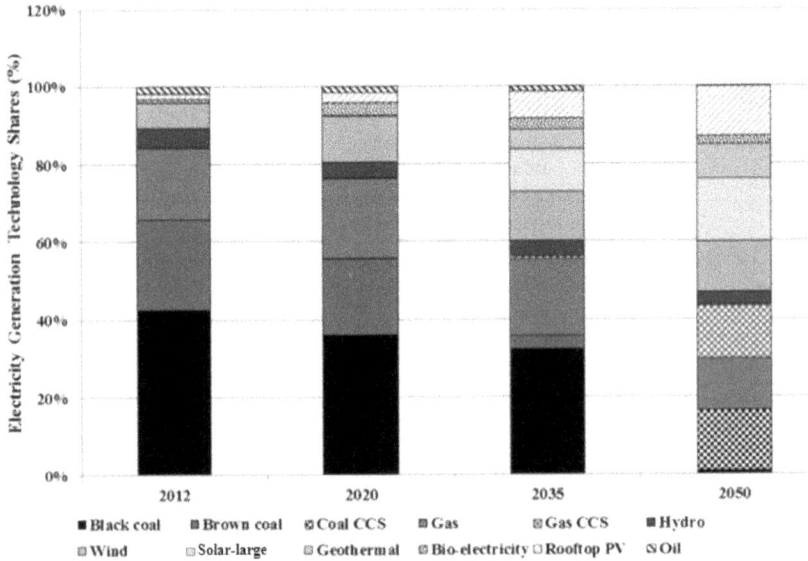

FIGURE 12.15   Electricity generation/production technology shares to 2050 [43].

South Wales, the government of Australia encourages customers to change over to solar power systems and LED lighting systems in place of the conventional lighting systems. Under this programme, incentives are given to such households and community establishments who commit toinstalla solar power system [40]. The country has a substantial renewable energy assets, especially solar energy. But it is felt that there is a need to rethink the existing policies of the country regarding this and create a balanced policy framework aiming for lower emission in the future [27]. If proper measures are taken to resolve the hindrances in developing the renewable energy, definitely Australia will be a leader in the contribution of renewable energy around the globe, especially in solar as the countryhas immense resources. Figure 12.15 sets out a likely production generation mix in 2050 according tothe Energy White Paper [43]. It is seen that renewable energy is a major component in future power generation, particularly solar energy.

12.3 CONCLUSION

The world energy system (WES) survey shows that industrial energy demand increases rapidly as global energy demand increases, while fossil fuels account for the largest share of the energy supply required by the industrial sectors. Hence, an important challenge in the coming years is to integrate renewable energy into the industrial process to reduce fossil fuel demand and greenhouse gas emissions.

Meanwhile, a significant share of energy demand in the industrial sectors is in the form of thermal energy for process heat applications at low and medium temperatures, which can be met by large-scale solar air and water heating systems in the form of solar thermal plants. Hence, in the last decade, solar thermal plants have been launched significantly around the world to provide the energy required by industrial sectors, with regard to regional climatic zones and policies.

Due to the importance of using renewable energy in reducing fossil fuel consumption, the trend of renewable energy demand in the coming years was predicted using the world energy model based on scenarios by considering a different set of basic assumptions about the evolution of the energy systems. Among the various scenarios, the Sustainable Development Scenario (SDS) and global roadmap scenario (REmap) comprehensively and separately predicted global energy consumption, fuel shares, and major types of renewable energy demand. According to the SDS, the share of solar heating systems to supply the heat energy required by industrial sectors will be increased up to 23% by 2030 compared to 2018, while REmap shows solar heating systems will meet 6% of the industry's heat demand. In SDS and REmap scenarios, the performance of the world energy system has been predicted without considering the undesirable obstacles and drawbacks, and in the meantime, only the goals have been identified, and the requirement has not been clearly defined. Hence, the Regional Climate Scenario (RCS) was developed by considering three indicators of regional policies, regional climate, and techno-economic potential for the development of integration of solar heating systems in industrial sectors. The RCS scenario emphasizes the expansion of solar thermal plants in arid and semiarid areas to improve the share of solar heating systems to supply the thermal energy needed by industry and achieve the goals of the SDS and REmap scenarios.

The technology that is used to generate solar energy and the market for the same in the global solar markets is increasing year on year. Not only is the revenue generated out of solar power generation and subsequent transmission of the same but also the technology and allied equipment for its implementation have contributed to a greater extent in fetching revenue in this sector. Researchers say that there is going to be a vast development and installation of solar photovoltaic electricity in countries like USA, Europe, Japan before 2030. Although there is a fall in the share of coal generation to 1.3% in 2020, its use and significance is not decreasing due to the marginal or very slow increase in renewable sources of power generation. Whatever increase we see in wind and solar energy generation, we still have along way to exit coal's dominance in the power sector globally.

REFERENCES

[1] IEA (2021). World Energy Model, IEA, Paris www.iea.org/reports/world-energy-model
[2] IEA (2021). World Energy Model, IEA, Paris www.iea.org/reports/world-energy-model
[3] IEA (2021). World Energy Model, IEA, Paris www.iea.org/reports/world-energy-model
[4] IEA, Industry direct CO2 emissions in the Sustainable Development Scenario, 2000–2030, IEA, Paris www.iea.org/data-and-statistics/charts/industry-direct-co2-emissions-in-the-sustainable-development-scenario-2000–2030
[5] IEA (2020). Tracking Industry 2020, IEA, Paris www.iea.org/reports/tracking-industry-2020
[6] IEA, Final energy consumption and fuel shares in the Sustainable Development Scenario, 2010–2030, IEA, Paris www.iea.org/data-and-statistics/charts/final-energy-consumption-and-fuel-shares-in-the-sustainable-development-scenario-2010–2030

[7] IEA, Renewable heat in industry in the Sustainable Development Scenario, 2010–2030, IEA, Paris www.iea.org/data-and-statistics/charts/renewable-heat-in-industry-in-the-sustainable-development-scenario-2010–2030

[8] IRENA (2018). Global Energy Transformation: A Roadmap to 2050, International Renewable Energy Agency, Abu Dhabi.

[9] Balasubramanian, A. (2013). World Climate Zones. Centre for Advanced Studies in Earth Science, University of Mysore, Mysore.

[10] Hakemzadeh, M.H. (2020). Investigation of photovoltaic system in Malaysian climate as a function of angle and orientation. Universiti Kebangsaan Malaysia. www.ukm.my/ptsl/portal/ethesis

[11] Climate change knowledge portal (https://climateknowledgeportal.worldbank.org/)

[12] Nicholas, Tasie., Israel-Cookey, Chigozie & Banyie, Ledum.(2018). The effect of relative humidity on the solar radiation intensity in Port Harcourt, Nigeria. International Journal of Research, 5: 128–136.

[13] Agbo, S.N., & Okoroigwe, E.C. (2007). Analysis of thermal losses in the flat-plate collector of a thermosyphon solar water heater. Research Journal of Physics, 1(1): 35–41.

[14] Zhenkui, W. et al. (2021). Heating performance of solar heat pump heating system with aluminum tube collector, IEEEAccess, 9: 26491–26501. doi: 10.1109/ACCESS.2021.3056121.

[15] Sukarno, Kartini, Teong, Khan Vun, Dayou, Jedol, Chee, Fueipien, Jackson, Chang. (2017). The monsoon effect on rainfall and solar radiation in Kota Kinabalu. Transactions on Science and Technology, 4(4): 460–465.

[16] Solangi, K.H., Islam, M.R., Saidura, R., Rahim, N.A., Fayaz, H., (2011). A review on global solar energy policy. Renewable and Sustainable Energy Reviews, 15(4): 2149–2163.

[17] www.un.org/en/climatechange/net-zer0-coalition

[18] www.bp.com/content/dam/bp/business-sites/en/global/corporate/pdfs/energy-economics/statistical-review/bp-stats-review-2021-full-report.pdf

[19] Statistical review of world energy 2021, 70th edition.

[20] www.eea.europa.eu/data-and-maps/figures/correlation-of-per-capita-energy.

[21] www.energy.gov/eere/solar/solar-energy-united-states.

[22] www.forbes.com/advisor/home-improvement/average-cost-of-solar-panels/.

[23] www.dataforma.com/blog/challenges-facing-the-solar-industry-right-now/.

[24] www.epa.gov/statelocalenergy/state-renewable-energy-resources, United States Environmental Protection Agency.

[25] The Power of Renewables: Opportunities and Challenges for China and the United States, (2010), The National Academies Press, Washington, D.C.

[26] https://chineseclimatepolicy.energypolicy.columbia.edu/en/solar-power#:~:text=panels%20are%20cleaned.-,Policies,power%20since%20at%20least% 202011.

[27] www.mordorintelligence.com/industry-reports/germany-solar-energy-market.

[28] https://taiyangnews.info/markets/japan-eyes-over-108-gw-solar-power-capacity-by-2030/

[29] https://en.wikipedia.org/wiki/Solar_power_in_Japan.

[30] https://solarify.in/blog/policies-regulations-solar-energy-india/#:~:text=The%20Indian%20government%20revised%20the,policies%20to%20promote%20solar%20energy.

[31] www.thehindu.com/sci-tech/energy-and-environment/indias-solar-capacity-milestones-and-challenges/article65227709.ece.

[32] www.iea.org/policies?country%5B0%5D=India&technology%5B0%5D=Solar%20PV&technology%5B1%5D=Solar.

[33] https://isolaralliance.org/about/background.

[34] www.aedb.org/images/Draft_ARE_Policy_2019_-_Version_2_July_21_2019.pdf

[35] https://blogs.worldbank.org/endpovertyinsouthasia/expanding-solar-and-wind-pakistan-requires-decisive-action.

[36] www.brecorder.com/news/40184919.

[37] Qudrat-Ullah, Hassan. (2022). A review and analysis of renewable energy policies and CO2 emissions of Pakistan. Energy, 238: 121849.

[38] www.energyfactsaustralia.org.au/explainers/energy-policy/.

[39] www.sciencedirect.com/science/article/abs/pii/S0960148113003170.

[40] https://australiangrants.org/renewable-energy-initiatives-australia/?gclid=EAIaIQobCh MIqvDP1en0-AIV0ZVLBR3h4g5-EAAYAiAAEgKb1fD_BwE.

[41] www.sta.org.au/explainers/energy-policy/.

[42] www.cleanenergycouncil.org.au/advocacy-initiatives/renewable-energy-target.

[43] Australian renewable energy policy: Barriers and challenges, Liam Byrnes a,b,*, Colin Brown a, John Foster b, Liam D. Wagner b, Renewable Energy, 60, December 2013, Pages 711–721.

Index

For Product Safety Concerns and Information please contact our EU
representative GPSR@taylorandfrancis.com
Taylor & Francis Verlag GmbH, Kaufingerstraße 24, 80331 München, Germany

www.ingramcontent.com/pod-product-compliance
Lightning Source LLC
Chambersburg PA
CBHW060803220326
41598CB00022B/2523

9 781032 203560